D1284143

Basic
Solid State Chemistry

Basic
Solid State Chemistry

Basic
Solid State Chemistry

ANTHONY R. WEST

Department of Chemistry
University of Aberdeen

JOHN WILEY & SONS

Chichester · New York · Brisbane · Toronto · Singapore

Copyright © 1984, 1988 by John Wiley & Sons Ltd.

All rights reserved.

No part of this book may be reproduced by any means, or
transmitted, or translated into a machine language
without the written permission of the publisher.

Library of Congress Cataloging-in-Publication Data:

West, Anthony R.
  Basic solid state chemistry.
  Rev. ed. of: Solid state chemistry and its
applications. c1984.
  Includes bibliographies and index.
  1. Solid state chemistry.   I. West, Anthony R.
Solid state chemistry and its applications.   II. Title.
III. Title: Solid state chemistry.
QD478.W47   1988   541'.0421   88-5625

ISBN 0 471 91797 4 (cloth)
ISBN 0 471 91798 2 (paper)

British Library Cataloguing in Publication Data:

West, Anthony R.
  Basic solid state chemistry.
  1. Solids.   Chemical bonding
  I. Title
  541'.0421

ISBN 0 471 91797 4 (cloth)
ISBN 0 471 91798 2 (paper)

Phototypesetting by Thomson Press (India) Limited, New Delhi
Printed in Great Britain by Anchor Brendon Ltd., Colchester, Essex.

Contents

Chapter 1

Crystal Structures

Many of the properties and applications of crystalline inorganic materials revolve around a surprisingly small number of structure types. Before describing these important structures, we must first consider some basic concepts and definitions of crystallography.

Definitions

Unit cells and crystal systems

Crystals are built up of regular arrangements of atoms in three dimensions; these arrangements can be represented by a repeat unit or motif called the *unit cell*. The unit cell is defined as *the smallest repeating unit which shows the full symmetry of the crystal structure*. Let us see exactly what this means, first in two dimensions. A section through the NaCl structure is shown in Fig. 1.1(a); possible repeat units are given in (b) to (e). In each, the repeat unit is a square and adjacent squares share edges and corners. Adjacent squares are identical, as they must be by definition; thus, all the squares in (b) have Cl^- ions at their corners and centres. The repeat units in (b), (c) and (d) are all of the same size and, in fact, differ only in their relative position. The choice of origin of the repeating unit is to a certain extent a matter of personal taste, even though the size and shape or orientation of the cell are fixed. The repeat unit of NaCl is usually chosen as either (b) or (c) rather than (d) because it is easier to drawn the unit and visualize the structure as a whole if the unit contains atoms or ions at special positions such as corners, edge centres, etc. Another guideline is that usually the origin is chosen so that the symmetry of the structure is evident (next section).

In the hypothetical case that two-dimensional crystals of NaCl could form, the repeat unit shown in (e), or its equivalent with chlorine at the corners and sodium in the middle, would be the correct unit. Comparing (e) and, for example, (b), both repeat units are square and show the symmetry of the two-dimensional structure; as the units in (e) are half the size of those in (b), (e) would be preferred according to the definition of the unit cell given above. In three dimensions, however, the unit cell of NaCl is based on (c) rather than (e) because only (c) shows the cubic symmetry of the structure (see later).

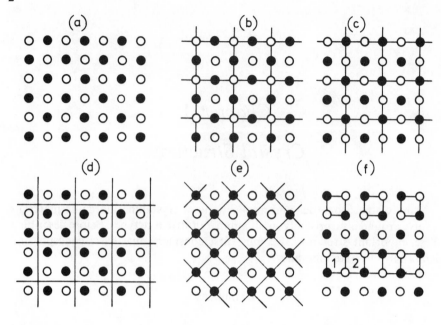

● Na   ○ Cl

Fig. 1.1   (a) Section through the NaCl structure, showing: (b) to (e) possible repeat units, and (f) incorrect units

In Fig. 1.1(f) are shown two examples of what is *not* a repeat unit. The top part of the diagram contains isolated squares whose area is one quarter of the squares in (b). It is true that each square is identical but it is not permissible to isolate unit cells or areas from each other, as appears to happen here. The bottom part of the diagram contains units that are not identical; thus square 1 has a sodium in its top right corner whereas 2 has a chlorine in this position.

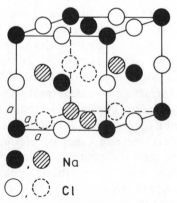

●, ▨ Na

○, ⬭ Cl

Fig. 1.2   Cubic unit cell of NaCl,
$a = b = c$

Table 1.1 *The seven crystal systems*

Crystal system	Unit cell shape†	Essential symmetry	Allowed lattices
Cubic	$a = b = c, \alpha = \beta = \gamma = 90°$	Four threefold axes	P, F, I
Tetragonal	$a = b \neq c, \alpha = \beta = \gamma = 90°$	One fourfold axis	P, I
Orthorhombic	$a \neq b \neq c, \alpha = \beta = \gamma = 90°$	Three twofold axes or mirror planes	P, F, I, A (B or C)
Hexagonal	$a = b \neq c, \alpha = \beta = 90°, \gamma = 120°$	One sixfold axis	P
Trigonal (a)	$a = b \neq c, \alpha = \beta = 90°, \gamma = 120°$	One threefold axis	P
(b)	$a = b = c, \alpha = \beta = \gamma \neq 90°$	One threefold axis	R
Monoclinic*	$a \neq b \neq c, \alpha = \gamma = 90°, \beta \neq 90°$	One twofold axis or mirror plane	P, C
Triclinic	$a \neq b \neq c, \alpha \neq \beta \neq \gamma \neq 90°$	None	P

*Two settings of the monoclinic cell are used in the literature, the one given here, which is most commonly used, and the other $a \neq b \neq c, \alpha = \beta = 90°, \gamma \neq 90°$.
†The symbol $\neq$ means not necessarily equal to. Sometimes, crystals possess *pseudo-symmetry* in which, say, the unit cell is geometrically cubic but does not possess the essential symmetry elements for cubic symmetry, and the symmetry is lower, perhaps tetragonal.

The unit cell of NaCl in three dimensions in shown in Fig. 1.2; it contains Na^+ ions at the corners and face centre positions with Cl^- ions at the edge centres and body centre. Each face of the unit cell looks like the unit area shown in Fig. 1.1 (c). As in the two-dimensional case, the choice of origin is somewhat arbitrary and an equally valid unit cell could be chosen in which the Na^+ and Cl^- ions were interchanged. The unit cell of NaCl is *cubic*. The three edges of the cell, a, b and c are equal in length. The three angles of the cell α (between b and c), β (between a and c) and γ (between a and b) are all equal to $90°$. A cubic unit cell also possesses certain symmetry elements, and these symmetry elements together with the shape define the cubic unit cell.

The seven *crystal systems* listed in Table 1.1 are the seven independent unit cell shapes that are possible in three-dimensional crystal structures. Each crystal system is governed by the presence or absence of symmetry in the structure and the essential symmetry for each is given in the third column. Let us deal next with symmetry because it is of fundamental importance in solid state chemistry and especially, in crystallography.

Symmetry

Symmetry is most easily defined by the use of examples. Consider the silicate tetrahedron shown in Fig. 1.3(a). If it is rotated about an axis passing along one of the Si—O bonds, say the vertical one, then every $120°$ the tetrahedron finds itself in an identical position. Effectively, the three basal oxygens change position with each other every $120°$. During a complete $360°$ rotation, the tetrahedron passes through three such identical positions. The fact that different (i.e. > 1) identical orientations are possible means that the SiO_4 tetrahedron possesses symmetry. The axis about which the tetrahedron may be rotated is called a *rotation axis*; it is an example of a *symmetry element*. The process of rotation is an example of a *symmetry operation*.

The symmetry elements that are important in crystallography are listed in Table 1.2. There are two nomenclatures for labelling symmetry elements, the Hermann–Mauguin system used in crystallography and the Schönflies system used in spectroscopy. It would be ideal if there was only one system which everybody used, but this is unlikely to come about since (a) both systems are very well established, (b) crystallographers require elements of *space symmetry* that

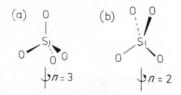

Fig. 1.3 (a) Threefold and (b) twofold rotation axes

Table 1.2 *Symmetry elements*

	Symmetry element	Hermann–Mauguin symbols (crystallography)	Schönflies symbols (spectroscopy)
Point symmetry	Mirror plane	m	σ_v, σ_h
	Rotation axis	$n(= 2, 3, 4, 6)$	$C_n(C_2, C_3, \text{etc.})$
	Inversion axis	$\bar{n} = (\bar{1}, \bar{2}, \text{etc.})$	—
	Alternating axis	—	$S_n(S_1, S_2, \text{etc.})$
	Centre of Symmetry	$\bar{1}$	i
Space symmetry	Glide plane	n, d, a, b, c	—
	Screw axis	$2_1, 3_1, \text{etc.}$	—

spectroscopists do not and vice versa, (c) spectroscopists use a more extensive range of *point symmetry* elements than crystallographers.

The symmetry element described above for the silicate tetrahedron is a *rotation axis*, with symbol n. Rotation about this axis by $360/n$ degrees gives an identical orientation and the operation is repeated n times before the original configuration is regained. In this case, $n = 3$ and the axis is a *threefold rotation axis*. The SiO_4 tetrahedron possesses four threefold rotation axes, one in the direction of each Si—O bond.

When viewed from another angle, SiO_4 tetrahedra possess *twofold rotation axes* (Fig. 1.3b) which pass through the central silicon and bisect the O—Si—O bonds. Rotation by $180°$ leads to an indistinguishable orientation of the tetrahedron. The tetrahedron possesses three of these twofold axes.

Crystals may display rotational symmetries of $2, 3, 4$ and 6. Others, such as $n = 5, 7$, are never observed. This is not to say that molecules which have pentagonal symmetry cannot exist in the crystalline state. They can, of course, but their fivefold symmetry cannot be exhibited by the crystal as a whole. This is shown in Fig. 1.4(a), where a fruitless attempt has been made to pack together pentagons to form a complete layer; for hexagons with sixfold rotation axes (b), a close packed layer is easily produced.

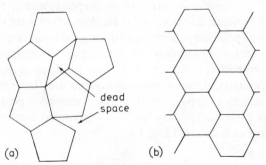

Fig. 1.4 (a) The impossibility of forming a close packed array of pentagons. (b) A close packed layer of hexagons

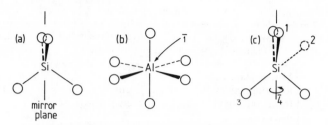

Fig. 1.5 Symmetry elements: (a) mirror plane, (b) centre of
symmetry, (c) fourfold inversion axis

A *mirror plane, m,* exists when two halves of, for instance, a molecule or complex
ion can be interconverted by carrying out the imaginary process of reflection
across the mirror plane. The silicate tetrahedron possesses six mirror planes, one
of which is shown in Fig. 1.5(a). The silicon and two oxygens lie on the mirror
plane and are unaffected by the process of reflection. The other two oxygens are
interchanged on reflection.

The *centre of symmetry,* $\bar{1}$, exists when any part of a molecule or ion can be
reflected through this centre of symmetry, which is a point, and an identical
arrangement found on the other side. An AlO_6 octahedron has a centre of
symmetry (b).

If a line is drawn from any oxygen, through the centre of symmetry and
extended an equal distance on the other side, it terminates at another oxygen. A
tetrahedron (e.g. SiO_4) does not have a centre of symmetry (located at Si).

The *inversion axis,* $\bar{n}$, is a combined symmetry operation involving rotation
(according to n) and inversion through the centre. A $\bar{4}$ (fourfold inversion) axis is
shown in Fig. 1.5(c). The first stage involves rotation by $360/4 = 90°$ and takes, for
example, oxygen 1 to position 2. This is followed by inversion through the centre,
at Si, and leads to the position of oxygen 3. Oxygens 1 and 3 are therefore related
by a $\bar{4}$ axis. Possible inversion axes in crystals are limited to $\bar{1}, \bar{2}, \bar{3}, \bar{4}$ and $\bar{6}$ for the
same reason that only certain pure rotation axes are allowed. The onefold
inversion axis is simply equivalent to the centre of symmetry and the twofold
inversion axis is the same as a mirror plane perpendicular to that axis.

The symmetry elements discussed so far are elements of *point symmetry*. For all
of them, at least one point stays unchanged during the symmetry operation, i.e. an
atom lying on a centre of symmetry, a rotation axis or a mirror plane does not
move during the respective symmetry operations. Finite-sized molecules can only
possess point symmetry elements, whereas crystals may have extra symmetries
that include translation steps as part of the symmetry operation. These are
elements of *space symmetry*; they are not considered further in this book,
although they are included in Table 1.2.

Symmetry and choice of unit cell

The geometric shapes of the various crystal systems (unit cells) are listed in
Table 1.1. These shapes do not *define* the unit cell; they are merely a consequence

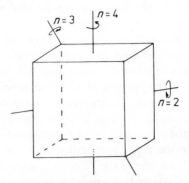

Fig. 1.6 Two-, three-, and four-fold axes of a cube

of the presence of certain symmetry elements. A cubic unit cell is defined as one having four threefold symmetry axes. These run parallel to the cube body diagonals (Fig. 1.6) and it is an automatic consequence of this condition that $a = b = c$ and $\alpha = \beta = \gamma = 90°$. The *essential* symmetry elements by which each crystal system is defined are listed in Table 1.1. In most crystal systems, other symmetry elements in addition to the essential ones are also present. For instance, cubic crystals have many others, including three fourfold axes passing through the centres of each pair of opposite cube faces (Fig. 1.6).

The *tetragonal* unit cell Table 1.1, is characterized by a single fourfold axis and is exemplified by the structure of CaC_2. This is related to the NaCl structure but, because the carbide ion is cigar-shaped rather than spherical, one of the cell axes becomes longer than the other two (Fig. 1.7a). A similar tetragonal cell may be drawn for NaCl by replacing Na for Ca and Cl for C_2; it occupies half the volume of the true, cubic unit cell (Fig. 1.7b). The choice of a tetragonal unit cell for NaCl is rejected, however, because it does not show the full cubic symmetry of the crystal (see Fig. 1.2 and the discussion with it).

The *trigonal* system is characterized by a single threefold axis. Its shape can be derived from that of a cube by stretching or compressing the cube along one of its

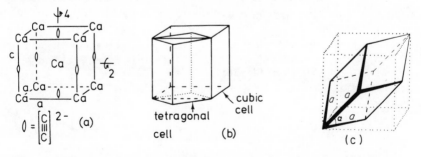

Fig. 1.7 (a) Tetragonal unit cell of CaC_2: note the cigar-shaped carbide ions are aligned parallel to c; (b) relation between 'tetragonal' and cubic cells for NaCl; (c) derivation of a trigonal unit cell from a cubic cell

body diagonals, Fig. 1.7(c). The threefold axis parallel to this body diagonal is retained but those along the other body diagonal directions are all destroyed. All three cell edges remain the same length and all three angles stay the same but are not equal to 90°. It is possible to describe such a trigonal cell for NaCl with $\alpha = \beta = \gamma = 60°$ where Na^+ ions are at the corners and a Cl^- ion is in the body centre, but this is again unacceptable because NaCl has symmetry higher than trigonal. $NaNO_3$ has a structure that may be regarded as a trigonal distortion of the NaCl structure: instead of spherical Cl^- ions, it has triangular nitrate groups, and the presence of these effectively causes a compression along one body diagonal (or rather an expansion in the plane perpendicular to the diagonal). All fourfold symmetry axes and all but one of the threefold axes are destroyed.

The *hexagonal* crystal system is discussed later (page 17) and a drawing of the hexagonal unit cell is given in Fig. 1.17.

The *orthorhombic* unit cell may be regarded as something like a shoebox in which the angles are all 90° but the sides are of unequal length. It usually possesses several mirror planes of symmetry and several twofold axes; the minimum requirement for orthorhombic symmetry is the presence of three mutually perpendicular mirror planes or twofold axes.

The *monoclinic* unit cell may be regarded as derived from our orthorhombic shoebox by a shearing action in which the top face is partially sheared relative to the bottom face and in a direction parallel to one of the box edges. As a consequence, one of the angles departs from 90° and most of the symmetry is lost, apart from a mirror plane and/or a single twofold axis.

The *triclinic* system possesses no symmetry at all and this is reflected in the shape of the unit cell: the sides are of unequal length and the angles differ from 90°.

Lattice, Bravais lattice

It is very useful to be able to represent the manner of repetition of atoms, ions or molecules in a crystal by an array of points, the array being called a *lattice* and the points *lattice points*. The section of the NaCl structure shown in Fig. 1.8(a)

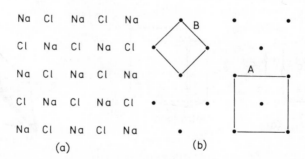

Fig. 1.8 (a) Representation of the NaCl structure in two dimensions by (b) an array of lattice points

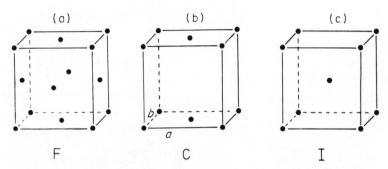

Fig. 1.9 (a) Face centred, (b) side centred and (c) body centred lattices

may be represented by an array of points (b); each point represents one Na$^+$ and one Cl$^-$ but whether the point is located at Na$^+$, at Cl$^-$ or in between is irrelevant. The unit cell may be constructed by linking up the lattice points; two ways of doing this, A and B, are shown in (b). A cell such as B which contains lattice points only at the corners is *primitive*, P, whereas a cell such as A which contains additional lattice points is *centred*. Several types of centred lattice are possible. The *face centred lattice*, F, contains additional lattice points in the centre of each face (Fig. 1.9a); NaCl is face centred cubic. A *side centred lattice* contains extra lattice points on only one pair of opposite faces, e.g. C-centred (b), whereas a *body centred lattice*, I, has an extra lattice point at the body centre (c). α-iron is *body centred cubic* because it has a cubic unit cell with iron atoms at the corner and body centre positions. CsCl is also cubic with caesium atoms at corners and chlorine atoms at the body centre (or vice versa), but it is primitive. This is because, in order for a lattice to be body centred, the atom or group of atoms which are located at or near the corner must be identical to those at or near the body centre position.

The combination of crystal system and lattice type gives the *Bravais lattice* of a structure. There are fourteen possible Bravais lattices. They are given in Table 1.2 by taking the different allowed combinations of crystal system, column 1, and lattice type, column 4, e.g. primitive monoclinic, C-centred monoclinic and primitive triclinic are three of the fourteen possible Bravais lattices. The lattice type plus unit cell combinations which are absent from the table either (a) would violate symmetry requirements, e.g. a C-centred lattice cannot be cubic because it would not have the necessary threefold axes; or (b) may be represented by a smaller, alternative cell, e.g. a face centred tetragonal cell can be redrawn as a body centred tetragonal cell; the symmetry is still tetragonal but the volume is halved (Fig. 1.7b).

Lattice planes and Miller indices

The concept of lattice planes causes considerable confusion because there are two separate ideas which can easily become mixed. Any close packed structure, such as metal structures or ionic structures may, in certain orientations, be

10

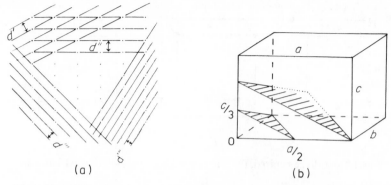

Fig. 1.10   (a) Lattice planes (in projection) (b) derivation of Miller indices

regarded as built up of layers or planes of atoms stacked to form a three-dimensional structure. These layers are often related in a simple manner to the unit cell of the crystal such that, for example, a unit cell face may coincide with a layer of atoms. The reverse is not necessarily true, however, especially in more complex structures and for example, unit cell faces or simple sections through the unit cell often do not coincide with layers of atoms in the crystal. *Lattice planes*, which are a concept introduced with Bragg's law of diffraction (Chapter 3) are defined purely from the shape and dimensions of the unit cell. Lattice planes are entirely imaginary and simply provide a reference grid to which the atoms in the crystal structure may be referred. Sometimes, a given set of lattice planes coincides with layers of atoms, but not usually.

Consider the two-dimensional array of lattice points shown in Fig. 1.10(a). This array of points may be divided up into many different sets of rows and for each set there is a characteristic perpendicular distance, *d*, between pairs of adjacent rows. In three dimensions, these rows become planes and adjacent planes are separated by the *interplanar d-spacing, d.* (Bragg's law treats X-rays as being diffracted from these various sets of lattice planes and the Bragg diffraction angle, *θ*, for each set is related to the *d*-spacing by Bragg's law).

Lattice planes are labelled by assigning three numbers known as *Miller indices* to each set. The derivation of Miller indices is illustrated in Fig. 1.10(b). The origin of the unit cell is at point O and two planes are shown which are parallel and pass obliquely through the unit cell. A third plane in this set must, by definition, pass through the origin. Each of these planes continues out to the surface of the crystal and in so doing cuts through many more unit cells; also, there are many more planes in this set parallel to the two shown, but which do not pass through this particular unit cell. In order to assign Miller indices to a set of planes, there are three stages:

1. Identify that plane which is adjacent to the one that passes through the origin.
2. Find the intersection of this plane on the three axes of the cell and write these intersections as fractions of the cell edges. The plane in question cuts the *x* axis

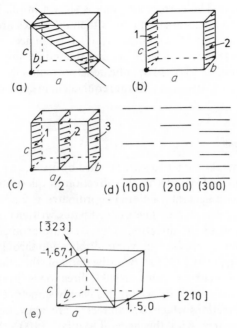

Fig. 1.11 Examples of Miller indices: (a) (101); (b) (100); (c) (200); (d) (h00); (e) indices of directions [210] and [$\bar{3}$23]

at $a/2$, the y axis at b at the z axis at $c/3$; the fractional intersections are $\frac{1}{2}$, 1, $\frac{1}{3}$.

3. Take reciprocals of these fractions; this gives (213).

These three integers are the Miller indices of the plane and all other planes that are parallel to it and are separated from adjacent planes by the same d-spacing.

Some other examples are shown in Fig. 1.11. In (a), the shaded plane cuts x, y and z at $1a$, ∞ b and $1c$, i.e. the plane is parallel to b. Taking reciprocals of 1, ∞ and 1 gives us (101) for the Miller indices. A Miller index of 0 means, therefore, that the plane is parallel to that axis. In (b), the planes of interest comprise opposite faces of the unit cell. We cannot determine directly the indices of plane 1 as it passes through the origin. Plane 2 has intercepts of $1a$, ∞ b and ∞ c and Miller indices of (100). Figure (c) is similar to (b) but there are now twice as many planes as in (b). To find the Miller indices, consider plane 2 which is the one that is closest to the origin but without passing through it. Its intercepts are $\frac{1}{2}$, ∞ and ∞ and the Miller indices are (200). A Miller index of 2 therefore indicates that the plane cuts the relevant axis at half the cell edge. This illustrates an important point. After taking reciprocals, do not divide through by the highest common factor. A common source of error is to regard, say, the (200) set of planes as those planes interleaved between the (100) planes, to give the sequence (100), (200),

(100), (200), (100),....The correct labelling is shown in (d). If extra planes are interleaved between adjacent (100) planes then *all* planes are labelled as (200).

The general symbol for Miller indices is (*hkl*). It is not necessary to use commas to separate the three letters or numbers and the indices are enclosed in curved brackets. The symbol { } is used to indicate sets of planes that are equivalent; for example, the sets (100), (010) and (001) are equivalent in cubic crystals and may be represented collectively as {100}.

Indices of directions

Directions in crystals and lattices are labelled by first drawing a line that passes through the origin and parallel to the direction in question. Let the line pass through a point with general fractional coordinates x, y, z; the line also passes through $2x, 2y, 2z; 3x, 3y, 3z$, etc. These coordinates, written in square brackets [x, y, z], are the indices of the direction; x, y and z are arranged to be the set of smallest possible integers, by division or multiplication throughout by a common factor. Thus $[\frac{1}{2}\frac{1}{2}0]$, [110], [330] all describe the same direction, but conventionally [110] is used. For cubic systems, an [*hkl*] direction is always perpendicular to the (*hkl*) plane of the same indices, but this is only sometimes true in non-cubic systems. Sets of directions which, by symmetry, are equivalent, e.g. cubic [100], [010], etc., are written with the general symbol $\langle 100 \rangle$. Some examples of directions and their indices are shown in Fig. 1.11(e).

d-spacing formulae

We have already defined the *d*-spacing of a set of planes as the perpendicular distance between any pair of adjacent planes in the set and it is this d value that appears in Bragg's law. For a cubic unit cell, the (100) planes simply have a *d*-spacing of a, the value of the cell edge (Fig. 1.11b). For (200) in a cubic cell, $d = a/2$, etc. For *orthogonal* crystals (i.e. $\alpha = \beta = \gamma = 90°$), the *d*-spacing for any set of planes is given by the formula,

$$\frac{1}{d^2_{hkl}} = \frac{h^2}{a^2} + \frac{k^2}{b^2} + \frac{l^2}{c^2} \tag{1.1}$$

The equation simplifies for tetragonal crystals, in which $a = b$, and still further for cubic crystals with $a = b = c$; *i.e.* for cubic crystals,

$$\frac{1}{d^2} = \frac{h^2 + k^2 + l^2}{a^2} \tag{1.2}$$

As a check, for cubic (200), $h = 2$, $k = l = 0$ and $1/d^2 = 4/a^2$. Therefore,

$$d = \frac{a}{2}$$

Monoclinic and especially triclinic crystals have much more complicated *d*-spacing formulae because each angle that is not equal to 90° becomes an additional variable. The relevant formulae for *d*-spacings and unit cell volumes of all the crystal systems are given in Appendix 1.

Unit cell contents and crystal densities

The unit cell, by definition, must contain at least one formula unit, whether it be an atom, ion pair, molecule, etc. In centred cells, and sometimes in primitive cells, the unit cell contains more than one formula unit. A simple relation may be derived between the cell volume, the number of formula units in the cell, the formula weight and the bulk crystal density. The density is given by

$$D = \frac{\text{mass}}{\text{volume}} = \frac{\text{formula weight}}{\text{molar volume}}$$

$$= \frac{\text{FW}}{\text{volume of formula unit} \times N}$$

where N is Avogadro's number. If the unit cell, of volume V, contains Z formula units, then

$$V = \text{volume of one formula unit} \times Z$$

Therefore,

$$D = \frac{\text{FW} \times Z}{V \times N} \qquad (1.3)$$

V is usually expressed as $Å^3$ and must be multiplied by 10^{-24} to give densities in the normal units of grams per cubic centimetre. Substituting for Avogadro's number, the above formula reduces to

$$D = \frac{\text{FW} \times Z \times 1.66}{V} \qquad (1.4)$$

If V is in $Å^3$ in equation (1.4), the units of D are in grams per cubic centimetre.

This simple formula has a number of uses, as shown by the following examples:

(a) It can be used to check that a given set of crystal data are consistent and that, for example, an erroneous formula weight has not been assumed.
(b) It can be used to determine any of the four variables if the other three are known. This is most common for Z (which must be a whole number) but is also used to determine FW and D.
(c) By comparison of D_{obs} (the experimental density of a material) and $D_{\text{X-ray}}$ (the density calculated from the above formula), information may be obtained on the presence of crystal defects such as vacancies or interstitials, the mechanisms of solid solution formation and the porosity of ceramic pieces.

Considerable confusion often arises in determining the value of the contents, Z, of a unit cell. This is because atoms or ions that lie on corners, edges or faces of the unit cell are also shared between adjacent cells, and this must be taken into consideration.

For example, α-Fe (Fig. 1.9c) has $Z = 2$. The corner iron atoms, of which there are eight, are each shared between eight neighbouring unit cells. Effectively, each contributes only $\frac{1}{8}$ to the particular cell in question, giving $8 \times \frac{1}{8} = 1$ net iron atom for the corners. The body centre iron atom lies entirely inside the unit cell and counts as one. Hence $Z = 2$.

For NaCl (Figs 1.2 and 1.9a), $Z = 4$, i.e. $4(Na^+Cl^-)$. The corner Na^+ ions again count as one. The face centre Na^+ ions, of which there are six, count as $\frac{1}{2}$ each, giving a total of $1 + 3 = 4\ Na^+$ ions. The Cl^- ions at the edge centres, of which there are twelve, count as $\frac{1}{4}$ each, which, together with Cl^- at the body centre, give a total of $12 \times \frac{1}{4} + 1 = 4\ Cl^-$ ions.

Description of crystal structures

Crystal structures may be described in various ways. The most common way, and one which gives all the necessary information, is to refer the structure to the unit cell. The structure is then given by the size and shape of the cell and the positions of the atoms inside the cell. However, a knowledge of the unit cell and atomic coordinates alone is often insufficient to give a revealing picture of the structure in three dimensions. This latter is obtained only by considering a larger part of the structure, comprising perhaps several unit cells, and by considering the arrangement of atoms relative to each other, their coordination numbers, interatomic distances, types of bonding, etc. From this more general view of structures, it is possible to find alternative ways of visualizing them and also to compare and contrast different types of structure.

Two of the most useful ways of describing structures are the *close packing* approach and the *space-filling polyhedron* approach. Neither of these can be applied to all types of structure and both have their advantages and limitations. They do, however, provide a much greater insight into crystal chemistry than is obtained using unit cells alone. Details of how to make models of crystal structures are given in Appendix 2 and notes on the geometry of octahedra and tetrahedra in Appendix 3.

Close packed structures—cubic and hexagonal close packing

Many metallic, ionic, covalent and molecular crystal structures can be described using the concept of close packing. The guiding factor is that structures are arranged so as to have the maximum density, although this needs some qualification when ionic structures are being discussed. The principles involved can be understood by considering the most efficient way of packing together equal-sized spheres in three dimensions.

The most efficient way of packing spheres in *two* dimensions is shown in Fig. 1.12. Each sphere, e.g. A, is surrounded by, and is in contact with, six others. By regular repetition, infinite sheets called *close packed layers* are formed. The

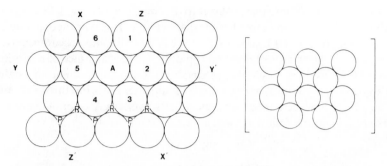

Fig. 1.12 A close packed layer of equal-sized spheres. Inset shows a non-close packed layer

coordination number of six is the maximum possible for a planar arrangement of contacting, equal-sized spheres. Lower coordination numbers are, of course, possible, as shown in the inset, but in such cases the layers are no longer close packed. Note also that within a close packed layer, three *close packed directions* occur. Thus, rows of spheres in contact occur in the directions XX', YY' and ZZ' and sphere A belongs to each of these rows.

The most efficient way of packing spheres in *three* dimensions is to stack close packed layers on top of each other to give *close packed structures*. There are two simple ways in which this may be done, resulting in *hexagonal close packed* and *cubic close packed* structures. These are derived as follows.

The most efficient way for two close packed layers A and B to be in contact is for each sphere of one layer to rest in a hollow between three spheres in the other layer, e.g. at P or R in Fig. 1.12. Two layers in such a position relative to each other are shown in Fig. 1.13. Atoms in the second layer may occupy either P or R positions, but not both together, nor a mixture of the two. Any B (dashed) sphere is therefore seated between three A (solid) spheres, and vice versa.

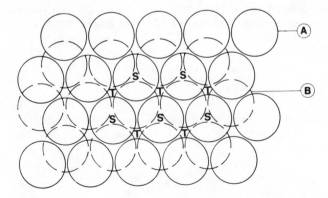

Fig. 1.13 Two close packed layers arranged in A and B positions. The B layer occupies the P positions shown in Fig. 1.12

Addition of a third close packed layer to the two shown in Fig. 1.13 can be done in two possible ways and herein lies the distinction between hexagonal and cubic close packing. In Fig. 1.13, suppose that the A layer lies underneath the B layer and we wish to place a third layer on top of B. There is a choice of positions for this third layer as there was for the second layer: the spheres can occupy either of the new sets of positions S or T but not both together nor a mixture of the two. If the third layer is placed at S, then it is directly over the A layer. As subsequent layers are added, the following sequence arises:

$$...ABABAB...$$

This is known as *hexagonal close packing*. If, however, the third layer is placed at T, then all three layers are staggered relative to each other and it is not until the fourth layer is positioned (at A) that the sequence is repeated. If the position of the third layer is called. C, this gives (Fig. 1.14):

$$...ABCABCABC...$$

This sequence is known as *cubic close packing*.

Hexagonal close packing (h.c.p.) and cubic close packing (c.c.p.) are the two simplest stacking sequences and are by far the most important in structural chemistry. Other more complex sequences with larger repeat units, e.g. ABCACB or ABAC, do occur in a few materials; some of these larger repeat units are related to the phenomenon of polytypism.

In a three-dimensional close packed structure each sphere is in contact with *twelve* others and this is the maximum coordination number possible for contacting and equal-sized spheres. (A common non-close packed structure is the body centred cube, e.g. in α-Fe, with a coordination number of eight; see Fig. 1.9(c).) Six of these neighbours are coplanar with the central sphere (Fig. 1.12) and from Figs 1.13 and 1.14 the remaining six are in two groups of three spheres, one in the plane above and one in the plane below (Fig. 1.15); h.c.p. and c.c.p. differ only in the relative orientations of these two groups of three neighbours.

The unit cells of cubic close packed and hexagonal close packed structures are given in Figs 1.16 and 1.17. The unit cell of a c.c.p. arrange-

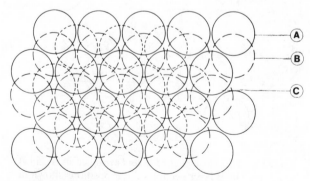

Fig. 1.14   Three close packed layer in c.c.p. sequence

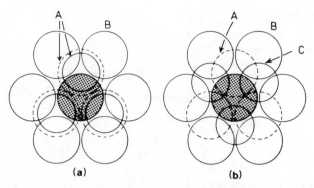

Fig. 1.15 Coordination number 12 of shaded sphere in
(a) h.c.p. and (b) c.c.p. structures

ment is the familiar face centred cubic (f.c.c.) unit cell (Fig. 1.9a) with spheres
at corner and face centre positions. The relation between c.c.p. and f.c.c. is not
immediately obvious since the faces of the f.c.c. unit cell do not correspond to
close packed (c.p.) layers. The c.p. layers are, instead, parallel to the {111} planes
of the f.c.c. unit cell. This is shown in Fig. 1.16(b) by removing sphere 1 from its
corner position in (a) to reveal part of a close packed layer underneath (compare
(b) with Fig. 1.12): the orientations of (a) and (b) are the same but the spheres in (b)
are shown larger and almost contacting. A similar arrangement to that shown in
(b) would be seen on removing any corner sphere and, therefore, in a c.c.p.
structure, c.p. layers occur in four orientations. These orientations are perpendicular to the body diagonals of the cube (the cube has eight corners but only
four body diagonals and hence, four different orientations of the c.p. layers).

The hexagonal unit cell of an h.c.p. arrangement of spheres (Fig. 1.17) is simpler
in that the *basal plane* of the cell coincides with a c.p. layer of spheres (b). The unit

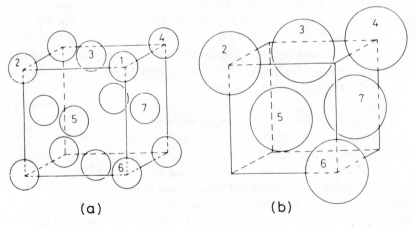

Fig. 1.16 Face centred cubic unit cell of a c.c.p. arrangement of spheres

18

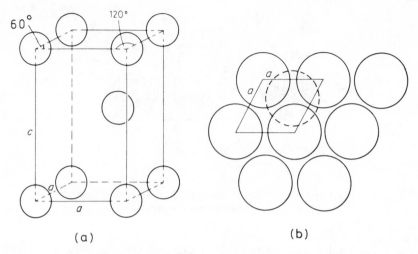

Fig. 1.17 Hexagonal unit cell of a h.c.p. arrangement of spheres

cell contains only two spheres, one at the origin (and hence at all corners) and one inside the cell at position $\frac{1}{3}, \frac{2}{3}, \frac{1}{2}$ (dashed circle in b). Close packed layers occur in only one orientation in a h.c.p. structure.

In c.p. structures, 74.05 per cent of the total volume is occupied by spheres. This is the maximum density possible in structures constructed of spheres of only one size. This density value may be calculated by considering the volume and contents of unit cells of c.p. structures. Let us do the calculation for a c.c.p. structure.

In a c.c.p. array of spheres, the unit cell is f.c.c. and contains four spheres, one at a corner and three at face centre positions, Fig. 1.16 (this is equivalent to the statement that a f.c.c. unit cell contains four lattice points). Close packed directions (XX', YY', ZZ' in Fig. 1.12), in which spheres are in contact, occur parallel to the face diagonals of the unit cell, e.g. spheres 2, 5 and 6 in Fig. 1.16(b) form part of a c.p. row. If the diameter of a sphere is $2r$, the length of the face diagonal is $4r$. The length of the cell edge is then $2\sqrt{2}r$ and the cell volume is $16\sqrt{2}r^3$. The volume of each sphere is $1.33\pi r^3$ and so the ratio of the total sphere volume to the unit cell volume is given by:

$$\frac{4 \times 1.33\pi r^3}{16\sqrt{2}r^3} = 0.7405$$

Similar results are obtained for hexagonal close packing by considering the contents and volume of the appropriate hexagonal unit cell (Fig. 1.17).

In non-close packed structures, densities lower than 0.7405 are obtained, e.g. the density of body centred cubic (b.c.c.) is 0.6802 (to calculate this it is necessary to know that the c.p. directions in b.c.c. are parallel to the body diagonals of the cube).

Materials that can be described as close packed

Metals

Most metals crystallize in one of the three arrangements, c.c.p., h.c.p. and b.c.c., the first two of which are c.p. structures. The distribution of structure type among the metals is irregular (Table 1.3) and no clear-cut trends are observed. It is still not well understood why particular metals prefer one structure type to another. Calculations reveal that the lattice energies of h.c.p. and c.c.p. metal structures are comparable and, therefore, the structure observed in a particular case probably depends on fine details of the bonding requirements or band structures of the metal. Structural details of elements not included in Table 1.3 are given in Appendix 4.

Alloys

Alloys are intermetallic phases or solid solutions and, as is the case for pure metals, many can be regarded as c.p. structures. As an example, copper and gold crystallize in c.c.p. structures, both as pure elements and when mixed together to form Cu–Au alloys. Thus, at high temperatures a complete range of *solid solutions* between copper and gold forms. In these the copper and gold atoms are distributed statistically over the lattice points of the f.c.c. unit cell and therefore the c.c.p. layers contain a random mixture of copper and gold atoms. On annealing the alloy compositions AuCu and $AuCu_3$ at lower temperatures, the gold and copper atoms *order* themselves; c.c.p. layers still occur but the arrangement of gold and copper atoms within the layers is no longer statistical. Such *order–disorder phenomena* are outside the scope of this introduction to crystal structures but occur commonly in both metallic and ionic structures.

Ionic structures

The structures of materials such as NaCl, Al_2O_3, Na_2O, ZnO, etc., in which the anion is somewhat larger than the cation, can be regarded as built of c.p. layers of

Table 1.3 *Structures and unit cell dimensions* (Å) *of some common metals*

Cubic close packed, a		Hexagonal close packed, a, c			Body centred cubic, a	
Cu	3.6147	Be	2.2856	3.5842	Fe	2.8664
Ag	4.0857	Mg	3.2094	5.2105	Cr	2.8846
Au	4.0783	Zn	2.6649	4.9468	Mo	3.1469
Al	4.0495	Cd	2.9788	5.6167	W	3.1650
Ni	3.5240	Ti	2.9506	4.6788	Ta	3.3026
Pd	3.8907	Zr	3.2312	5.1477	Ba	5.019
Pt	3.9239	Ru	2.7058	4.2816		
Pb	4.9502	Os	2.7353	4.3191		
		Re	2.760	4.458		

20

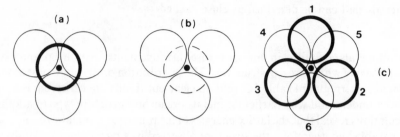

Fig. 1.18 Interstitial sites in a c.p. structure. Heavy circles are above and the dashed circles below the plane of the paper: (a) T_+ site, (b) T_- site (c) O site

anions with the cations placed in interstitial sites. A considerable variety of structure types are possible in which the variables are the stacking sequence of the anions, either h.c.p. or c.c.p., and the number and type of interstitial sites that are occupied by cations. The cations are, however, often too large for the prescribed interstitial sites and the structure can accommodate them only by expanding the anion array. Consequently, the arrangement of anions is the same as in a c.p. array of spheres, but the anions may not be in contact. O'Keeffe has suggested the term *eutactic* for structures such as these in which the arrangement of ions is the same as that in c.p. array but in which the ions are not necessarily contacting. In the subsequent discussion of ionic structures, use of the terms h.c.p. and c.c.p. for the anion arrays does not necessarily imply that the anions are in contact but rather that the structures are eutactic.

Two types of interstitial sites, tetrahedral and octahedral, are present in c.p. structures (Fig. 1.18). These may be seen by considering the nature of the interstitial space between any pair of adjacent c.p. layers. For the tetrahedral sites, three anions that form the base of the tetrahedron belong to one c.p. layer with the apex of the tetrahedron either in the layer above (a) or below (b). This gives two types of tetrahedral sites, T_+ and T_-, in which the apex is up and down, respectively. Because the centre of gravity of a tetrahedron is nearer to the base than to the apex, cations in tetrahedral sites are not located midway between adjacent anion layers but are nearer to one layer than the other. Octahedral sites, on the other hand, are coordinated to three anions in each layer (c) and are placed midway between the anion layers. A more common way of regarding octahedral coordination is as four coplanar atoms with one atom at each apex above and below the plane. In (c), atoms 1, 2, 4 and 6 are coplanar; 3 and 5 form apices of the octahedron. Alternatively, atoms 2, 3, 4, 5 and 1, 3, 5, 6 are coplanar.

The distribution of interstitial sites between any two adjacent layers of c.p. anions is shown in Fig. 1.19. Counting up the numbers of each, for every anion there is one octahedral site and two tetrahedral sites, one T_+ and one T_-. It is rare that all the interstitial sites in a c.p. structure are occupied; often one set is fully or partly occupied and the remaining two sets are empty. A selection of c.p. ionic structures, classified according to the anion layer stacking sequence and the occupancy of the interstitial sites, is given in Table 1.4. Individual structures are

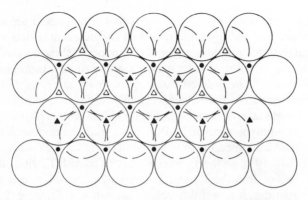

● octahedral sites
▲ T+ tetrahedral sites
△ T- tetrahedral sites

Fig. 1.19 Distribution of interstitial sites between two c.p. layers. Dashed circles are below the plane of the paper

described in more detail later; here we note how a wide range of diverse structures can be grouped into one large family and this helps to bring out the similarities and differences between them. For example:

(a) The rock salt and nickel arsenide structures both have octahedrally co-ordinated cations and differ only in the anion stacking sequence. A similar relation exists between the structures of olivine and spinel.

Table 1.4 *Some close packed structures*

| Anion arrangement | Interstitial sites | | | Examples |
	T_+	T_-	Oct	
c.c.p.	—	—	1	NaCl, rock salt
	1	—	—	ZnS, blende or sphalerite
	$\frac{1}{8}$	$\frac{1}{8}$	$\frac{1}{2}$	$MgAl_2O_4$, spinel
	—	—	$\frac{1}{2}$	$CdCl_2$
	1	—	—	$CuFeS_2$, chalcopyrite
	—	—	$\frac{1}{3}$	$CrCl_3$
	1	1	—	K_2O, antifluorite
h.c.p.	—	—	1	NiAs
	1	—	—	ZnS, wurtzite
	—	—	$\frac{1}{2}$	CdI_2
	—	—	$\frac{1}{2}$	TiO_2^*, rutile
	—	—	$\frac{2}{3}$	Al_2O_3, corundum
	$\frac{1}{8}$	$\frac{1}{8}$	$\frac{1}{2}$	Mg_2SiO_4, olivine
	1	—	—	β-Li_3PO_4
	$\frac{1}{2}$	$\frac{1}{2}$	—	γ-$Li_3PO_4^*$
c.c.p. 'BaO$_3$' layers	—	—	$\frac{1}{4}$	$BaTiO_3$, perovskite

*The h.c.p. oxide layers in rutile and γ-Li_3PO_4 are not planar but are buckled. The oxide ion arrangement in these may alternatively be described as tetragonal packed (t.p.).

(b) Rutile, TiO_2 and CdI_2 both have hexagonal close packed anions (although the layers are buckled in rutile) with half the octahedral sites occupied by cations but they differ in the manner of occupancy of these octahedral sites. In rutile, half the octahedral sites between any pair of c.p. anion layers are occupied by Ti^{4+}, in ordered fashion; in CdI_2, layers of fully occupied octahedral sites alternate with layers in which all sites are empty and this gives CdI_2 an obvious layered structure which is reflected in its physical properties.

(c) Both β and γ polymorphs of Li_3PO_4 have h.c.p. oxide ions (although the layers are buckled, especially in γ) with half the tetrahedral sites occupied by cations; in β-Li_3PO_4, all the T_+ sites are occupied with T_- sites empty, but in γ-Li_3PO_4 half of both the T_+ and T_- sites are occupied, in ordered fashion.

(d) In a few structures, it is useful to regard the *cation* as forming the c.p. layers with the anions occupying the interstitial sites. The most common example is the fluorite structure, CaF_2, which may be regarded as cubic close packed Ca^{2+} ions with all the T_+ and T_- tetrahedral sites occupied by F^- ions. The antifluorite structure of, for example, K_2O is the exact inverse of fluorite and is included in Table 1.4.

(e) The close packing concept may be extended to include structures in which a mixture of anions and large cations form the packing layers and smaller cations occupy the interstitial sites. In perovskite, $BaTiO_3$, c.c.p. layers of composition 'BaO_3' occur and one quarter of the octahedral sites between these layers are occupied by titanium although it occupies only those octahedral sites in which all six corners are O^{2-} ions.

(f) Some structures may be regarded as anion-deficient c.p. structures in which the anions form an essentially c.p. array but some anions are missing. The ReO_3 structure may be regarded as having c.c.p. oxide layers with one quarter of the O^{2-} ion sites empty. It is analogous to the perovskite structure just described in which the Ti^{4+} ions are replaced by Re^{6+} and the Ba^{2+} ions are removed and their sites left vacant. The structure of β-alumina, nominally of formula $NaAl_{11}O_{17}$, contains c.p. oxide layers in which every fifth layer has approximately three quarters of the O^{2-} ions missing.

Covalent network structures

Materials such as diamond and silicon carbide, which have very strong, directional, covalent bonds, can also be described as c.p. structures or eutactic structures; many have the same structures as ionic compounds. Thus, one polymorph of SiC has the wurtzite structure and it is immaterial whether silicon or carbon is regarded as the packing atom since the net result—a three-dimensional framework of corner sharing tetrahedra—is the same. Diamond can be regarded as a sphalerite structure in which half the carbon atoms form a c.c.p. array and the other half occupy T_+ interstitial sites, but again the two types of atom are equivalent. Classification of diamond as a eutactic structure is useful since in diamond all the atoms are of the same size and it is unrealistic to distinguish between packing atoms and interstitial atoms.

Many structures have mixed ionic–covalent bonding, e.g. ZnS and $CrCl_3$ in Table 1.4, and one advantage of describing their structures in terms of close packing is that this can be done, if necessary, without reference to the type of bonding that is present.

Molecular structures

Since c.p. structures provide an efficient means of packing atoms together, many molecular compounds crystallize as c.p. structures even though the bonding forces between adjacent molecules are weak van der Waals forces. If the molecules are roughly spherical in shape or become spherical because they can rotate or occupy different orientations at random, then simple h.c.p. or c.c.p. structures result, e.g. in crystalline H_2, CH_4 and HCl. Non-spherical molecules, especially if they are built of tetrahedra and octahedra, can also fit into a c.p. arrangement. For example, Al_2Br_6 is a dimeric molecule which can be regarded as two $AlBr_4$ tetrahedra sharing a common edge (Fig. 1.20b). In crystalline Al_2Br_6, the bromine atoms form an h.c.p. arrangement and aluminium atoms occupy $\frac{1}{6}$ of the available tetrahedral sites. One molecule, with the bromine atoms

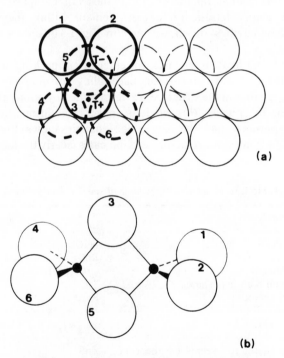

Fig. 1.20 Hexagonal close packing arrangement of bromine atoms in crystalline Al_2Br_6 molecules; aluminium atoms occupy T_+ and T_- sites. Dashed circles are below the plane of the paper

in heavy outline is shown in Fig. 1.20(a); aluminium atoms occupy one pair of adjacent T_+ and T_- sites. Bromine atoms 3 and 5 are common to both tetrahedra and are the bridging atoms in (b). Adjacent Al_2Br_6 molecules are arranged so that each bromine in the h.c.p. array belongs to only one molecule. $SnBr_4$ is a tetrahedral molecule and also crystallizes with an h.c.p. bromine array, but only one eighth of the tetrahedral sites are occupied.

Structures built of space-filling polyhedra

This approach to describing crystal structures emphasizes the coordination number of the cations and the way that structures may be regarded as built of polyhedra, formed by the cations with their immediate anionic neighbours, linked together by sharing corners, edges or faces. For example, in NaCl, each Na^+ ion has six Cl^- ions as its nearest neighbours, arranged octahedrally; this is represented as an octahedron with Cl^- ions at the corners and Na^+ at the centre. A three-dimensional overview of the structure is obtained by looking at the way in which neighbouring octahedra are linked to each other. In NaCl, each octahedron edge is shared between two octahedra, resulting in an infinite framework of edge-sharing octahedra. Although the polyhedra may link up to form a three-dimensional framework, not all the available space is usually filled; e.g. in NaCl, empty tetrahedral cavities remain within the framework of octahedra. These cavities are shown later in Fig. 1.34(d) for a single layer of octahedra.

Although it is legitimate to estimate the efficiency of packing of spheres in c.p. structures, it is incorrect to do this for space-filling polyhedra since the anions, usually the largest ions in the structure, are represented by points at the corners of the polyhedra. In spite of this obvious misrepresentation, the space-filling polyhedron approach has the advantage that it shows the topology or connectivity of a framework structure and indicates clearly the location of empty

Table 1.5 *Some structures built of space-filling polyhedra*

Octahedra only	
12 edges shared	NaCl
6 corners shared	ReO_3
3 edges shared	$CrCl_3$, BiI_3
2 edges and 6 corners shared	TiO_2
4 corners shared	$KAlF_4$
Tetrahedra only	
4 corners shared (between 4 tetrahedra)	ZnS
4 corners shared (between 2 tetrahedra)	SiO_2
1 corner shared (between 2 tetrahedra)	$Si_2O_7^{6-}$
2 corners shared (between 2 tetrahedra)	$(SiO_3)_n^{2n-}$, chains or rings

interstitial sites. Some examples of structures that can be viewed as built of space-filling polyhedra are given in Table 1.5. A variety of polyhedra occur in inorganic structures, although tetrahedra and octahedra appear to be the most common.

A complete scheme for classifying polyhedral structures has been given by Wells and others in which the initial problem is a geometrical one—what types of network built of linked polyhedra are possible? The variables to be considered are the following. Polyhedra may be tetrahedra, octahedra, trigonal prisms, etc. Polyhedra may share some or all of their corners, edges and faces with adjacent polyhedra, which may or may not be of the same type. The corners and edges may be common to more than two polyhedra (obviously only two polyhedra can share a common face). An enormous number of structures are feasible, at least theoretically, and it is an interesting exercise to categorize real structures on this basis.

The topological problem of what kinds of structure may be generated by arranging polyhedra in various ways takes no account of the bonding forces between atoms or ions. Such information must come from elsewhere. Also the description of structures in terms of polyhedra does not necessarily imply that such entities exist in the structure as separate species. Thus, in NaCl, the bonding is mainly ionic, and physically distinct $NaCl_6$ octahedra do not occur. Similarly, SiC has a covalent network structure and separate SiC_4 tetrahedral entities do not exist. Polyhedra *do* have a separate existence in structures of (a) molecular materials, e.g. Al_2Br_6 contains pairs of edge-sharing tetrahedra, and (b) compounds that contain complex ions, e.g. the silicate structures are built up of SiO_4 tetrahedra which link up to form complex anions ranging in size from isolated monomeric units to infinite chains, sheets and three-dimensional frameworks.

In considering the types of polyhedral linkage that are likely to occur in crystal structures, Pauling's third rule for the structures of complex ionic crystals (see next chapter for other rules) provides a very useful guideline. This rule states that the presence of shared polyhedron edges and, especially, shared faces decreases the stability of a structure. The effect is particularly important for cations of high valence and small coordination number, i.e. for small polyhedra, especially tetrahedra, that contain highly charged cations. When polyhedra share edges and, particularly, faces, the cation–cation distances (i.e. distances between the centres of the polyhedra) decrease and the cations repel each other electrostatically. In Fig. 1.21 are shown pairs of (a) corner-sharing and (b) edge-sharing octahedra. The cation–cation distance is clearly less in the latter and for face-sharing octahedra, not shown, the cations are even closer. Comparing edge-shared tetrahedra and edge-shared octahedra, the cation–cation distance M–M (relative to the cation–anion distance M–X) is less between the edge-shared tetrahedra because the MXM bridging angle is (c) 71° for the tetrahedra and (b) 90° for the octahedra. A similar effect occurs for face-sharing tetrahedra and octahedra. In Table 1.6, the M–M distances for the various polyhedral linkages are given as a function of the distance M–X. The M–M distance is greatest for corner sharing of either tetrahedra or octahedra and is least for face-sharing

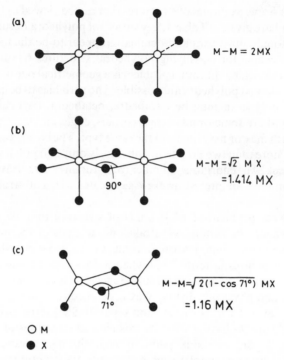

Fig. 1.21 Cation–cation separation in octahedra which share (a) corners and (b) edges and (c) in tetrahedra which share edges

tetrahedra. For the sharing of corners and edges, the M–M distances given are the maximum possible distances. Reduced distances occur if the polyhedra are rotated about the shared corner or edge (e.g. so that the M–X–M angle between corner-shared polyhedra is less than 180°).

From Table 1.6, the M–M distance in face-shared tetrahedra is considerably *less than* the M–X bond distance. This represents an unstable situation, therefore, because of the large cation–cation repulsions which would arise, and does not normally occur. A good example of the non occurrence of face-sharing tetrahedra

Table 1.6 *The distance M–M between centres of regular* MX_4 *or* MX_6 *groups sharing X atom(s)*

	Corner* sharing	Edge* sharing	Face sharing
Two tetrahedra	2.00 MX(tet.)	1.16 MX(tet.)	0.67 MX(tet.)
Two octahedra	2.00 MX(oct.)	1.41 MX(oct.)	1.16 MX(oct.)

*Maximum possible value.

is provided by the absence of any crystal whose structure is the h.c.p. equivalent of the fluorite (or antifluorite) structure. In, for example, Na_2O which has the antifluorite structure, the oxide ions are in a c.c.p. array and the NaO_4 tetrahedra share edges. However, in a structure with an h.c.p. anion array and all the tetrahedral cation sites occupied, the MX_4 tetrahedra would share faces. Edge sharing of tetrahedra, in which the M–M separation is only 16 per cent larger than the M–X bond distance Table 1.6, appears to be energetically acceptable in principle, as shown by the common occurrence of compounds with a fluorite type of structure, but face sharing of tetrahedra appears to be generally unacceptable.

For tetrahedra containing cations of high charge, edge sharing appears to be energetically unacceptable and only corner sharing occurs. For example, in silicate structures which are built of SiO_4 tetrahedra, edge sharing of SiO_4 tetrahedra never occurs. (*Note*: In an ideally ionic structure the charge on silicon would be $4+$, but the actual charges are considerably less due to partial covalency of the Si—O bonds).

An additional factor to be taken into account when comparing linkages of tetrahedra and octahedra (Table 1.6) is that, for a given M and X, the M–X distance varies, depending on the nature of the polyhedra. In a c.p. structure, the tetrahedral sites are smaller than the octahedral sites, i.e.

$$\frac{\text{M–X in } MX_4(\text{tet.})}{\text{M–X in } MX_6(\text{oct.})} = \frac{\sqrt{3}}{2} = 0.866$$

This may be shown by, for example, calculating and comparing the M–X distances present in the f.c.c. cells of NaCl and CaF_2 structures (see Table 1.10).

Some important structure types

Rock salt (NaCl), zinc blende or sphalerite (ZnS), fluorite (CaF$_2$), antifluorite (Na$_2$O)

These structures are considered together because they have in common a cubic close packed (i.e. a face centred cubic) arrangement of anions and differ only in the positions of the cations. In Fig. 1.22 are shown the anions (●) in a f.c.c. unit cell with the possible octahedral O, tetrahedral T_+ and tetrahedral T_- sites for the cations. Octahedral sites occur at the edge centre 1, 2, 3 and body centre 4 positions. In order to see the T_+ and T_- sites clearly it is convenient to divide the unit cell into eight minicubes by bisecting each cell edge (dashed lines). These minicubes contain anions at only four of the eight corners and in the middle of each is a tetrahedral site. There are two orientations for the tetrahedral sites, corresponding to T_+ and T_- in Fig. 1.18, as shown in Fig. 1.23. Note this useful way to represent tetrahedra, as a cube with alternate corners missing, and which permits ready calculation of bond distances, bond angles etc., in tetrahedra. A tetrahedral site occurs in the middle of each minicube and parallel to any of the cell axis, x, y and z, T_+ and T_- sites alternate.

The c.p. anion layers are oriented parallel to the {111} planes of the unit cell.

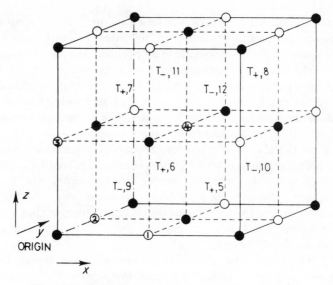

Fig. 1.22 Available cation sites in an f.c.c. anion array

Thus, in Fig. 1.23(a) the three anions P, Q and R, belong to one layer, forming one face of the tetrahedron, with anion S in the adjacent layer, forming an apex to the tetrahedron.

The *fractional coordinates* of the four anions in the unit cell, Fig. 1.22, are

$$000, \quad \tfrac{1}{2}\tfrac{1}{2}0, \quad \tfrac{1}{2}0\tfrac{1}{2}, \quad 0\tfrac{1}{2}\tfrac{1}{2}$$

Only one corner atom is included in this list at the origin, i.e. at 000, since the other seven corner atoms, at 100, 110, etc., are equivalent and can be regarded as the corner atoms of adjacent unit cells. Likewise for each pair of opposite faces,

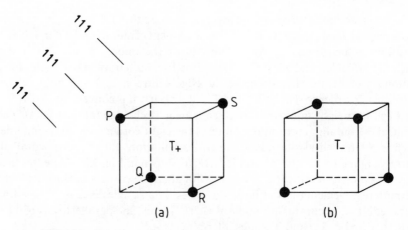

Fig. 1.23 Tetrahedral sites T_+, T_- and their relation to a cube

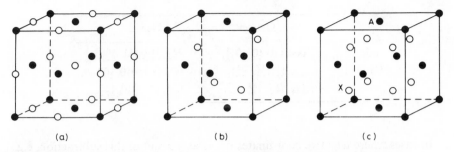

Fig. 1.24 Unit cell of (a) NaCl, (b) ZnS, sphalerite, and (c) Na_2O. Open circles are cations; closed circles anions

e.g. $\frac{1}{2}\frac{1}{2}0$ and $\frac{1}{2}\frac{1}{2}1$, only one of these is specified since the other is equivalent and is generated by translation of one unit cell dimension in the z direction.

The various cation positions have the following coordinates:

Octahedral: $1 - \frac{1}{2}00,$ $2 - 0\frac{1}{2}0,$ $3 - 00\frac{1}{2},$ $4 - \frac{1}{2}\frac{1}{2}\frac{1}{2}$

Tetrahedral T_+: $5 - \frac{3}{4}\frac{1}{4}\frac{1}{4},$ $6 - \frac{1}{4}\frac{3}{4}\frac{1}{4},$ $7 - \frac{1}{4}\frac{1}{4}\frac{3}{4},$ $8 - \frac{3}{4}\frac{3}{4}\frac{3}{4}$

Tetrahedral T_-: $9 - \frac{1}{4}\frac{1}{4}\frac{1}{4},$ $10 - \frac{3}{4}\frac{3}{4}\frac{1}{4},$ $11 - \frac{1}{4}\frac{3}{4}\frac{3}{4},$ $12 - \frac{3}{4}\frac{1}{4}\frac{3}{4}$

Note that there are four of each type of cation site, O, T_+, T_-, in the unit cell together with four anions. By having different sites occupied, different structures are generated, as follows:

Rock salt: O sites occupied by cations; T_+, T_- empty

Zinc blende: T_+ (or T_-) sites occupied; O, T_- (or T_+) empty

Antifluorite: T_+, T_- occupied; O empty.

Unit cells of these three structures are shown in Fig. 1.24(a), (b) and (c). In rock salt, both anions and cations are octahedrally coordinated, whereas in zinc blende, both are tetrahedrally coordinated. In antifluorite, cations are tetrahedrally coordinated and anions are eight-coordinated.

A general rule regarding coordination numbers is that in any structure of formula A_xX_y, the coordination numbers of A and X must be in the ratio of $y:x$. In both rock salt and zinc blende, $x = y$ and therefore, in each, anions and cations have the same coordination number. In antifluorite, of formula A_2X, the coordination numbers of cation and anion must be in the ratio of 1:2. Since the cations occupy tetrahedral sites, the anion coordination number must therefore be eight.

In order to see the anion coordination number of eight, it is convenient to redefine the origin of the unit cell so that the origin coincides with a cation rather than an anion. This may be done by displacing the unit cell along one of the body diagonals and by one quarter of the length of the diagonal. The cation at X in Fig. 1.24(c), with coordinates $\frac{1}{4}\frac{1}{4}\frac{1}{4}$, may be chosen as the new origin of the unit cell. The coordinates of all the atoms in the new cell are given by subtracting $\frac{1}{4}\frac{1}{4}\frac{1}{4}$ from their coordinates in the old cell, as in the following table.

	Old cell	New cell
Anion	$000, \frac{1}{2}\frac{1}{2}0, \frac{1}{2}0\frac{1}{2}, 0\frac{1}{2}\frac{1}{2},$	$\frac{333}{444}, \frac{113}{444}, \frac{131}{444}, \frac{311}{444},$
Cations	$\frac{111}{444}, \frac{113}{444}, \frac{131}{444}, \frac{311}{444},$	$000, 00\frac{1}{2}, 0\frac{1}{2}0, \frac{1}{2}00,$
	$\frac{133}{444}, \frac{313}{444}, \frac{331}{444}, \frac{333}{444},$	$0\frac{1}{2}\frac{1}{2}, \frac{1}{2}0\frac{1}{2}, \frac{1}{2}\frac{1}{2}0, \frac{1}{2}\frac{1}{2}\frac{1}{2},$

In cases where negative coordinates occur as a result of this subtraction, e.g. $-\frac{1}{4}-\frac{1}{4}-\frac{1}{4}$, the position lies outside the new unit cell and it is necessary to find an equivalent position within the unit cell. In this particular case, 1 is added to each coordinate, giving $\frac{3}{4}\frac{3}{4}\frac{3}{4}$. Addition of 1 to, say, the x coordinate is equivalent to moving to a similar position in the next unit cell in the x direction. The new unit cell of the antifluorite structure with its origin at cation X, is shown in Fig. 1.25(a). It contains cations at corners, edge centres, face centres and body centre. In order to see the anion coordination more clearly, the unit cell may be imagined as divided into eight smaller cubes (as in Fig. 1.22). Each of these smaller cubes has cations at all eight corners and hence at the centre of each is an eight-coordinate site. Anions occupy four of these eight smaller cubes such that parallel to the cell axes the eight coordinate sites are alternately occupied and empty. The eightfold coordination for one anion, A, is shown in Fig. 1.25(b).

In the antifluorite structure, the effect of changing the origin from an anion position to a cation position is to show the structure in a completely different light. This does not happen with the rock salt and zinc blende structures, however. In these the cation and anion positions are interchangeable and it is immaterial whether the origin coincides with an anion or a cation.

So far the NaCl, ZnS and Na_2O structures have been described in two ways: (a) as c.p. structures and (b) in terms of their unit cells. A third way is to regard them as built of space-filling polyhedra. Each ion and its nearest neighbours may be represented by the appropriate polyhedron, e.g. in zinc blende a tetrahedron

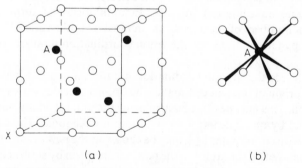

(a)   (b)

Fig. 1.25 Alternative view of the antifluorite structure

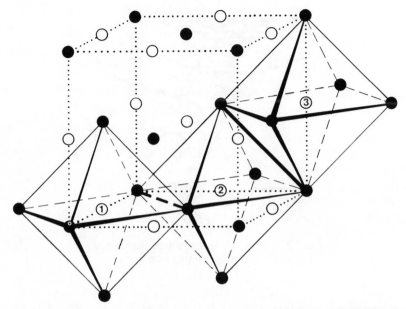

Fig. 1.26   Unit cell of the rock salt structure showing edge-sharing octahedra

represents a zinc ion and its four sulphide neighbours (or vice versa). It is then necessary to consider how neighbouring polyhedra are linked—whether by sharing corners, edges or faces.

In the rock salt structure, $NaCl_6$ octahedra share common edges. Each octahedron has twelve edges and each edge is shared between two octahedra. In Fig. 1.26, the unit cell of NaCl is outlined (dotted) and is in the same orientation as in Fig. 1.24(a). Octahedra centred on sodium ions in edge centre positions 1, 2 and 3 are outlined. Octahedra 1 and 2 share a common edge, indicated by the thick dashed line, and octahedra 2 and 3 share a common edge, shown by the thick solid line. Since each octahedron is linked by its edges to twelve other octahedra it is difficult to represent this satisfactorily in a drawing.

A schematic drawing of the rock salt structure as an array of octahedra is shown in Fig. 1.27. Also shown are tetrahedral interstices (arrowed) which are normally empty in NaCl. In this structure, each octahedron face is parallel to layers of c.p. anions. This is emphasized in the drawing in that coplanar octahedron faces are numbered or shaded. Parts of four different sets of faces are shown, corresponding to the four c.p. orientations in a c.c.p. array.

A large number of AB compounds possess the rock salt structure. A selection is given in Table 1.7 together with values of the a dimension of the cubic unit cell. Most halides and hydrides of the alkali metals and Ag^+ have this structure, as do also a large number of the chalcogenides (oxides, sulphides, etc.) of divalent metals such as alkaline earths and divalent transition metals. Many of these are ionic but others, e.g. TiO, have a metallic character.

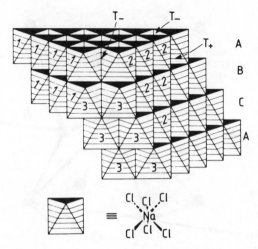

Fig. 1.27 The rock salt structure depicted as
an array of edge-sharing octahedra

The zinc blende structure contains ZnS_4 tetrahedra (and/or SZn_4 tetrahedra) which are linked at their corners. Each corner is common to four such tetrahedra; this must be so since anion and cation have the same coordination number in ZnS. The unit cell of zinc blende (Fig. 1.24b) is shown again in Fig. 1.28(a), but in terms of corner-sharing ZnS_4 tetrahedra. The faces of the tetrahedra are parallel to the c.p. anion layers, i.e. the {111} planes, and this is emphasized in a more extensive model of the structure (b), in which the model is oriented so that one set of tetrahedron faces is approximately horizontal. Conventionally, the ZnS structure is regarded as built of c.p. layers of sulphide anions with the smaller zinc

Table 1.7 *Some compounds with the NaCl structure*

	$a(\text{Å})$		$a(\text{Å})$		$a(\text{Å})$		$a(\text{Å})$
MgO	4.213	MgS	5.200	LiF	4.0270	KF	5.347
CaO	4.8105	CaS	5.6948	LiCl	5.1396	KCl	6.2931
SrO	5.160	SrS	6.020	LiBr	5.5013	KBr	6.5966
BaO	5.539	BaS	6.386	LiI	6.00	KI	7.0655
TiO	4.177	αMnS	5.224	LiH	4.083	RbF	5.6516
MnO	4.445	MgSe	5.462	NaF	4.64	RbCl	6.5810
FeO	4.307	CaSe	5.924	NaCl	5.6402	RbBr	6.889
CoO	4.260	SrSe	6.246	NaBr	5.9772	RbI	7.342
NiO	4.1769	BaSe	6.600	NaI	6.473	AgF	4.92
CdO	4.6953	CaTe	6.356	NaH	4.890	AgCl	5.549
SnAs	5.7248	SrTe	6.660	ScN	4.44	AgBr	5.7745
TiC	4.3285	BaTe	7.00	TiN	4.240		
UC	4.955	LaN	5.30	UN	4.890		

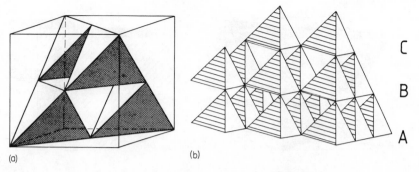

Fig. 1.28 The sphalerite (zinc blende) structure showing (a) the unit cell contents and (b) a more extended network of corner-sharing tetrahedra

cations in tetrahedral sites. Since the same structure is generated by interchanging the Zn^{2+} and S^{2-} ions, the structure could also be described as a c.p. array of Zn^{2+} ions with S^{2-} ions occupying one set of tetrahedral sites. A third, equivalent description is as an array of c.c.p. ZnS_4 (or SZn_4) tetrahedra.

Some compounds which exhibit the zinc blende structure are listed in Table 1.8 together with their cubic lattice parameter, a. The bonding in these compounds is less ionic than in corresponding AB compounds which have the rock salt structure. Thus, oxides do not usually have the zinc blende structure (ZnO, not included in Table 1.8, is an exception; it is dimorphic with zinc blende and wurtzite structure polymorphs). Chalcogenides (S, Se, Te) of the alkaline earth

Table 1.8(a) *Some compounds with the zinc blende (sphalerite) structure (in ångstroms)*

CuF	4.255	BeS	4.8624	β-CdS	5.818	BN	3.616	GaP	5.448
CuCl	5.416	BeSe	5.07	CdSe	6.077	BP	4.538	GaAs	5.6534
γ-CuBr	5.6905	BeTe	5.54	CdTe	6.481	BAs	4.777	GaSb	6.095
γ-CuI	6.051	β-ZnS	5.4060	HgS	5.8517	AlP	5.451	InP	5.869
γ-AgI	6.495	ZnSe	5.667	HgSe	6.085	AlAs	5.662	InAs	6.058
β-MnS, red	5.600	ZnTe	6.1026	HgTe	6.453	AlSb	6.1347	InSb	6.4782
β-MnSe	5.88	β-SiC	4.358						

Table 1.8(b) *Elements with the diamond structure*

Element	a (Å)
C	3.5667
Si	5.4307
Ge	5.6574
α-Sn (gray)	6.4912

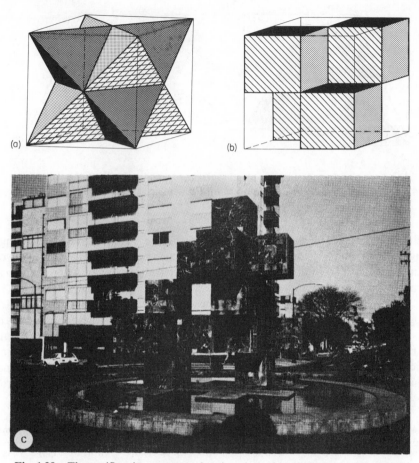

Fig. 1.29 The antifluorite structure showing the unit cell in terms of (a) NaO_4 tetrahedra or (b) ONa_8 cubes. A more extended array of cubes is shown in (c); this model resides on a roundabout in Mexico City

metals (but not Be) have the rock salt structure whereas the more covalent corresponding compounds of Be, Zn, Cd and Hg have the zinc blende structure. Most metal(I) halides are like rock salt, with the exception of the more covalent copper(I) halides and γ-AgI which have the zinc blende structure.

The antifluorite structure contains tetrahedrally coordinated cations and eight-coordinate anions (Figs 1.24c and 1.25). This leads to two ways of describing the structure, either as a three-dimensional network of tetrahedra or as a three-dimensional network of cubes. The results of these two methods are shown in Fig. 1.29. The unit cell contains either (a) eight NaO_4 tetrahedra or (b) four ONa_8 cubes. In (a), tetrahedra share edges and each edge is common to two tetrahedra. In (b), cubes share corners, but each corner is common to only four cubes (up to a maximum number of eight cubes may share a common corner).

Also, the cubes share edges, but each edge is common to only two cubes (the maximum possible is four). A more extended network of corner- and edge-sharing cubes is shown in (c). This must surely rate as one of the world's largest models of the antifluorite structure!

The two ways of describing the arrangement of polyhedra in the antifluorite structure coincide with the two classes of compound which possess this structure. So far, we have considered the *antifluorite* structure which is shown by a large number of oxides and other chalcogenides of the alkali metals (Table 1.9), i.e. compounds of the general formula $A_2^+ X^{2-}$. A second group of compounds, which includes fluorides of large, divalent cations and oxides of large tetravalent cations, has the inverse, *fluorite* structure, i.e. $M^{2+}F_2$ and $M^{4+}O_2$. In the fluorite structure, the cations are eight coordinate and the anions are four coordinate, which is the inverse of the antifluorite structure. The arrangement of cubes in Fig. 1.29(b) and (c) also show the MF_8 and MO_8 coordination in the fluorite structure.

A common alternative way of describing the fluorite structure is as a primitive cubic array of anions in which the eight-coordinate sites at the cube body centres are alternately empty and occupied by a cation: this description is consistent with (b) and (c). It should be stressed, however, that the true lattice type of the fluorite structure is face centred cubic and not primitive cubic, since the primitive cubes represent only a small part (one eighth) of the f.c.c. unit cell. Description of the fluorite structure as a primitive cubic array of anions with alternate cube body centres occupied by cations shows a similarity to the CsCl structure (next section). This also has a primitive cubic array of anions, but, instead, cations occupy all the cube body centre sites.

It is very often desirable to be able to make calculations of bond and other interatomic distances in crystal structures. This is usually a straightforward exercise, especially for crystals which have orthogonal unit cells (i.e. $\alpha = \beta = \gamma = 90°$), and involves simple trigonometric calculations. For example, in the rock salt structure, the anion–cation distance is $a/2$ and the anion–anion distance is

Table 1.9 *Some compounds with fluorite and antifluorite structure*

Fluorite structure				Antifluorite structure			
	$a(\text{Å})$		$a(\text{Å})$		$a(\text{Å})$		$a(\text{Å})$
CaF_2	5.4626	PbO_2	5.349	Li_2O	4.6114	K_2O	6.449
SrF_2	5.800	CeO_2	5.4110	Li_2S	5.710	K_2S	7.406
$SrCl_2$	6.9767	PrO_2	5.392	Li_2Se	6.002	K_2Se	7.692
BaF_2	6.2001	ThO_2	5.600	Li_2Te	6.517	K_2Te	8.168
$BaCl_2$	7.311	UO_2	5.372	Na_2O	5.55	Rb_2O	6.74
CdF_2	5.3895	NpO_2	5.4334	Na_2S	6.539	Rb_2S	7.65
HgF_2	5.5373	CmO_2	5.3598	Na_2Se	6.823		
EuF_2	5.836	PuO_2	5.386	Na_2Te	7.329		
$\beta\text{-}PbF_2$	5.940	AmO_2	5.376				

Table 1.10 *Calculation of interatomic distances in some simple structures*

Structure type	Distance	Number of such distances	Magnitude of distance in terms of unit cell dimensions
Rock salt (cubic)	Na–Cl	6	$a/2 = 0.5a$
	Cl–Cl	12	$a/\sqrt{2} = 0.707a$
	Na–Na	12	$a/\sqrt{2} = 0.707a$
Zinc blende (cubic)	Zn–S	4	$a\dfrac{\sqrt{3}}{4} = 0.433a$
	Zn–Zn	12	$a/\sqrt{2} = 0.707a$
	S–S	12	$a/\sqrt{2} = 0.707a$
Fluorite (cubic)	Ca–F	4 or 8	$a\dfrac{\sqrt{3}}{4} = 0.433a$
	Ca–Ca	12	$a/\sqrt{2} = 0.707a$
	F–F	6	$a/2 = 0.5a$
Wurtzite* (hexagonal)	Zn–S	4	$a\sqrt{\dfrac{3}{8}} = 0.612a = \dfrac{3c}{8} = 0.375c$
	Zn–Zn	12	$a = 0.612c$
	S–S	12	$a = 0.612c$
Nickel arsenide* (hexagonal)	Ni–As	6	$a/\sqrt{2} = 0.707a = 0.433c$
	As–As	12	$a = 0.612c$
	Ni–Ni	2	$c/2 = 0.5c = 0.816a$
	Ni–Ni	6	$a = 0.612c$
Caesium chloride (cubic)	Cs–Cl	8	$a\dfrac{\sqrt{3}}{2} = 0.866a$
	Cs–Cs	6	a
	Cl–Cl	6	a
Cadmium iodide (hexagonal)	Cd–I	6	$a/\sqrt{2} = 0.707a = 0.433c$
	I–I	12	$a = 0.612c$
	Cd–Cd	6	$a = 0.612c$

*These formulae do not necessarily apply when c/a is different from the ideal value of 1.633.

$a/\sqrt{2}$. As an aid to rapid calculation of interatomic distances, formulae are given in Table 1.10 for the more important structure types. These may be used in conjunction with the tables of unit cell dimensions (Table 1.7, etc.) to make calculations on specific compounds. Values of bond distances (to oxygen and fluorine) are given for all the elements in Appendix 4; where relevant a selection of values, for different coordination numbers and oxidation states is given.

Consideration of the anion and cation arrangements in the three structure types described above shows that the concept of c.p. anions with cations in interstitial sites begins to break down in the fluorite structure. Thus, while the antifluorite structure of, for example, Na_2O may be regarded as containing cubic close packed O^{2-} ions with Na^+ ions in tetrahedral sites, in the fluorite structure of, for example, CaF_2, it is necessary to regard the Ca^{2+} ions as forming the c.c.p. array with the F^- ions occupying tetrahedral interstitial sites. In CaF_2, although the Ca^{2+} ions have the same arrangement as in a c.c.p. arrangement of spheres, the Ca^{2+} ions are well separated from each other; from Tables 1.9 and 1.10, Ca–Ca = 3.86 Å, which is much larger than the diameter of a Ca^{2+} ion (depending on which table of ionic radii is consulted, the diameter of a Ca^{2+} ion is in the range ~ 2.2 to 2.6 Å). Therefore, CaF_2 is a good example of a eutactic structure.

The fluorine–fluorine distance in CaF_2 is 2.73 Å, which indicates that the fluorines are approximately contacting ($r_{F^-} = 1.2$ to 1.4 Å). Although the primitive cubic array of F^- ions in CaF_2 is not a c.p. arrangement, the anions are approximately in contact and this is perhaps a more realistic way of describing the structure than as containing a c.c.p. array of Ca^{2+} ions.

Diamond

This structure type, so important to the semiconductor industry, has in fact already been described, in the form of the zinc blende or sphalerite structure, Figs 1.24 and 1.28. The diamond structure is obtained when the two elements in the zinc blende, ZnS, structure are identical, as in C. The diamond structure may therefore be described as a c.c.p. array of carbon atoms, with one set of tetrahedral sites (either T_+ or T_-) occupied also by carbon atoms. It is, however, rather artificial to make a distinction between packing and interstitial atoms since structurally they are identical. Most of the Group IV elements crystallize with the diamond structure and their lattice parameters are given in Table 1.8(b).

Wurtzite (ZnS) and nickel arsenide (NiAs)

These structures have in common a hexagonal close packed arrangement of anions and differ only in the positions of the cations, as follows:

Wurtzite: T_+(or T_-) sites occupied; T_-(or T_+), O empty
Nickel arsenide: O sites occupied; T_+, T_- empty.

These structures are the hexagonal close packed analogues of the cubic close packed, sphalerite and rock salt structures respectively. Note that there is no hexagonal equivalent of the fluorite and antifluorite structures.

Both wurtzite and nickel arsenide have hexagonal symmetry and unit cells. A unit cell containing hexagonal close packed anions is shown in Fig. 1.30(a). It is less easy to visualize and draw on paper than a cubic cell because of the γ angle of 120°. The unit cell contains two anions, one at the origin and one inside the cell.

38

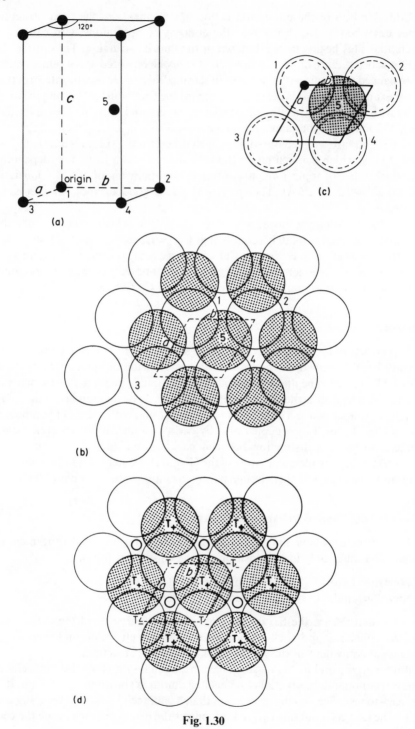

Fig. 1.30

$$T_- \; : \; 0,0, \; \tfrac{3}{8} \quad ; \quad \tfrac{1}{3}, \tfrac{2}{3}, \tfrac{7}{8}$$

$$T_+ \; : \; \tfrac{1}{3}, \tfrac{2}{3}, \tfrac{1}{8} \quad ; \quad 0,0, \; \tfrac{5}{8}$$

$$O \quad : \; \tfrac{2}{3}, \tfrac{1}{3}, \tfrac{1}{4} \quad ; \quad \tfrac{2}{3}, \tfrac{1}{3}, \tfrac{3}{4}$$

$$ANION \; : \; 0,0,0 \quad \quad \tfrac{1}{3}, \tfrac{2}{3}, \tfrac{1}{2}$$

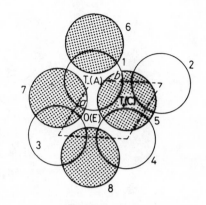

(e)

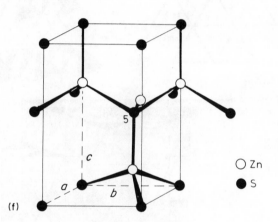

(f)

○ Zn
● S

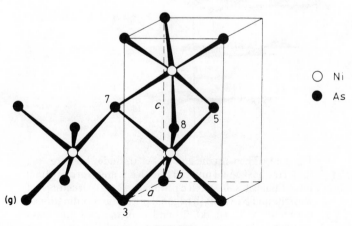

(g)

○ Ni
● As

Fig. 1.30

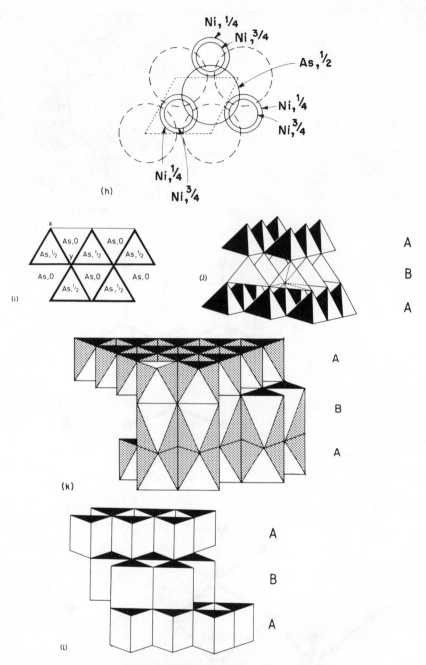

Fig. 1.30 The wurtzite and nickel arsenide structures: (a), (b), (c) the hexagonal unit cell of an h.c.p. anion array; (d), (e) interstitial sites in an h.c.p. array; (f), (g) structures of wurtzite and NiAs; (h), (i) trigonal prismatic coordination of arsenic in NiAs; (j), (k), (l) models of the ZnS and NiAs structures showing the arrangement and linkages of the polyhedra

Their coordinates are:

$$0,0,0; \qquad \tfrac{1}{3},\tfrac{2}{3},\tfrac{1}{2}$$

In Fig. 1.30(b) is shown a projection down c of the same structure. Close packed layers occur in the basal plane, i.e. at $c = 0$ (open circles) and at $c = \tfrac{1}{2}$ (shaded circles). The layer arrangement at $c = 0$ is repeated at $c = 1$ and therefore the stacking sequence is hexagonal... ABABA.... Atoms 1 to 4 in the basal plane outline the base of the unit cell (dashed).

The contents of one unit cell are shown in Fig. 1.30(c). Dashed circles represent atoms at the top four corners of the unit cell, i.e. at $c = 1$.

In metals which have hexagonal close packed structures, metal atoms are in contact, e.g. 1 and 2, 1 and 4, 1 and 5. In eutactic close packed ionic structures, however, the anions may be pushed apart by the cations in the interstitial sites and anion–anion contacts do not usually occur. Assuming for the moment that the anions are in contact, then the hexagonal unit cell has a definite shape given by the ratio $c/a = 1.633$. This is because a is equal to the shortest distance X–X, i.e. the diameter of an anion, and c is equal to twice the vertical height of a tetrahedron comprising four anions. The ratio c/a may then be calculated by geometry (Appendix 3).

The interstitial sites available for cations in a hexagonal close packed anion array are shown in Fig. 1.30(d). Since the cell contains two anions, there must be two of each type of interstitial site, T_+, T_- and O. Thus, in Fig. 1.30(e), a T_- site, A, occurs along the c edge of the cell at height $\tfrac{3}{8}$ above anion 1 at the origin. This T_- site is coordinated to three anions, shaded (5 to 7), at $c = \tfrac{1}{2}$ and to one anion (1) at the corner, $c = 0$. The tetrahedron so formed therefore points downwards. The position of the interstitial site inside this tetrahedron is at the centre of gravity of the tetrahedron, i.e. at one quarter of the vertical distance from base to apex (Appendix 3). Since the apex and base are at $c = 0$ and $c = \tfrac{1}{2}$, this T_- site is at $c = \tfrac{3}{8}$. In practice the occupation of this T_- site in the wurtzite structure may not be at exactly $0.375c$. For those structures which have been studied accurately (Table 1.11), values range from 0.345 to 0.385; the letter u is used to represent the fractional c value.

Table 1.11 *Some compounds with the wurtzite structure.* (Data taken from Wyckoff, 1971, Vol. 1)

	a(Å)	c(Å)	u	c/a		a(Å)	c(Å)	u	c/a
ZnO	3.2495	5.2069	0.345	1.602	AgI	4.580	7.494		1.636
ZnS	3.811	6.234		1.636	AlN	3.111	4.978	0.385	1.600
ZnSe	3.98	6.53		1.641	GaN	3.180	5.166		1.625
ZnTe	4.27	6.99		1.637	InN	3.533	5.693		1.611
BeO	2.698	4.380	0.378	1.623	TaN	3.05	4.94		1.620
CdS	4.1348	6.7490		1.632	NH_4F	4.39	7.02	0.365	1.600
CdSe	4.30	7.02		1.633	SiC	3.076	5.048		1.641
MnS	3.976	6.432		1.618	MnSe	4.12	6.72	1.631	

The three anions (5 to 7) at $c = \frac{1}{2}$ that form the base of this T_- site, A, also form the base of a T_+ site, B (not shown), centred at $0, 0, \frac{5}{8}$. The apex of the latter tetrahedron is the anion at the top corner with coordinates $0, 0, 1$. Another T_+ site, C, is shown in (e) at $\frac{1}{3}, \frac{2}{3}, \frac{1}{8}$. It is coordinated to anions 1, 2 and 4 in the basal plane at three corners of the unit cell and to anion 5 at $\frac{1}{3}, \frac{2}{3}, \frac{1}{2}$. The triangular base of this site, at $c = 0$, is shared with a T_- site underneath (not shown) at $\frac{1}{3}, \frac{2}{3}, -\frac{1}{8}$. The equivalent T_- site that lies inside the unit cell, D (not shown) is at $\frac{1}{3}, \frac{2}{3}, \frac{7}{8}$.

Octahedral site E, in (e) is coordinated to anions 1, 3 and 4 at $c = 0$ and to anions 5, 7 and 8 at $c = \frac{1}{2}$. The centre of gravity of the octahedron lies midway between these two groups of anions and has coordinates $\frac{2}{3}, \frac{1}{3}, \frac{1}{4}$. The second octahedral site, F (not shown) in the cell lies above E at $c = \frac{3}{4}$ and has coordinates $\frac{2}{3}, \frac{1}{3}, \frac{3}{4}$. The three anions 5, 7 and 8 are therefore common to the two octahedra.

The coordination environments of the cations in wurtzite and NiAs are emphasized in (f) and (g). Zinc is shown in T_+ sites and forms ZnS_4 tetrahedra, (f), which link up at their corners to form a three-dimensional network, as in (j). A similar structure results on considering the tetrahedra formed by four zinc atoms around a sulphur and the manner in which these SZn_4 tetrahedra are linked. The tetrahedral environment of sulphur (5) is shown in (f). The SZn_4 tetrahedron which it forms points downwards, in contrast to the ZnS_4 tetrahedra which all point upwards; on turning the SZn_4 tetrahedra upside down, however, the same structure results.

Comparing larger-scale models of the zinc blende (Fig. 1.28b) and wurtzite (Fig. 1.30j) structures, they are clearly very similar and both can be regarded as networks of tetrahedra. In zinc blende, layers of tetrahedra form an ABC stacking sequence and the orientation of the tetrahedra within each layer is identical. In wurtzite, however, the layers form an AB sequence and alternate layers are rotated by $180°$ about c relative to each other.

The $NiAs_6$ octahedra in NiAs are shown in Fig. 1.30(g). They share one pair of opposite faces (e.g. the face formed by arsenic ions 5, 7 and 8) to form chains of face-sharing octahedra that run parallel to c. In the ab plane, however, the octahedra share only common edges: arsenic ions 3 and 7 are shared between two octahedra such that chains of edge-sharing octahedra form parallel to b. Similarly, chains of edge-sharing octahedra form parallel to a (not shown). A more extended view of the octahedra and their linkages is shown in (k).

The NiAs structure is unusual in that the anions and cations have the same coordination number but do not have the same coordination environment. In the other AB structures—rock salt, sphalerite, wurtzite and CsCl—anions and cations are interchangeable and have the same coordination number and environment. Since the cation:anion ratio is 1:1 in NiAs and the nickel coordination is octahedral, the arsenic ions must also be six coordinate. However, the six nickel neighbours are arranged as in a trigonal prism and not octahedrally. This is shown for arsenic at $c = \frac{1}{2}$ in (h), which is coordinated to three nickel ions at $c = \frac{1}{4}$ and to three at $c = \frac{3}{4}$. The two sets of nickel ions are superposed in projection down c and give the trigonal prismatic coordination for arsenic. (Note that in a similar projection for octahedral coordination, the two sets of three outer atoms appear to be staggered relative to each other, as in (e) for octahedral site E.)

The NiAs structure may also be regarded as built of $AsNi_6$ trigonal prisms, therefore, which link up by sharing edges to form a three-dimensional array. In (i) each triangle represents a prism in projection down c. The prism edges that run parallel to c, i.e. those formed by nickel ions at $c = \frac{1}{4}$ and $\frac{3}{4}$ in (h), are shared between three prisms. Thus the vertical edge at y (in projection it is a point) is shared between three prisms and therefore three arsenic ions at $c = \frac{1}{2}$. Prism edges that lie in the ab plane are shared between only two prisms, however. In (i), the edge xy is shared between arsenic ions at $c = \frac{1}{2}$ and $c = 0$.

The structure may be regarded as built up of layers of prisms; two layers are shown in (i) centred at $c = \frac{1}{2}$ and $c = 0$ and are rotated by $180°$ about c relative to each other. The next layer of prisms, at $c = 1$, is in the same orientation as the layer at $c = 0$ and hence, an ... ABABA... hexagonal stacking sequence arises. This is shown further in (1).

A selection of compounds which have wurtzite and NiAs structures is given in Tables 1.11 and 1.12 with values of their hexagonal cell parameters a and c. The wurtzite structure is formed mainly by chalcogenides of some divalent metals and may be regarded as a fairly ionic structure. The NiAs structure is a more metallic structure and is adopted by a variety of intermetallic compounds and by some transition metal chalcogenides (S, Se, Te). The value of the ratio c/a is approximately constant in the family of wurtzite structures but varies considerably between the different compounds with the NiAs structure. This is associated with the presence of metallic bonding which arises from metal–metal interactions in the c direction, as follows:

Table 1.12 *Some compounds with the NiAs structure.* (Data taken from Wyckoff, 1971, Vol. 1)

	$a(\text{Å})$	$c(\text{Å})$	c/a		$a(\text{Å})$	$c(\text{Å})$	c/a
NiS	3.4392	5.3484	1.555	CoS	3.367	5.160	1.533
NiAs	3.602	5.009	1.391	CoSe	3.6294	5.3006	1.460
NiSb	3.94	5.14	1.305	CoTe	3.886	5.360	1.379
NiSe	3.6613	5.3562	1.463	CoSb	3.866	5.188	1.342
NiSn	4.048	5.123	1.266	CrSe	3.684	6.019	1.634
NiTe	3.957	5.354	1.353	CrTe	3.981	6.211	1.560
FeS	3.438	5.880	1.710	CrSb	4.108	5.440	1.324
FeSe	3.637	5.958	1.638	MnTe	4.1429	6.7031	1.618
FeTe	3.800	5.651	1.487	MnAs	3.710	5.691	1.534
FeSb	4.06	5.13	1.264	MnSb	4.120	5.784	1.404
δ'-NbN*	2.968	5.549	1.870	MnBi	4.30	6.12	1.423
PtB*	3.358	4.058	1.208	PtSb	4.130	5.472	1.325
PtSn	4.103	5.428	1.323	PtBi	4.315	5.490	1.272

*Anti-NiAs structure.

Let us consider the environment of nickel and arsenic ions.
Each arsenic is surrounded by (Table 1.10):
  6 nickel ions (in a trigonal prism) at distance $0.707a$
  12 arsenic ions (h.c.p. arrangement) at distance a
Each nickel is surrounded by:
  6 arsenic ions (octahedrally) at distance $0.707a$
  2 nickel ions (linearly, parallel to c) at distance $0.816a$ (i.e. $c/2$)
  6 nickel ions (hexagonally, in ab plane) at distance a

The main effect of changing the value of the c/a ratio is to alter the nickel–nickel distance parallel to c. Thus, in FeTe, $c/a = 1.49$, and hence the iron–iron distance parallel to c is reduced to $0.745a$ (i.e. $c/2 = \frac{1}{2}(1.49a)$), thereby bringing these iron atoms into closer contact and increasing the metallic bonding in the c direction. Quantitative calculations of the effect of changing the c/a ratio are more difficult to make since it is not readily possible to distinguish between, for example, an increase in a and a decrease in c, either of which could cause the same effect on the c/a ratio.

The non-occurrence of AX_2 compounds whose structures are the hexagonal equivalent of the cubic fluorite and antifluorite structures has already been mentioned. This may be understood by considering the various interatomic distances that would be present in such a structure. A hexagonal fluorite-like structure, AX_2, would have hexagonal packed A cations with X anions fully occupying T_+ and T_- sites. From Fig. 1.30(d), X ions would occupy, for example, sites at $0, 0, \frac{3}{8}(T_-)$ and $0, 0, \frac{5}{8}(T_+)$, thereby giving an X–X distance of $c/4 = 0.25c$. This compares with the A–X distance (as in wurtzite, Table 1.10) of $0.375c$. Since the shortest interatomic separations in ionic structures are almost always cation–anion contacts, it is most unlikely that a structure would exist in which either anion–anion distances or cation–cation distances were much shorter than anion–cation distances.

Caesium chloride (CsCl)

The unit cell of CsCl is shown in Fig. 1.31. It is primitive cubic, containing Cl^- ions at the corners and a Cs^+ ion at the body centre, or vice versa (note that it is *not* body centred cubic since there are different ions at corner and body centre positions). The coordination numbers of both Cs^+ and Cl^- are eight with interatomic distances of $0.866a$ (Table 1.10). The CsCl structure cannot be regarded as close packed. In a c.p. structure, e.g. NaCl, each Cl^- has twelve other Cl^- ions as next nearest neighbours, which is a characteristic feature of c.p. (or eutactic c.p.) structures. In CsCl, however, each Cl has only six Cl^- ions as next nearest neighbours (arranged octahedrally). Some compounds which exhibit the CsCl structure are given in Table 1.13. They fall into two groups, halides of large monovalent elements and a variety of intermetallic compounds.

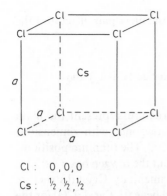

Cl : 0 , 0 , 0

Cs : ½ , ½ , ½

Fig. 1.31 The primitive cubic
unit cell of CsCl

Table 1.13 *Some compounds with the CsCl structure*

	a(Å)		a(Å)
CsCl	4.123	CuZn	2.945
CsBr	4.286	CuPd	2.988
CsI	4.5667	AuMg	3.259
CsCN	4.25	AuZn	3.19
NH_4Cl	3.8756	AgZn	3.156
NH_4Br	4.0594	LiAg	3.168
TlCl	3.8340	AlNi	2.881
TlBr	3.97	LiHg	3.287
TlI	4.198	MgSr	3.900

Other AX structures

There are five main AX structure types: rock salt, CsCl, NiAs, sphalerite and wurtzite, each of which is found in a large number of compounds. There is also a considerable number of less common AX structures. Some may be regarded as distorted variants of one of the main structure types, e.g.:

(a) FeO at low temperatures, < 90 K, has a rock salt structure which has undergone a slight rhombohedral distortion (the α angle is increased from 90 to 90.07° by a slight compression along one threefold axis). This rhombohedral distortion is associated with magnetic ordering in FeO at low temperatures (Chapter 8).

(b) TlF has a rock salt related structure in which the f.c.c. cell is distorted into a face centred orthorhombic cell by changing the lengths of all three cell axes by different amounts.

(c) NH_4CN has a distorted CsCl structure (as in NH_4Cl) in which the CN^- ions do not assume spherical symmetry but are oriented parallel to face diagonals. This distorts the symmetry to tetragonal and effectively increases *a* relative to *c*.

Other AX compounds have completely different structures, e.g.:

(a) Compounds of the d^8 ions, Pd and Pt (in PdO, PtS, etc.) often have a square planar coordination for the cation with tetragonal or orthorhombic symmetry for the structures as a whole; d^9 ions show this effect as well, e.g. Cu in CuO.

(b) Compounds of heavy p-block ions in their lower oxidation states (Pb^{2+}, Bi^{3+}, etc.) often have distorted polyhedra in which the cation exhibits the *inert pair effect*. Thus PbO and SnO have structures in which the M^{2+} ion has four O^{2-} neighbours to one side giving a square pyramidal arrangement; the O^{2-} coordination is a more regular tetrahedron of M^{2+} ions. InBi has a similar structure in which Bi^{3+} is the ion with the inert pair effect and the irregular coordination.

Rutile (TiO$_2$), cadmium iodide (CdI$_2$), cadmium chloride (CdCl$_2$) and caesium oxide (Cs$_2$O)

The title structures, together with the fluorite structure, represent the main AX$_2$ structure types. The unit cell of rutile is tetragonal with dimensions $a = b = 4.594$ Å, $c = 2.958$ Å, and is shown in Fig. 1.32(a). The titanium positions, two per cell, are fixed at the corner and body centre, but the oxygen has a variable parameter, x, whose value must be determined experimentally. Crystal structure determination and refinement gave the x values listed beneath Fig. 1.32(a) for the four oxygens in the unit cell.

The body centre titanium at $(\frac{1}{2}, \frac{1}{2}, \frac{1}{2})$ is coordinated octahedrally to six oxygens. Four of these—two at 0 and two at 1 directly above the two at 0—are coplanar with titanium. Two oxygens at $z = \frac{1}{2}$ are collinear with titanium and form the axes of the octahedron. The corner titaniums are also octahedrally coordinated but the orientation of their octahedra is different (b). The oxygens are coordinated trigonally to three titaniums: e.g. oxygen at 0 in (a) is coordinated to titanium atoms at the corner, at the body centre and at the body centre of the cell below.

Since the oxygen atoms form the corners of TiO$_6$ octahedra this means that each corner oxygen is shared between three octahedra. The octahedra are linked by sharing edges and corners to form a three-dimensional framework. Consider the TiO$_6$ octahedron in the centre of the cell in (b); a similar octahedron in identical orientation occurs in the cells above and below such that octahedra in adjacent cells share edges to form infinite chains parallel to c. For example, titanium atoms at $z = +\frac{1}{2}$ and $-\frac{1}{2}$ in adjacent cells are both coordinated to two oxygens at $z = 0$. Chains of octahedra are similarly formed by the octahedra centred at the corners of the unit cell. The two types of chains, which differ in

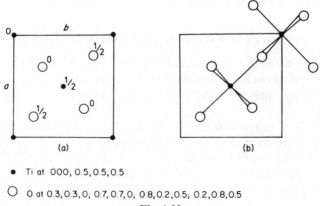

(a)   (b)

●   Ti at 000; 0.5, 0.5, 0.5

○   O at 0.3, 0.3, 0; 0.7, 0.7, 0; 0.8, 0.2, 0.5; 0.2, 0.8, 0.5

Fig. 1.32

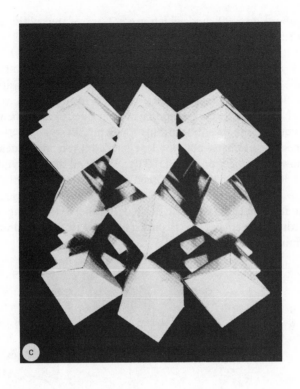

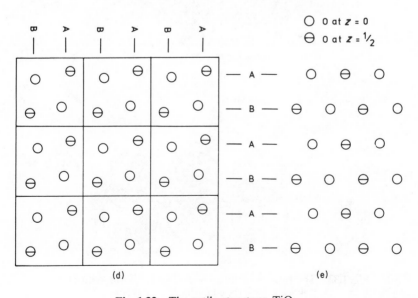

Fig. 1.32 The rutile structure, TiO_2

48

orientation about z by $90°$ and which are $c/2$ out of step with each other, are linked by their corners to form a three-dimensional framework (c).

The rutile structure is also commonly described as a distorted hexagonal close packed oxide array with half the octahedral sites occupied by titanium. A 3×3 block of unit cells is shown in (d) with only the oxygen positions marked. Corrugated close packed layers occur, both horizontally and vertically. This contrasts with the undistorted hexagonal close packed arrangement (e), in which the layers occur in one orientation only (horizontally).

The octahedral sites between two close packed layers in an ideal h.c.p. anion array are shown in projection in Fig. 1.33(a). While all these sites are occupied in NiAs (Fig. 1.30h), only half are occupied in rutile and in such a manner that alternate rows of octahedral sites are full and empty. The orientation of the tetragonal unit cell in rutile is shown dashed. Parallel to the tetragonal c axis, the TiO_6 octahedra share edges. This is shown in (b) for two octahedra with oxygens 1 and 2 forming the common edge.

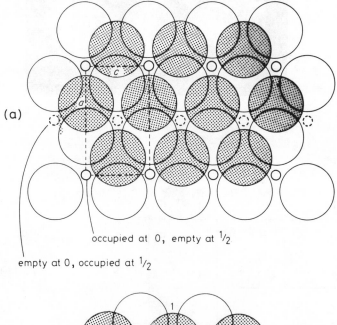

occupied at 0, empty at $1/2$

empty at 0, occupied at $1/2$

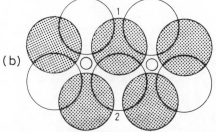

Fig. 1.33 The rutile structure: (a) in idealized form, with planar oxide layers; (b) edge-sharing octahedra

Recently, an alternative way of describing the packing arrangement of oxide ions in rutile has been proposed. The oxide ion arrangement is a slightly distorted version of a new type of packing called *primitive tetragonal packing* (p.t.p.), which is characterized by fourfold symmetry and a sphere coordination number of *eleven.* This contrasts with hexagonal and cubic close packing which have a packing sphere coordination number of twelve and body centred tetragonal

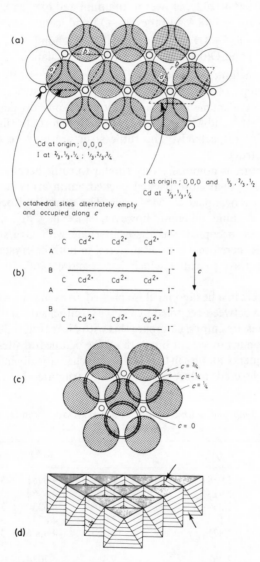

(a)

Cd at origin; 0,0,0
I at $^2/_3, ^1/_3, ^1/_4$; $^1/_3, ^2/_3, ^3/_4$

I at origin ; 0,0,0 and $^1/_3, ^2/_3, ^1/_2$
Cd at $^2/_3, ^1/_3, ^1/_4$

octahedral sites alternately empty
and occupied along c

(b)

(c)

(d)

Fig. 1.34 The CdI$_2$ structure: (a) the unit cell, with two possible choices of origin; (b) the layer stacking sequence; (c) the coordination environment of I; (d) a layer of close packed octahedra

packing which has a coordination number of ten. More details are given in the reading list.

The bond lengths in TiO_2 may be calculated by geometry; e.g. for the Ti—O bond length between titanium at $(\frac{1}{2}, \frac{1}{2}, \frac{1}{2})$ and oxygen at $(0.3, 0.3, 0)$ the difference in a and b parameters of titanium and oxygen is $(\frac{1}{2} - 0.3)a = 0.92$ Å. From a right-angled triangle calculation, the titanium–oxygen distance in projection down c (Fig. 1.32a) is $\sqrt{0.92^2 + 0.92^2}$. However, titanium and oxygen have a difference in c height of $(\frac{1}{2} - 0)c = 1.48$ Å and the Ti—O bond length is therefore equal to $\sqrt{0.92^2 + 0.92^2 + 1.48^2} = 1.97$ Å. The axial Ti—O bond length between, for example, $Ti(\frac{1}{2}, \frac{1}{2}, \frac{1}{2})$ and $O(0.8, 0.2, 0.5)$ is easier to calculate because both atoms are at the same c height. It is equal to $\sqrt{2(0.3 \times 4.594)^2} = 1.95$ Å.

Two main groups of compounds exhibit the rutile structure (Table 1.14): oxides of some tetravalent metal ions and fluorides of some divalent metal ions. In both cases, these M^{4+} and M^{2+} ions are too small to form the fluorite structure with O^{2-} and F^-, respectively. The rutile structure may be regarded as an essentially ionic structure.

The CdI_2 structure is nominally very similar to rutile because it also may be described as a hexagonal close packed packed anion array in which half the octahedral sites are occupied by M^{2+} ions. The manner of occupancy of the octahedral sites is quite different, however, since in CdI_2, entire layers of octahedral sites are occupied and these alternate with layers of empty sites (Fig. 1.34). CdI_2 is therefore a layered material in both its crystal structure and properties, in contrast to rutile which has a more rigid, three-dimensional character.

Two iodide layers in a hexagonal close packed array are shown in (a) with the octahedral sites in between occupied by Cd^{2+}. To either side of the iodide layers, the octahedral sites are empty. Compare this with NiAs (Fig. 1.30d and h) which has the same anion arrangement but with all the octahedral sites occupied. The layer stacking sequence along c in CdI_2 is shown schematically in Fig. 1.34(b) and emphasizes the layered nature of the CdI_2 structure. I^- layers form an

Table 1.14 *Some compounds with the rutile structure.* (Data taken from Wyckoff, 1971, Vol. 1)

	a(Å)	c(Å)	x		a(Å)	c(Å)	x
TiO_2	4.5937	2.9581	0.305	CoF_2	4.6951	3.1796	0.306
CrO_2	4.41	2.91		FeF_2	4.6966	3.3091	0.300
GeO_2	4.395	2.859	0 307	MgF_2	4.623	3.052	0.303
IrO_2	4.49	3.14		MnF_2	4.8734	3.3099	0.305
β-MnO_2	4.396	2.871	0.302	NiF_2	4.6506	3.0836	0.302
MoO_2	4.86	2.79		PdF_2	4.931	3.367	
NbO_2	4.77	2.96		ZnF_2	4.7034	3.1335	0.303
OsO_2	4.51	3.19		SnO_2	4.7373	3.1864	0.307
PbO_2	4.946	3.379		TaO_2	4.709	3.065	
RuO_2	4.51	3.11		WO_2	4.86	2.77	

...ABABA... stacking sequence. Cd^{2+} ions occupy octahedral sites which may be regarded as the C positions relative to the AB positions for I^-. The CdI_2 structure may be regarded as a sandwich structure in which Cd^{2+} ions are sandwiched between layers of I^- ions and adjacent sandwiches are held together by weak van der Waals bonds between the layers of I^- ions. In this sense, CdI_2 has certain similarities to molecular structures. For example, solid CCl_4 has strong C—Cl bonds within the molecule but only weak Cl—Cl bonds between adjacent molecules. Because the intermolecular forces are weak, CCl_4 is volatile with a low melting and boiling point. In the same way, CdI_2 may be regarded as an infinite sandwich 'molecule' in which there are strong Cd—I bonds within the molecule but weak van der Waals bonds between adjacent molecules.

The coordination of the I^- ion in CdI_2 is shown in (c). An I^- ion at $c = \frac{1}{4}$ (shaded) has three close Cd^{2+} neighbours to one side at $c = 0$. The next nearest neighbours are the twelve I^- ions that form the h.c.p. array.

The layered nature of CdI_2 is emphasized further by constructing a model from polyhedra: CdI_6 octahedra link up at their edges to form infinite sheets (d), but there are no direct polyhedral linkages between adjacent sheets. A self-supporting, three-dimensional model of octahedra cannot be made for CdI_2, therefore, unlike rutile, for example.

Some compounds which have the CdI_2 structure are listed in Table 1.15. This structure occurs mainly in transition metal iodides and also with some bromides, chlorides and hydroxides.

The cadmium chloride structure is closely related to that of CdI_2 and differs only in the nature of the anion packing: Cl^- ions are cubic close packed in $CdCl_2$ whereas I^- ions are hexagonal close packed in CdI_2. $CdCl_2$ and CdI_2 form a pair of closely related structures, therefore, in the same way that wurtzite and zinc blende or rock salt and nickel arsenide differ only in the anion stacking sequence.

Table 1.15 *Some compounds with the CdI_2 structure.* (Data taken from Wyckoff, 1971, Vol. 1)

	$a(\text{Å})$	$c(\text{Å})$		$a(\text{Å})$	$c(\text{Å})$
CdI_2	4.24	6.84	VBr_2	3.768	6.180
CaI_2	4.48	6.96	$TiBr_2$	3.629	6.492
CoI_2	3.96	6.65	$MnBr_2$	3.82	6.19
FeI_2	4.04	6.75	$FeBr_2$	3.74	6.17
MgI_2	4.14	6.88	$CoBr_2$	3.68	6.12
MnI_2	4.16	6.82	$TiCl_2$	3.561	5.875
PbI_2	4.555	6.977	VCl_2	3.601	5.835
ThI_2	4.13	7.02	$Mg(OH)_2$	3.147	4.769
TiI_2	4.110	6.820	$Ca(OH)_2$	3.584	4.896
TmI_2	4.520	6.967	$Fe(OH)_2$	3.258	4.605
VI_2	4.000	6.670	$Co(OH)_2$	3.173	4.640
YbI_2	4.503	6.972	$Ni(OH)_2$	3.117	4.595
$ZnI_2(I)$	4.25	6.54	$Cd(OH)_2$	3.48	4.67

The CdCl$_2$ structure may be represented by a hexagonal unit cell, although a smaller rhombohedral cell can also be chosen. The base of the hexagonal cell is of similar size and shape to that in CdI$_2$ but the c axis of CdCl$_2$ is three times as long as c in CdI$_2$. This is because in CdCl$_2$, the Cd^{2+} positions, and the CdCl$_6$ octahedra, are staggered along c and give rise to a three-layer repeat for Cd^{2+} ions (CBA) and a six-layer repeat for Cl$^-$ ions (ABCABC) (Fig. 1.35). In contrast, in CdI$_2$, the Cd^{2+} positions and the CdI$_6$ octahedra are stacked on top of each other and the c repeat contains only two I$^-$ layers (AB) and one Cd^{2+} layer (C).

The unit cell of CdCl$_2$ in projection down c is shown in Fig. 1.35(b). Chloride layers occur at $c = 0(A)$, $\frac{2}{12}(B)$ and $\frac{4}{12}(C)$ and this sequence is repeated at $c = \frac{6}{12}$, $\frac{8}{12}$ and $\frac{10}{12}$. Between those Cl$^-$ layers at 0 and $\frac{2}{12}$ are Cd^{2+} ions in the octahedral sites at $\frac{1}{12}$. However, the octahedral sites between Cl$^-$ layers at $\frac{2}{12}$ and $\frac{4}{12}$ are empty (these sites, at $c = \frac{3}{12}$, are directly below the Cd^{2+} ions at $\frac{9}{12}$).

The CdCl$_2$ structure is a layered structure, similar to CdI$_2$, and many of the comments made about structure and bonding in CdI$_2$ apply equally well to CdCl$_2$. Some compounds which have the CdCl$_2$ structure are given in Table 1.16; it occurs with a variety of transition metal halides.

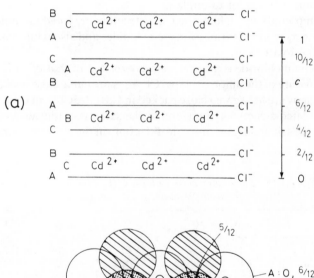

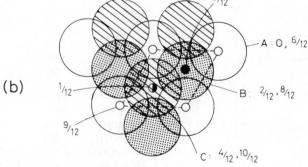

Fig. 1.35 The CdCl$_2$ structure

Table 1.16 *Some compounds with the CdCl$_2$ structure.* (Data taken from Wyckoff, 1971, Vol. 1)

	a(Å)	c(Å)		a(Å)	c(Å)
CdCl$_2$	3.854	17.457	NiCl$_2$	3.543	17.335
CdBr$_2$	3.95	18.67	NiBr$_2$	3.708	18.300
CoCl$_2$	3.544	17.430	NiI$_2$	3.892	19.634
FeCl$_2$	3.579	17.536	ZnBr$_2$	3.92	18.73
MgCl$_2$	3.596	17.589	ZnI$_2$	4.25	21.5
MnCl$_2$	3.686	17.470	Cs$_2$O*	4.256	18.99

*Cs$_2$O has an anti-CdCl$_2$ structure.

The structure of Cs$_2$O is unusual since it may be regarded as an anti-CdCl$_2$ structure (as in fluorite and antifluorite structures). Cs$^+$ ions form cubic close packed layers and oxide ions occupy the octahedral sites between alternate pairs of caesium layers. This raises some interesting questions because caesium is the most electropositive element and caesium salts are usually regarded as highly ionic. However, the structure of Cs$_2$O clearly shows that Cs$^+$ ions are not surrounded by anions, as expected for an ionic structure, but have only three oxide neighbours, all located at one side. The structure is held together, in three dimensions, by bonding between caesium ions in adjacent layers.

It may be that the structure of Cs$_2$O does not reflect any peculiar type of bonding but rather that it is the only structural arrangement which is feasible for a compound of this formula and for ions of this size. Thus, from the formula, the coordination numbers of Cs$^+$ and O^{2-} must be in the ratio of 1:2; since Cs$^+$ is considerably larger than O^{2-}, the maximum possible coordination number of oxygen by caesium may be six, which then leads to a coordination number of three for Cs$^+$.

A related question arises with the structures of the other alkali oxides, in particular K$_2$O and Rb$_2$O. These have the antifluorite structure with coordination numbers of four and eight for M$^+$ and O^{2-}, respectively. These compounds are unusual since Rb$^+$ is normally far too large a cation to enter into tetrahedral coordination with oxygen. However, if there is no feasible alternative structure, then perhaps Rb$^+$ ions have no choice but to enter the tetrahedral sites. With Cs$_2$O, tetrahedral coordination of Cs$^+$ by O^{2-} is probably impossible and hence it adopts the anti-CdCl$_2$ structure rather than the antifluorite structure. Thermodynamic data support these observations since neither Cs$_2$O nor Rb$_2$O are very stable: they oxidize readily to give peroxides, M$_2$O$_2$, and superoxides, MO$_2$, which contain much larger anions.

Perovskite (SrTiO$_3$)

This very important structure type, of general formula ABX$_3$, has a primitive cubic unit cell, shown in Fig. 1.36 as a projection down one axis (a) and as an oblique projection (b). It contains Ti at the cube corners (coordinates 0, 0, 0), Sr at

54

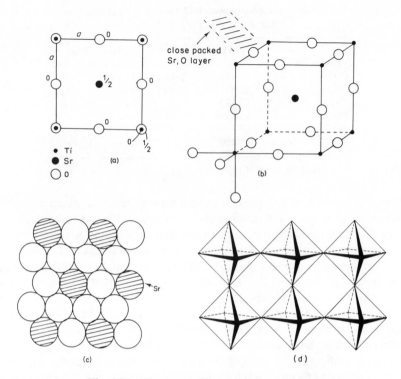

Fig. 1.36 The perovskite structure of $SrTiO_3$

the body centre (1/2, 1/2, 1/2) and oxygen at the edge centres (1/2, 0, 0; 0, 1/2, 0; 0, 0, 1/2). The coordination environment of each atom may be seen in (b) and interatomic distances calculated by simple geometry. The octahedral coordination by oxygen of one of the corner titaniums is shown. The Ti—O bond length $= a/2 = 1.953$ Å. The strontium in the cube centre is equidistant from all twelve oxygens at the centres of the edges of the unit cell. The strontium–oxygen distance is equal to half the diagonal of any cell face, i.e. $a/\sqrt{2}$ or 2.76 Å (from the geometry of triangles, the length of a cell face diagonal is equal to $\sqrt{a^2 + a^2}$).

Each oxygen has two titaniums as its nearest cationic neighbours, at a distance of 1.953 Å, and four strontium atoms, coplanar with the oxygen at a distance of 2.76 Å. However, eight other oxygens are at the same distance, 2.76 Å, as the four strontiums. It is debatable whether the oxygen coordination number is best regarded as two (linear) or as six (a grossly squashed octahedron with two short and four long distances) or as fourteen (six cations and eight oxygens). No firm recommendation is made!

Having arrived at the unit cell of $SrTiO_3$, the atomic coordinates, coordination numbers and bond distances, we now wish to view the structure on a rather larger scale. There are several questions which may be asked. Does the structure have a close packed anion arrangement? Can the structure be regarded as some kind of framework with atoms in the interstices? Answers to these questions are as follows.

Perovskite does not contain close packed oxide ions as such but the oxygens and strontiums, considered together, do form a cubic close packed array with the layers parallel to the $\{111\}$ planes (Fig. 1.36b and c). To see this, compare the perovskite structure with that of NaCl (Fig. 1.2). The latter contains Cl^- ions at the edge centre and body centre positions of the cell and is cubic close packed. By comparison, perovskite contains O^{2-} ions at the edge centres and Sr^{2+} at the body centre. The structure of the mixed Sr, O close packed layers in perovskite is such that one quarter of the atoms are strontium, arranged in a regular fashion (Fig. 1.36c).

It is quite common for fairly large cations, such as Sr^{2+} ($r = 1.1$ Å) to play apparently different roles in different structures, i.e. as twelve coordinate packing ions in perovskite or as octahedrally coordinated cations within a close packed oxide array, as in SrO (rock salt structure).

The formal relation between rock salt and perovskite also includes the Na^+ and Ti^{4+} cations as both are in octahedral sites. Whereas in NaCl all octahedral sites are occupied (corners and face centres), in perovskite only one quarter (the corner sites) are occupied.

Perovskite may also be regarded as a framework structure constructed from corner-sharing (TiO_6) octahedra and with Sr^{2+} ions placed in twelve-coordinate interstices. The octahedral coordination of one titanium is shown in Fig. 1.36b; each oxygen of this octahedron is shared with one other octahedron, such that the Ti–O–Ti arrangement is linear. In this way, octahedra are linked at their corners to form sheets (d), and neighbouring sheets are linked similarly to form a three-dimensional framework.

Several hundred oxides and halides form the perovskite structure; a selection is given in Table 1.17. The oxides contain two cations, whose combined oxidation state is six; several cation combinations are therefore possible.

As well as the cubic perovskite structure, described above, a variety of

Table 1.17 *Some compounds with the perovskite structure*

Compound	a(Å)	Compound	a(Å)
$KNbO_3$	4.007	$SrTiO_3$	3.9051
$KTaO_3$	3.9885	$SrZrO_3$	4.101
KIO_3	4.410	$SrHfO_3$	4.069
$NaNbO_3$	3.915	$SrSnO_3$	4.0334
$NaWO_3$	3.8622	$SrThO_3$	
$LaCoO_3$	3.824	$CsCaF_3$	4.522
$LaCrO_3$	3.874	$CsCdBr_3$	5.33
$LaFeO_3$	3.920	$CsCdCl_3$	5.20
$LaGaO_3$	3.875	$CsHgBr_3$	5.77
$LaVO_3$	3.99	$CsHgCl_3$	5.44

distorted, non-cubic structures exist, with tetragonal, rhombohedral, monoclinic, etc., symmetry. These lower-symmetry structures form on cooling the high-temperature cubic structures. Their structures are similar to the cubic perovskite structure, but the framework of octahedra may be slightly twisted or distorted. Some have interesting and useful properties, such as ferroelectricity (Chapter 7).

Rhenium trioxide (ReO_3) and tungsten bronzes

These structures are closely related to perovskite described above. The cubic ReO_3 structure is the same as perovskite, $SrTiO_3$, but without the body centre Sr atoms. Its unit cell contains Re at corners with oxygen at edge centres. The ReO_6 octahedra link up at their corners to form a three-dimensional framework, similar to that formed by the TiO_6 octahedra in perovskite, but now the 12-coordinate cavities are empty.

A few oxides and halides form the ReO_3 structure, Table 1.18, together with an example of the anti-ReO_3 structure in Cu_3N.

The *tungsten bronze* structure is intermediate between that of ReO_3 and perovskite. It occurs in the series, Na_xWO_3. It comprises a three-dimensional framework of WO_6 octahedra, as in ReO_3, but with some $(0 < x < 1)$ of the 12-coordinate sites occupied by Na. To accommodate this variation in stoichiometry, x, the oxidation state of tungsten is a mixture of V and VI. The formula of the bronzes may be written more completely as

$$Na_x W_x^V W_{1-x}^{VI} O_3$$

The tungsten bronzes have interesting electrical properties. At low x, the materials are pale green/yellow in colour and are semiconducting. As x rises and electrons begin to occupy the $5d$ band of tungsten, the materials become electrically conducting and exhibit metallic lustre, hence the name 'bronze'.

A wide variety of monovalent cations enter this structure; similar series occur with MoO_3 in the *molybdenum bronzes*.

Table 1.18 *Some compounds with the ReO_3 structure*

Compound	$a(\text{Å})$
ReO_3	3.734
UO_3	4.156
MoF_3	3.8985
NbF_3	3.903
TaF_3	3.9012
Cu_3N	3.807

Spinel

Several of the commercially important magnetic oxides have the spinel structure. The parent spinel is $MgAl_2O_4$. It has an essentially cubic close packed array of oxide ions with Mg^{2+}, Al^{3+} in tetrahedral and octahedral interstices, respectively. There are well over a hundred compounds with the spinel structure reported to date. Most are oxides. Some are sulphides, selenides and tellurides. A few are halides. Many different cations may be introduced into the spinel structure and several different charge combinations are possible, viz.:

2, 3	as in	$MgAl_2O_4$
2, 4	as in	Mg_2TiO_4
1, 3, 4	as in	$LiAlTiO_4$
1, 3	as in	$Li_{0.5}Al_{2.5}O_4$
1, 2, 5	as in	$LiNiVO_4$
1, 6	as in	Na_2WO_4

Similar cation combinations occur with sulphides, e.g. $2, 3:ZnAl_2S_4$ and $2, 4:Cu_2SnS_4$. With halide spinels, cations are limited to charges of 1 and 2, in order to give an overall cation: anion ratio of 3:4, e.g. Li_2NiF_4.

$MgAl_2O_4$ has a large, cubic unit cell with $a = 8.08$ Å. The cell contents are eight formula units ($Z = 8$) corresponding to '$Mg_8Al_{16}O_{32}$'. The structure is complex, therefore, and is not reproduced here. However, the structure may be described rather simply as a cubic close packed structure with certain of the interstitial sites occupied by the cations. Specifically, one half of the octahedral sites are occupied by Al and one eighth of the tetrahedral sites, both T_+ and T_-, are occupied by Mg.

A complicating factor in some other compounds with the spinel structure is that the cation distribution may vary. Two extreme types of behaviour may be distinguished. In *normal* spinels, the cations are in those sites given by the formula

$$[A]^{tet}[B_2]^{oct}O_4$$

i.e. with A in tetrahedral sites and B in octahedral sites. Examples of normal spinels are $MgAl_2O_4$ and $MgTi_2O_4$. In *inverse* spinels, half of the B ions are in tetrahedral sites, leaving the remaining B ions and all the A ions in octahedral sites, i.e.

$$[B]^{tet}[A, B]^{oct}O_4$$

Usually the two types of cations A and B are disordered over the octahedral sites. Examples of inverse spinels are $MgFe_2O_4$ and Mg_2TiO_4.

As well as the two extreme types of behaviour exhibited by the normal and inverse spinels, the complete range of intermediate cation distributions is possible and in some cases, the distribution changes with temperature. The cation distribution may be quantified in a simple way by using a parameter, γ, which

Table 1.19(a) *Some compounds with the spinel structure*

Crystal	Type	$a(\text{Å})$	Structure
$MgAl_2O_4$	2, 3	8.0800	Normal
$CoAl_2O_4$	2, 3	8.1068	Normal
$CuCr_2S_4$	2, 3	9.629	Normal
$CuCr_2Se_4$	2, 3	10.357	Normal
$CuCr_2Te_4$	2, 3	11.051	Normal
$MgTi_2O_4$	2, 3	8.474	Normal
Co_2GeO_4	2, 4	8.318	Normal
Fe_2GeO_4	2, 4	8.411	Normal
$MgFe_2O_4$	2, 3	8.389	Inverse
$NiFe_2O_4$	2, 3	8.3532	Inverse
$MgIn_2O_4$	2, 3	8.81	Inverse
$MgIn_2S_4$	2, 3	10.708	Inverse
Mg_2TiO_4	2, 4	8.44	Inverse
Zn_2SnO_4	2, 4	8.70	Inverse
Zn_2TiO_4	2, 4	8.467	Inverse
$LiAlTiO_4$	1, 3, 4	8.34	Li in 8a
$LiMnTiO_4$	1, 3, 4	8.30	Li in 8a
$LiZnSbO_4$	1, 2, 5	8.55	Li in 8a
$LiCoSbO_4$	1, 2, 5	8.56	Li in 8a

Table 1.19(b) *The spinel structure: crystallographic data*

Atom	Fractional coordinates
O	uuu; $\bar{u}\bar{u}u$; $u\bar{u}\bar{u}$; $\bar{u}u\bar{u}$ $\frac{1}{4}-u, \frac{1}{4}-u, \frac{1}{4}-u$; $\frac{1}{4}+u, \frac{1}{4}-u, \frac{1}{4}+u$; $\frac{1}{4}-u, \frac{1}{4}+u, \frac{1}{4}+u$; $\frac{1}{4}+u, \frac{1}{4}+u, \frac{1}{4}-u$ $+$ face centring
Al	$\frac{5}{8}\frac{5}{8}\frac{5}{8}$; $\frac{5}{8}\frac{7}{8}\frac{7}{8}$; $\frac{7}{8}\frac{5}{8}\frac{7}{8}$; $\frac{7}{8}\frac{7}{8}\frac{5}{8}$; $+$ face centring
Mg	000; $\frac{1}{4}\frac{1}{4}\frac{1}{4}+$ face centring

Coordination numbers:
Mg   tetrahedral, MgO_4
Al   octahedral, AlO_6
O   tetrahedral, $OMgAl_3$

corresponds to the fraction of A ions on the octahedral sites:

Normal:	$[A]^{tet}[B_2]^{oct}O_4,$	$\gamma = 0$
Inverse:	$[B]^{tet}[A, B]^{oct}O_4,$	$\gamma = 1$
Random:	$[B_{0.67}A_{0.33}]^{tet}[A_{0.67}B_{1.33}]^{oct}O_4$	$\gamma = 0.67$

The cation distribution in spinels and the degree of inversion, γ, have been studied in considerable detail. Several factors influence γ, including the site preferences of ions in terms of size, covalent bonding effects and crystal field stabilization

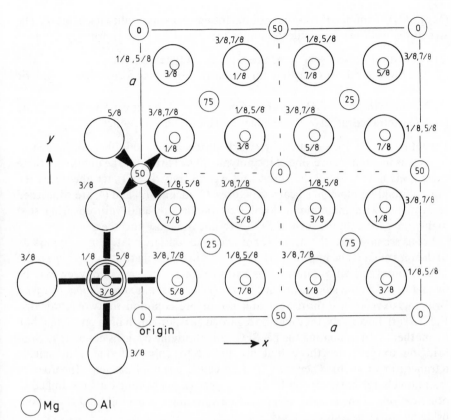

○ Mg   ○ Al

Fig. 1.37 The spinel structure: atom heights are given as either fractions or percentages of c

energies (Chapter 2). The actual γ value in any particular spinel is given by the net effect of these various parameters taken together. Some compounds with the spinel structure are given in Table 1.19a with crystallographic data in Table 1.19b; a projection of the structure is given in Fig. 1.37.

Silicate structures—some tips to understanding them

Silicates, especially many minerals, often have very complicated formulae. The purpose of this section is not to give a review of the crystal structures of silicates but simply to show that a considerable amount of structural information may be obtained from their chemical formulae. Using certain guidelines one can appreciate, without the necessity of remembering a large number of complex formulae, whether a particular silicate is a three-dimensional framework structure, whether it is sheet-like or chain-like, etc.

It is common practice to regard many silicate structures as composed of cations and silicate anions. Various types of silicate anion are possible, ranging from the extremes of isolated SiO_4^{4-} tetrahedra in orthosilicates such as olivine

(Mg_2SiO_4), to infinite three-dimensional frameworks, as in silica itself (SiO_2). The structures of the various silicate anions are based on certain principles:

(1) Almost all silicate structures are built of SiO_4 tetrahedra.
(2) The SiO_4 tetrahedra may link up by sharing corners to form larger polymeric units.
(3) No more than two SiO_4 tetrahedra may share a common corner (i.e. oxygen).
(4) SiO_4 tetrahedra never share edges or faces with each other.

Exceptions to (1) are structures in which silicon is octahedrally coordinated to oxygen as in, for instance, one of the polymorphs of SiP_2O_7. The number of these exceptions is very small, however, and we can regard SiO_4 tetrahedra as the normal building block in silicate structures. Guidelines (3) and (4) are concerned respectively with maintaining local electroneutrality and with ensuring that highly charged cations, such as Si^{4+}, are not too close together.

Let us see now how the structures of silicate anions are related, in a simple way to their formulae. The important factor in relating formula to structure type is the silicon to oxygen ratio. This ratio is variable since two types of oxygen may be distinguished in the silicate anions: *bridging oxygens* and *non-bridging oxygens*. Bridging oxygens are those that link up, or are common to, two tetrahedra, Fig. 1.38(a). Effectively, they may be regarded as belonging half to one Si and half to another Si. In evaluating the net Si:O ratio, bridging oxygens count as $\frac{1}{2}$. Non-bridging oxygens are those that are linked to only one silicon or silicate tetrahedron as in (b). They may also be called terminal oxygens. In order to maintain charge balance, non-bridging oxygens must of course also be linked to other cations in the crystal structure. In evaluating the overall Si:O ratio, non-bridging oxygens count as 1.

The overall Si:O ratio in a silicate crystal structure depends on the relative number of bridging and non-bridging oxygens. Some examples are given in Table 1.20. In these, the alkali and alkaline earth cations do not form part of the silicate anion whereas in certain other cases, e.g. in aluminosilicates, cations such as Al^{3+} may substitute for silicon in the silicate anion. The examples given in the table are all straightforward and one may deduce the type of silicate anion directly from the chemical formula.

Many other more complex examples could be given. In these, although the detailed structure cannot be deduced from the formula, one can at least get an approximate idea of the type of silicate anion. For example, in the phase

(a) (b)

Fig. 1.38 Silicate anions with (a) bridging and (b) non-bridging oxygens

Table 1.20 *Relation between chemical formula and silicate anion structure*

Si:O ratio	Number of oxygens per Si		Type of silicate anion	Examples
	bridging	non-bridging		
1:4	0	4	isolated SiO_4^{4-}	Mg_2SiO_4 olivine, Li_4SiO_4
1:3.5	1	3	dimer $Si_2O_7^{6-}$	$Ca_3Si_2O_7$ rankinite, $Sc_2Si_2O_7$ thortveite
1:3	2	2	chains $(SiO_3)_n^{2n-}$	Na_2SiO_3, $MgSiO_3$ pyroxene
			rings, e.g. $Si_3O_9^{6-}$	$CaSiO_3^*$, $BaTiSi_3O_9$ benitoite
			$Si_6O_{18}^{12-}$	$Be_3Al_2Si_6O_{18}$ beryl
1:2.5	3	1	sheets $(Si_2O_5)_n^{2n-}$	$Na_2Si_2O_5$
1:2	4	0	3D framework	$SiO_2^\dagger$

*$CaSiO_3$ is dimorphic. One polymorph has $Si_3O_9^{6-}$ rings. The other polymorph has infinite $(SiO_3)_n^{2n-}$ chains.
†The three main polymorphs of silica: quartz, tridymite and cristobalite each have a different kind of 3D framework structure.

$Na_2Si_3O_7$, the Si:O ratio is 1:2.33. This corresponds to a structure in which, on average, two-thirds of an oxygen per SiO_4 tetrahedron are non-bridging. Clearly, therefore, some of the SiO_4 tetrahedra in this structure must be composed entirely of bridging oxygens whereas others contain one non-bridging oxygen. The structure of the silicate anion would therefore be expected to be something between an infinite sheet and a 3D framework. In fact, the structure contains an infinite, double-sheet silicate anion in which two thirds of the silicate tetrahedra have one non-bridging oxygen.

The relation between formula and anion structure becomes more complex in cases where ions such as Al^{3+} may substitute for Si^{4+} in the silicate anion. Examples are as follows.

The plagioclase feldspars are a family of aluminosilicates typified by albite, $NaAlSi_3O_8$ and anorthite, $CaAl_2Si_2O_8$. In both of these Al is partly replacing Si in the silicate anion. It is therefore appropriate to consider the overall ratio $(Si + Al):O$. In both cases this ratio is 1:2 and therefore, a 3D framework structure is expected. Framework structures also occur in orthoclase, $KAlSi_3O_8$, kalsilite, $KAlSiO_4$, eucryptite, $LiAlSiO_4$ and spodumene, $LiAlSi_2O_6$.

Substitution of Al for Si occurs in many sheet structures such as micas and the clay minerals. The mineral talc has the formula $Mg_3(OH)_2Si_4O_{10}$ and, as expected for an Si:O ratio of 1:2.5, the structure contains infinite silicate sheets. In the mica phlogopite, one quarter of the Si atoms in talc are effectively replaced by Al and extra K^+ ions are added to preserve electroneutrality. Hence, phlogopite has the formula $KMg_3(OH)_2(Si_3Al)O_{10}$. In talc and phlogopite, Mg^{2+} ions occupy octahedral sites between silicate sheets; K^+ ions occupy 12-coordinate sites.

Further complications arise in other aluminosilicates in which Al^{3+} ions may occupy octahedral sites as well as tetrahedral ones. In such cases, one has to have information on the coordination number of aluminium since this cannot be deduced from the formula. One example is the mica muscovite, $KAl_2(OH)_2(Si_3Al)O_{10}$. This is structurally similar to phlogopite, with infinite sheets of constitution $(Si_3Al)O_{10}$ and a $(Si + Al):O$ ratio of $1:2.5$. However, the other two Al^{3+} ions replace the three Mg^{2+} ions of phlogopite and occupy octahedral sites. By convention, only ions that replace Si in tetrahedral sites are included as part of the complex anion. Hence octahedral Al^{3+} ions are formally regarded as cations in much the same way as alkali and alkaline earth cations, Table 1.20.

Chapter 2

Bonding in Solids

Crystalline materials exhibit the complete spectrum of bond types, ranging from ionic to covalent, van der Waals and metallic. Sometimes more than one type of bonding is present in a particular material, as in salts of complex anions, e.g. Li_2SO_4, which have both ionic and covalent bonds. Commonly, the bonding may be a blend of the different types, as in TiO which is ionic/metallic or CdI_2 which is ionic/covalent/van der Waals. In discussing structures, it is often convenient to ignore temporarily the complexities of mixed bond types and to treat bonds as though they were purely ionic.

Comparing ionic and covalent bonding, ionic bonding leads to structures with high symmetry and in which the coordination numbers are as high as possible. Consequently the net electrostatic attractive force which holds crystals together (and hence the lattice energy) is maximized. Covalent bonding, on the other hand, gives highly directional bonds in which one or all of the atoms present have a definite preference for a certain coordination environment, irrespective of the other atoms that are present. The coordination numbers in covalently bonded structures are usually small and may be less than those in corresponding ionic structures which contain atoms of similar size to those in the covalent structure.

The type of bonding that occurs in a compound correlates fairly well with the position of the component atoms in the periodic table and, especially, with their electronegativity. Alkali and alkaline earth elements usually form essentially ionic structures (beryllium is sometimes an exception), especially in combination with smaller, more electronegative anions such as O^{2-} and F^-. Covalent structures occur especially with: (a) small atoms of high valency which, in the cationic state, would be highly polarizing, e.g. B^{3+}, Si^{4+}, P^{5+}, S^{6+}, etc.; and to a lesser extent with (b) large atoms which in the anionic state are highly polarizable, e.g. I^-, S^{2-}.

Most non-molecular compounds have bonding which is a mixture of ionic and covalent and, as discussed later, it is becoming possible to make quantitative assessments of the *ionicity* of a particular bond, i.e. the percentage of ionic character in the bond. An additional factor in some transition metal compounds especially is the occurrence of metallic bonding.

Ionic bonding

Purely ionic bonding in crystalline compounds is an idealized or extreme form of bonding which is rarely attained in practice. Even in structures that are regarded as essentially ionic, e.g. NaCl and CaO, there is usually a certain amount of covalent bonding between cation and anion which acts to reduce the charge on each. The degree of covalent bonding increases with increasing valence of the ions to the extent that ions with a *net* charge greater than $+1$ or -1 appear unlikely to exist. Thus, while NaCl may reasonably be represented as Na^+Cl^-, TiC (which also has the NaCl structure) certainly does not contain Ti^{4+} and C^{4-} ions and the main bonding type in TiC must be non-ionic. This brings us to a dilemma. Do we continue to use the ionic model in the knowledge that for many structures, e.g. Al_2O_3, $CdCl_2$, a large degree of covalent bonding must be present? If not, we must find an alternative model for the bonding. In this section, ionic bonding is given considerable prominence because of its apparent wide applicability and its usefulness as a *starting point* for describing structures which in reality often have a considerable amount of covalent bonding. In later sections methods of assessing the degree of covalent character in 'ionic structures' are discussed.

Ions and ionic radii

It is difficult to imagine discussing crystal chemistry without having information available on the sizes of ions in crystals. However, crystal chemistry is currently undergoing a minor revolution in that the long-established tables of ionic radii of Pauling (1928), Goldschmidt and others are now thought to be seriously in error; at the same time, our concepts of ions and ionic structures are also undergoing revision. In more recent compilations of ionic radii, e.g. of Shannon and Prewitt (1969, 1970), cations are shown as being larger and anions smaller than previously thought. For example, Pauling radii of Na^+ and F^- are 0.98 and 1.36 Å, respectively, whereas Shannon and Prewitt give values of 1.14 to 1.30 Å, depending on the coordination number, for Na^+, and 1.19 Å for F^-.

These changes have arisen largely because, with modern, high-quality X-ray diffraction work it is possible to obtain fairly accurate maps of the distribution of electron density throughout ionic crystals. Thus, one can effectively 'see' ions and tell something about their size, shape and nature. In Fig. 2.1 is shown an electron density 'contour map' for LiF for a section passing through the structure parallel to one unit cell face. The map therefore passes through the centres of Li^+ and F^- ions located on the (100) planes. In Fig. 2.2 is shown the variation of electron density with distance along the line that connects adjacent Li^+ and F^- ions. From Figs 2.1 and 2.2 and similar diagrams for other structures, the following conclusions about ions in crystals may be drawn:

(a) Ions are essentially spherical.
(b) Ions may be regarded as composed of two parts: a central core in which most of the electron density is concentrated and an outer sphere of influence which contains very little electron density.

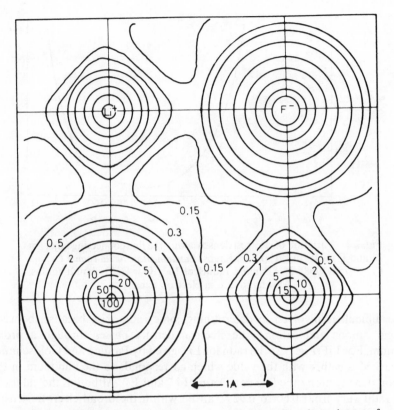

Fig. 2.1 Electron density contour map of LiF: a section through part of the unit cell face. The electron density (electrons Å^{-3}) is constant along each of the contour lines. (From Krug, Witte and Wolfel, *Zeit Phys. Chem.*, *Frankfurt*, **4**, 36, 1955)

(c) Assignment of radii to ions is difficult; even for ions which are supposedly in contact, it is not obvious (Fig. 2.2) where one ion ends and another begins.

Conclusion (b) is in contrast to the oft-stated assumption that 'ions can be treated as charged, incompressible, non-polarizable spheres'. Certainly, ions are charged, but they cannot be regarded as spheres with a clearly defined radius. Their electron density does not decrease abruptly to zero at a certain distance from the nucleus but decreases only gradually with increasing radius. Instead of being incompressible, ions are probably quite elastic, by virtue of flexibility in the outer sphere of influence of an ion while the inner core remains unchanged. This flexibility is necessary in order to explain variations of apparent ionic radii with coordination number and environment (see later). Within limits, ions can therefore expand or contract as the situation demands.

From Figs 2.1 and 2.2, most of the electron density is concentrated close to the nuclei of the ions; in a crystal, therefore, most of the total volume is essentially free space and contains relatively little electron density.

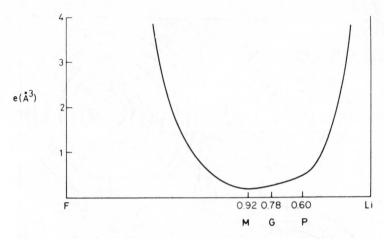

Fig. 2.2 Variation of electron density along the line connecting adjacent Li and F nuclei in LiF (From Krebs, *Fundamentals of Inorganic Crystal Chemistry*, 1968). P = Pauling radius of Li^+, G = Goldschmidt radius, M = minimum in electron density

The difficulties involved in determining ionic radii arise because, between adjacent anions and cations, the electron density passes through a broad minimum. For LiF(Fig. 2.2), the radii for Li^+ given by Pauling and Goldschmidt are marked together with the value which corresponds to the minimum in the electron density along the line connecting Li^+ and F^-. Although the values of these radii vary from 0.60 to 0.92 Å, all lie within the broad electron density minimum of Fig. 2.2.

The many methods that have been used in the past to estimate ionic radii will not be discussed here. In spite of the difficulties involved in determining absolute radii, it is necessary to have a set of radii for reference. Fortunately, most sets of radii are *additive* and self-consistent; provided one does not mix radii from different tabulations it is possible to use any set of radii to evaluate interionic distances in crystals with reasonable confidence. Shannon and Prewitt give two sets of radii: one is based on $r_{O^{2-}} = 1.40$ Å and is similar to Pauling, Goldschmidt, etc.; the other is based on $r_{F^-} = 1.19$ Å (and $r_{O^{2-}} = 1.26$ Å) and is related to the values determined from X-ray electron density maps. Both sets are comprehensive for cations in their different coordination environments but only pertain to oxides and fluorides. We choose here to use the Shannon and Prewitt set based on $r_{F^-} = 1.19$ Å and $r_{O^{2-}} = 1.26$ Å. Cation radii for some ions M^+ to M^{4+} are shown graphically in Fig. 2.3 as a function of cation coordination number; it should be stressed that the more highly charged ions are unlikely to exist as such but probably have their positive charge reduced by polarization of the anion and consequent partial covalent bonding between cation and anion. Further data, in the form of bond distances, are given in Appendix 4.

The following trends in ionic radii, with position in the periodic table, formal charge and coordination number, occur:

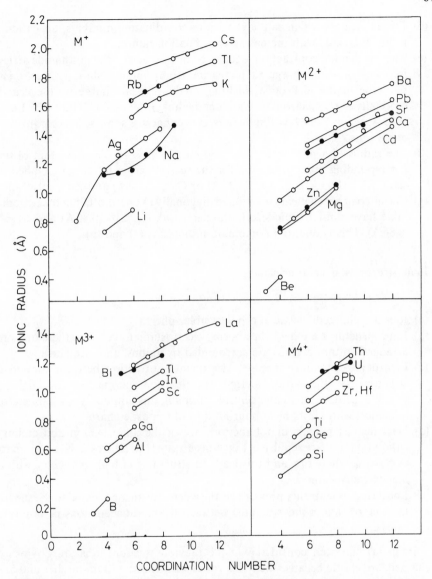

Fig. 2.3 Ionic radii as a function of coordination number for cations M^+ to M^{4+}. (From Shannon and Prewitt, *Acta Cryst.*, **B25**, 725, 1969; **B26**, 1046, 1970. Data based on $r_{F^-} = 1.19$ Å and $r_{O^{2-}} = 1.26$ Å)

(a) For the *s*- and *p*-block elements, radii increase with atomic number for any vertical group, e.g. octahedrally coordinated alkali ions.

(b) For any isoelectronic series of cations, the radius decreases with increasing charge, e.g. Na^+, Mg^{2+}, Al^{3+} and Si^{4+}.

(c) For any element which can have variable oxidation states, the cation radius decreases with increasing oxidation state, e.g. V^{2+}, V^{3+}, V^{4+}, V^{5+}.

(d) For an element which can have various coordination numbers, the cationic radius increases with increasing coordination number.

(e) The 'lanthanide contraction' occurs as follows: across the lanthanide series, ions with the same charge but increasing atomic number show a reduction in size (due to the ineffective shielding of the nuclear charge by the d and, especially, f electrons), e.g. octahedral radii, La^{3+}(1.20 Å)... Eu^{3+} (1.09 Å)... Lu^{3+} (0.99 Å). Similar effects occur across some series of transition metal ions.

(f) The radius of a particular transition metal ion is smaller than that of the corresponding main group ion for the reasons given in (e), e.g. octahedral radii, Rb^+ (1.63 Å) and Ag^+ (1.29 Å) or Ca^{2+} (1.14 Å) and Zn^{2+} (0.89 Å).

(g) Certain pairs of elements positioned diagonally to one another in the periodic table have similar ionic size (and chemistry), e.g. Li^+ (0.88 Å) and Mg^{2+} (0.86 Å). This is due to a combination of effects (a) and (b).

Ionic structures–general principles

Consider the following as a guide to ionic structures:

(a) Ions are charged, elastic and polarizable spheres.

(b) Ionic structures are held together by electrostatic forces and, therefore, are arranged so that cations are surrounded by anions, and vice versa.

(c) In order to maximize the net electrostatic attraction between ions in a structure (i.e. the lattice energy), coordination numbers are as high as possible, provided that the central ion maintains contact (via its sphere of influence) with all its neighbouring ions of opposite charge.

(d) Next nearest neighbour interactions are of the anion–anion and cation–cation type and are repulsive. Like ions arrange themselves to be as far apart as possible, therefore, and this leads to structures of high symmetry with a maximized volume.

(e) Local electroneutrality prevails, i.e. the valence of an ion is equal to the sum of the electrostatic bond strengths between it and adjacent ions of opposite charge.

Point (a) has been considered in the previous section; ions are obviously charged, are elastic because their size varies with coordination number and are polarizable when departures from purely ionic bonding occurs. For example, the electron density map for LiF (Fig. 2.1) shows a small distortion from spherical shape in the outer part of the sphere of influence of the Li^+ ion, and this may be attributed to the occurrence of a small amount of covalent bonding between Li^+ and F^-.

Points (b)–(d) imply that the forces which hold ionic crystals together and the net energy of interaction between the ions are the same as would be obtained by regarding the crystal as a three-dimensional array of point charges and considering the net coulombic energy of the array. From Coulomb's law, the force F between two ions of charge Z_+e and Z_-e, separated by distance r, is given by

$$F = \frac{(Z_+ e)(Z_- e)}{r^2} \tag{2.1}$$

A similar equation applies to each pair of ions in the crystal and evaluation of the resulting force between all the ions leads to the lattice energy of the crystal (later).

Point (c) includes the proviso that nearest neighbour ions should be 'in contact'. Given the nature of electron density distributions in ionic crystals (Figs 2.1 and 2.2), it is hard to quantify what is meant by 'in contact'. It is nevertheless an important factor since, although the apparent size of ions varies with coordination number, most ions, smaller ones especially, appear to have a maximum coordination number; for Be^{2+} this is four and for Li^+ it is usually six. Ions are flexible, therefore, but expand or contract within fairly narrow limits.

The idea of *maximizing* the volume of ionic crystals, point (d), is rather unexpected since one is accustomed to regarding ionic structures and derivative close packed structures, especially, as having *minimum* volume. There is no conflict, however. The prime bonding force in ionic crystals is the nearest neighbour cation–anion *attractive* force and this force is maximized at a small cation–anion separation (when ions become too close, additional repulsive forces come into play, thereby reducing the net attractive force). Superposed on this is the effect of next nearest neighbour *repulsive* forces between like ions. With the constraints that (a) cation–anion distances be as short as possible and (b) coordination numbers be as large as possible, like ions arrange themselves to be as far apart as possible in order to reduce their mutual repulsion. This leads to regular and highly symmetrical arrays of like ions. It has been shown that such regular arrays of like ions tend to have maximized volumes whereas, by distorting the structures, a reduction in volume may be possible, at least in principle.

An excellent example of a structure whose volume is maximized is rutile. The buckling of the oxide layers (Fig. 1.32d) causes the next nearest neighbour coordination number of oxygen (by oxygen) to be reduced from 12 (as in h.c.p.) to 11 (as in p.t.p.). The coordination of titanium by oxygen, and vice versa, is unaffected by this distortion but the overall volume of the structure increases by 2 to 3 per cent. Hence the oxide layers are buckled so that the volume can be maximized.

Point (e) is Pauling's electrostatic valence rule, the second of a set of rules formulated by Pauling for ionic crystals. Basically, the rule means that the charge on a particular ion, e.g. an anion, must be balanced by an equal and opposite charge on the immediately surrounding cations. However, since these cations are also shared with other anions, it is necessary to estimate the amount of positive charge that is effectively associated with each cation–anion bond. For a cation M^{m+} surrounded by n anions, X^{x-}, the *electrostatic bond strength* (e.b.s.) of the cation–anion bonds is defined as

$$\text{e.b.s.} = \frac{m}{n} \tag{2.2}$$

For each anion, the sum of the electrostatic bond strengths of the surrounding cations must balance the negative charge on the anion, i.e.

$$\sum \frac{m}{n} = x \qquad (2.3)$$

For example:

(a) Spinel, $MgAl_2O_4$, contains octahedral Al^{3+} and tetrahedral Mg^{2+} ions; each oxygen is surrounded tetrahedrally by three Al^{3+} ions and one Mg^{2+} ion. We can check that this must be so, as follows:

For Mg^{2+}: e.b.s. $= \frac{2}{4} = \frac{1}{2}$

For Al^{3+}: e.b.s. $= \frac{3}{6} = \frac{1}{2}$

Therefore,

$$\sum \text{e.b.s.}(3\,Al^{3+} + 1\,Mg^{2+}) = 2$$

(b) We can show that three SiO_4 tetrahedra *cannot* share a common corner in silicate structures:

For Si^{4+}: e.b.s. $= \frac{4}{4} = 1$

Therefore, for an oxygen that bridges two SiO_4 tetrahedra, $\sum$e.b.s. $= 2$, which is acceptable. However, three tetrahedra sharing a common oxygen would give $\sum$e.b.s. $= 3$ for that oxygen, which is quite unacceptable.

This rule of Pauling's provides an important guide to the kinds of polyhedral linkages that are and are not possible in crystal structures. In Table 2.1 is given a list of some common cations with their formal charge, coordination number and electrostatic bond strength. In Table 2.2 are listed some allowed and unallowed combinations of polyhedra about a common oxide ion, together with some examples of the allowed combinations. Many other combinations are possible and the reader may like to deduce some, bearing in mind that there are also

Table 2.1 *Electrostatic bond strengths of some cations*

Cation with formal charge	Coordination number(s)	Electrostatic bond strength(s)
Li^+	4, 6	$\frac{1}{4}, \frac{1}{6}$
Na^+	6, 8	$\frac{1}{6}, \frac{1}{8}$
Be^{2+}	3, 4	$\frac{2}{3}, \frac{1}{2}$
Mg^{2+}	4, 6	$\frac{1}{2}, \frac{1}{3}$
Ca^{2+}	8	$\frac{1}{4}$
Zn^{2+}	4	$\frac{1}{2}$
Al^{3+}	4, 6	$\frac{3}{4}, \frac{1}{2}$
Cr^{3+}	6	$\frac{1}{2}$
Si^{4+}	4	1
Ge^{4+}	4, 6	$1, \frac{2}{3}$
Ti^{4+}	6	$\frac{2}{3}$
Th^{4+}	8	$\frac{1}{2}$

Table 2.2 *Allowed and unallowed combinations of corner-sharing oxide polyhedra*

Allowed	Example	Unallowed
$2SiO_4$tet.	Silica	$> 2SiO_4$tet.
$1MgO_4$tet. $+ 3AlO_6$oct.	Spinel	$3AlO_4$tet.
$1SiO_4$tet. $+ 3MgO_6$oct.	Olivine	$1SiO_4$tet. $+ 2AlO_4$tet.
$8LiO_4$tet.	Li_2O	$4TiO_6$oct.
$2TiO_6$oct. $+ 4CaO_{12}$dod.	Perovskite	
$3TiO_6$oct.	Rutile	

topological restrictions on the number of possible polyhedral combinations; thus the maximum number of octahedra that can share a common corner is six (as in rock salt), etc.

Pauling's third rule is concerned with the topology of polyhedra and has been considered in the previous chapter. Pauling's first rule states: 'A coordinated polyhedron of anions is formed about each cation, the cation–anion distance being determined by the radius sum and the coordination number of the cation by the radius ratio.' The idea that cation–anion distances are determined by the radius sum is implicit to every tabulation of ionic radii since a major objective of such tabulations is to be able to predict, correctly, interatomic distances. Let us now consider coordination numbers and *the radius ratio rules.*

The radius ratio rule

In ideally ionic crystal structures, the coordination numbers of the ions are determined largely by electrostatic considerations. Cations surround themselves with as many anions as possible, and vice versa. This maximizes the electrostatic attractions between neighbouring ions of opposite charge and hence maximizes the lattice energy of the crystal (see later). This requirement led to the formulation of the *radius ratio rule* for *ionic structures* in which the ions and the structure adopted for a particular compound depend on the *relative* sizes of the ions. This rule is very useful qualitatively but, as we shall see, cannot usually be applied rigorously to explain or predict coordination numbers. There are two guidelines to be followed in using this rule. First, a cation must be in contact with its anionic neighbours. This places a *lower* limit on the size of cation which may occupy a particular site, since a situation in which a cation may 'rattle' inside its anion polyhedron is assumed to be unstable. Second, neighbouring anions may or may not be in contact.

Using these guidelines, one may calculate the range of cation sizes that can occupy the various interstitial sites in an anion array. Let us first calculate the minimum radius for an octahedral cation site, i.e. one in which both anion–anion and anion–cation contacts occur. This then gives us the lower limiting radius ratio for octahedral coordination. In such an octahedral site, the cation is in contact with all six of its anion neighbours; in Fig. 2.4 is a section through an octahedron,

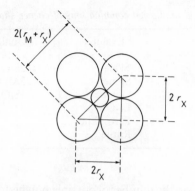

Fig. 2.4 Radius ratio calculation for octahedral coordination

showing the central cation with four coplanar anions. Applying the Pythagoras theorem,

$$(2r_X)^2 + (2r_X)^2 = [2(r_M + r_X)]^2 \tag{2.4}$$

i.e.

$$2r_X\sqrt{2} = 2(r_M + r_X)$$

and

$$r_M/r_X = \sqrt{2} - 1 = 0.414 \tag{2.5}$$

For radius ratios < 0.414, the cation is too small for an octahedral site and instead, should occupy a site of smaller coordination number. For radius ratios > 0.414, the cation would push the anions apart and this happens increasingly up to a radius ratio of 0.732. At and above this value, the cation is sufficiently large to have eight anionic neighbours, all of which are in contact with the cation.

Let us now make the calculation for the minimum size of cation that can occupy such an eight coordinate site (as in CsCl, Fig. 1.31). Here it is necessary to recognize that cations and anions are in contact along the body diagonal of the cube, i.e.

$$2(r_M + r_X) = \text{(cube body diagonal)} \tag{2.6}$$

In addition, anions are in contact along the cell edge, a, i.e.

$$a = 2r_X \tag{2.7}$$

Therefore,

$$2(r_M + r_X) = 2r_X\sqrt{3}$$

and

$$r_M/r_X = \sqrt{3} - 1 = 0.732 \tag{2.8}$$

In order to evaluate the minimum radius ratio for tetrahedral coordination, it is necessary to recognize that, by regarding a tetrahedron as a cube with alternate corners missing, Fig. 1.23, anions and cations are in contact along the cube *body* diagonals, equation (2.6), but that anions are in contact along the cube *face* diagonals, i.e.

$$2r_X = (\text{face diagonal}) \tag{2.9}$$

Combining equations (2.6) and (2.9) and rearranging gives

$$(2r_X)^2 + (\sqrt{2}r_X)^2 = [2(r_M + r_X)]^2$$

i.e.

$$2(r_M + r_X) = \sqrt{6}r_X$$

and

$$r_M/r_X = \frac{\sqrt{6}-2}{2} = 0.225 \tag{2.10}$$

The minimum radius ratios for various coordination numbers are given in Table 2.3. Note that CN = 5 is absent from the table; in close packed structures, at least, it is not possible to have a coordination number of five in which all M—X bonds are of the same length.

The radius ratio rules have had a limited amount of success in predicting trends in coordination number and structure type and at best can be used only as a general guideline. Radius ratios depend very much on which table of ionic radii is consulted and there appears to be no clear advantage in using either one of the more traditional sets or the modern set of values based on X-ray diffraction results. For example, for RbI, $r+/r- = 0.69$ or 0.80, according to the tables based on $r_{O^{2-}} = 1.40$ and 1.26 Å, respectively. Thus one value would predict six-coordination (rock salt), as observed, but the other predicts eight-coordination (CsCl). On the other hand, LiI has $r+/r- = 0.28$ and 0.46, according to the same tables; one value predicts tetrahedral coordination and the other octahedral (as observed). For the larger cations, especially caesium, $r+/r- > 1$, and it is perhaps more realistic to consider instead the inverse ratio, $r-/r+$, in for instance, CsF.

A more convincing example of the relevance of radius ratio rules is provided by oxides and fluorides of general formula MX_2. Possible structure types, with their cation coordination numbers, are silica (4), rutile (6) and fluorite (8). A selection of oxides in each group is given in Table 2.4, together with the radius ratios

Table 2.3 *Minimum radius ratios for different cation coordination numbers*

Coordination	Minimum $r_M:r_X$
Linear, 2	—
Trigonal, 3	0.155
Tetrahedral, 4	0.225
Octahedral, 6	0.414
Cubic, 8	0.732

Table 2.4 *Structures and radius ratios of oxides, MO_2*

Oxide	Calculated radius ratio*		Observed structure type	
CO_2	~ 0.1	(CN = 2)	Molecular	(CN = 2)
SiO_2	0.32	(CN = 4)	Silica	(CN = 4)
GeO_2	$\begin{cases} 0.43 \\ 0.54 \end{cases}$	(CN = 4) (CN = 6)	$\begin{cases} \text{Silica} \\ \text{Rutile} \end{cases}$	(CN = 4) (CN = 6)
TiO_2	0.59	(CN = 6)	Rutile	(CN = 6)
SnO_2	0.66	(CN = 6)	Rutile	(CN = 6)
PbO_2	0.73	(CN = 6)	Rutile	(CN = 6)
HfO_2	$\begin{cases} 0.68 \\ 0.77 \end{cases}$	(CN = 6) (CN = 8)	Fluorite	(CN = 8)
CeO_2	$\begin{cases} 0.75 \\ 0.88 \end{cases}$	(CN = 6) (CN = 8)	Fluorite	(CN = 8)
ThO_2	0.95	(CN = 8)	Fluorite	(CN = 8)

*Since cation radii vary with coordination number, as shown in Fig. 2.3, radius ratios may be calculated for different coordination numbers. The coordination numbers used here are shown in parentheses. Calculations are based on $r_{O^{2-}} = 1.26$ Å.

calculated from Fig. 2.3 (based on $r_{O^{2-}} = 1.26$ Å). Changes in coordination number are expected to occur at radius ratios of 0.225, 0.414 and 0.732. Bearing in mind that the calculated radius ratio values depend on the particular table of radii that is consulted, the agreement between theory and practice is reasonable. For example, GeO_2 is polymorphic and can have both silica and rutile structures; the radius ratio calculated for tetrahedral coordination of germanium is borderline between the values predicted for CN = 4 and 6.

Borderline radius ratios and distorted structures

The structural transition from CN = 4 to 6 which occurs with increasing cation size is often clear-cut. A good example is provided by GeO_2 which has a borderline radius ratio and which also exhibits polymorphism. Both polymorphs have highly symmetric structures; one has a silica-like structure with CN = 4 and the other has a rutile structure with CN = 6. Polymorphs with CN = 5 do not occur with GeO_2.

In some other cases of borderline radius ratios, however, distorted polyhedra and/or coordination numbers of 5 are observed. Thus V^{5+} (radius ratio = 0.39 for CN = 4 or radius ratio = 0.54 for CN = 6) has an environment in one polymorph of V_2O_5 which is a gross distortion of octahedral; five V—O bonds are of reasonable length, in the range 1.5 to 2.0 Å, but the sixth bond is much longer, 2.8 Å, and the coordination is better regarded as distorted square pyramidal. It appears that V^{5+} is rather small to happily occupy an octahedral site and that, instead, a structure occurs which is transitional between tetrahedral and octahedral. Similar types of distortions occur between CN = 6 and 8. Thus,

ZrO_2 has a borderline radius ratio (0.68 for $CN = 6$; 0.78 for $CN = 8$) and although it may have the fluorite structure at very high temperatures ($> 2000\,°C$), with a CN of 8 for zirconium, in its normal form at room temperature as the mineral baddleyite, zirconium has a CN of 7.

Less severe distortions occur in cases where a cation is only slightly too small for its anion environment. The regular anion coordination is maintained but the cation may rattle or undergo small displacements within its polyhedron. In, for example, $PbTiO_3$ (radius ratio for $Ti = 0.59$ for $CN = 6$), titanium may undergo displacement by $\sim 0.2\,\text{Å}$ off the centre of its octahedral site towards one of the corner oxygens. The direction of displacement may be reversed under the action of an applied electric field and this gives rise to the important property of ferroelectricity (Chapter 7).

The concept of 'maximum contact distance' has been proposed by Dunitz and Orgel (1960). If the metal–anion distance increases above this distance then the cation is free to rattle. If the metal–anion distance decreases the metal ion is subjected to compression. However, the maximum contact distance does not correspond to the sum of ionic radii, as they are usually defined, and this is a difficult concept to quantify.

Lattice energy of ionic crystals

Ionic crystals may be regarded as regular three-dimensional arrangements of point charges. The forces that hold the crystals together are entirely electrostatic in origin and may be calculated by summing all the electrostatic repulsions and attractions in the crystal. The *lattice energy*, U, of a crystal is defined as the net potential energy of the arrangement of charges that forms the structure. It is equivalent to the energy required to sublime the crystal and convert it into a collection of gaseous ions, e.g.

$$NaCl(s) \rightarrow Na^+(g) + Cl^-(g), \qquad \Delta H = U$$

The value of U depends on the crystal structure that is adopted, the charge on the ions and the internuclear separation between the anion and cation.

Two principal kinds of force are involved in determining the crystal structure of ionic materials:

(a) The electrostatic forces of attraction and repulsion between ions. Two ions M^{z+} and X^{z-} separated by a distance, r, experience an attractive force, F, given by Coulomb's law:

$$F = \frac{Z_+ Z_- e^2}{r^2} \tag{2.11}$$

and their coulombic potential energy, V, is given by

$$V = \int_\infty^r F\,dr = -\frac{Z_+ Z_- e^2}{r} \tag{2.12}$$

(b) Short-range repulsive forces which are important when atoms or ions are so close together that their electron clouds begin to overlap. Born suggested that this repulsive energy has the form:

$$V = \frac{B}{r^n} \qquad (2.13)$$

where B is a constant and the Born exponent, n, has a value in the range 5 to 12. Because n is large, V falls rapidly to zero with increasing r.

The lattice energy, U, of a crystal may be calculated by combining the net electrostatic attraction and the Born repulsion energies and finding the internuclear separation, r_e, which gives the maximum U value. The procedure is as follows.

Consider the NaCl structure (Fig. 1.24a). Between each pair of ions in a crystal there is an electrostatic interaction given by equation (2.11). We wish to sum all such interactions which occur in the crystal and calculate the net attractive energy. Let us first consider one particular ion, e.g. Na^+ in the body centre of the unit cell (Fig. 1.24a), and calculate the interaction between it and its neighbours. Its nearest neighbours are six Cl^- ions in the face centre positions and at a distance r ($2r$ is the value of the unit cell edge). The attractive potential energy is given by

$$V = -6\frac{e^2 Z_+ Z_-}{r} \qquad (2.14)$$

The next nearest neighbours to the Na^+ ion are twelve Na^+ ions arranged at the edge centre positions of the unit cell, i.e. at a distance $\sqrt{2}r$; this gives a repulsive potential energy term

$$V = +12\frac{e^2 Z_+ Z_-}{\sqrt{2}r} \qquad (2.15)$$

The third nearest neighbours are eight Cl^- ions at the cube corners and distance $\sqrt{3}r$; these are attracted to the central Na^+ ion according to

$$V = -8\frac{e^2 Z_+ Z_-}{\sqrt{3}r} \qquad (2.16)$$

The net attractive energy between our Na^+ ion and all other ions in the crystal is given by an infinite series:

$$V = -\frac{e^2 Z_+ Z_-}{r}\left(6 - \frac{12}{\sqrt{2}} + \frac{8}{\sqrt{3}} - \frac{6}{\sqrt{4}} + \cdots\right) \qquad (2.17)$$

This summation is repeated for each ion in the crystal, i.e. for $2N$ ions per mole of NaCl. Since each ion pair interaction is thereby counted twice it is necessary to divide the final value by 2, giving

$$V = -\frac{e^2 Z_+ Z_-}{r} NA \qquad (2.18)$$

Table 2.5   *Madelung constants for some simple structures*

Structure type	A
Rock salt	1.748
CsCl	1.763
Wurtzite	1.641
Sphalerite	1.638
Fluorite	2.520
Rutile	2.408

where the *Madelung constant*, A, is the numerical value of the infinite series summation in parentheses in equation (2.17). The Madelung constant depends only on the geometrical arrangement of point charges. It has the same value, 1.748, for all compounds with the rock salt structure. Values of A for some other simple structure types are given in Table 2.5.

If equation (2.18) represented the only factor involved in the lattice energy, the structure would collapse in on itself since $V \propto -1/r$ (Fig. 2.5). This catastrophe is avoided by the mutual repulsion between ions, of whatever charge, when they become too close and which is given by equation (2.13). The dependence of this

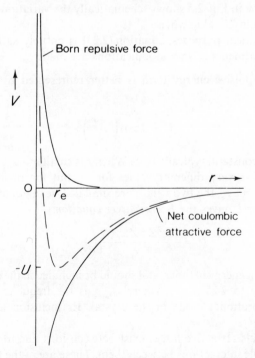

Fig. 2.5 Lattice energy (dashed line) of ionic crystals as a function of internuclear separation

repulsive force on r is shown schematically in Fig. 2.5. The total energy of the crystal, the lattice energy U, is given by summing equations (2.18) and (2.13) and differentiating with respect to r to find the maximum U value and equilibrium interatomic distance, r_e; i.e.

$$U = -\frac{e^2 Z_+ Z_- N A}{r} + \frac{BN}{r^n} \tag{2.19}$$

Therefore,

$$\frac{dU}{dr} = \frac{e^2 Z_+ Z_- N A}{r^2} - \frac{nBN}{r^{n+1}} \tag{2.20}$$

When

$$\frac{dU}{dr} = 0 \tag{2.21}$$

then

$$B = \frac{e^2 Z_+ Z_- A r^{n-1}}{n} \tag{2.22}$$

and therefore

$$U = -\frac{e^2 Z_+ Z_- N A}{r_e}\left(1 - \frac{1}{n}\right) \tag{2.23}$$

The dashed line in Fig. 2.5 shows schematically the variation of U with r and gives the minimum U value when $r = r_e$.

For most practical purposes, equation (2.23) is entirely satisfactory, but in more refined treatments certain modifications are made:

(a) The Born repulsive energy term is better represented by an exponential function:

$$V = B\exp\left(\frac{-r}{\rho}\right) \tag{2.24}$$

where ρ is a constant, typically 0.35. When r is small ($r \ll r_e$), equations (2.13) and (2.24) give very different values for V, but for realistic interatomic distances, i.e. $r \simeq r_e$, the two values are similar. Use of equation (2.24) in the expression for U gives the *Born–Mayer equation*:

$$U = -\frac{e^2 Z_+ Z_- A N}{r_e}\left(1 - \frac{\rho}{r_e}\right) \tag{2.25}$$

(b) The zero point energy of the crystal should be included in the calculation of U. This is equal to $2.25\, h v_{O_{max}}$, where $v_{O_{max}}$ is the frequency of the highest occupied vibrational mode in the crystal. Its inclusion leads to a small reduction in U.

(c) Van der Waals attractive forces exist between ions due to induced dipole-induced dipole interactions between them. These are of the form NC/r^6 and lead to an increase in U.

A more complete equation for U, after correcting for these factors, is

$$U = -\frac{Ae^2Z_+Z_-N}{r} + BNe^{-r/\rho} - CNr^{-6} + 2.25Nhv_{O_{max}} \qquad (2.26)$$

Typical values for these four terms, in kilojoules per mole, are (from Greenwood):

Substance	$NAe^2Z_+Z_-r^{-1}$	$NBe^{-r/\rho}$	NCr^{-6}	$2.25Nhv_{O_{max}}$	U
NaCl	−859.4	98.6	−12.1	7.1	−765.8
MgO	−4631	698	−6.3	18.4	−3921

from which it can be seen that the Born repulsive term contributes 10 to 15 per cent to the value of U whereas the zero point vibrational and van der Waals terms contribute about 1 per cent each and, being of opposite sign, tend to cancel each other out. For most purposes, therefore, we can use the simplified equation (2.23). Let us now consider each of the terms in equation (2.23) and evaluate their relative significance.

The magnitude of U depends on six parameters A, N, e, Z, n and r_e, four of which are constants for a particular ionic structure type. This leaves just two variables, the charge on the ions, Z_+, Z_-, and the internuclear separation, r_e. Of the two, the charge is by far the most important since the value of the product (Z_+Z_-) is capable of much larger variation than is r_e. For instance, a material with divalent ions should have a lattice energy that is four times as large as an isostructural crystal with the same r_e but containing monovalent ions (compare the lattice energies of SrO and LiCl, Table 2.6, which have similar r_e, Table 1.7). For a series of isostructural phases with the same Z values but increasing r_e, a decrease in U is expected (e.g. alkali fluorides, alkaline earth oxides with NaCl structure). A selection of lattice energies for materials with the rock salt structure and showing these two trends is given in Table 2.6.

Since the lattice energy of a crystal is equivalent to its heat of dissociation, a correlation exists between U and the melting point of the crystal (a better correlation may be sought between U and the sublimation energy, but such data are not so readily available). The effect of (Z_+Z_-) on the melting point is shown by the refractoriness of the alkaline earth oxides (m.p. of CaO = 2572 °C) compared with the alkali halides (m.p. of NaCl = 800 °C). The effect of r_e on

Table 2.6 *Some lattice energies in kJ mol⁻¹*. (Data from Ladd and Lee, *Progr. Solid State Chem.*, **1**, 37–82, 1963; **2**, 378–413, 1965)

MgO	3938	LiF	1024	NaF	911
CaO	3566	LiCl	861	KF	815
SrO	3369	LiBr	803	RbF	777
BaO	3202	LiI	744	CsF	748

Table 2.7   *Thermochemical radii* (Å) *of complex anions.* (Data from Kapustinskii, *Quart. Rev.*, 283–294, 1956)

BF_4^-	2.28	CrO_4^{2-}	2.40	IO_4^-	2.49
SO_4^{2-}	2.30	MnO_4^-	2.40	MoO_4^{2-}	2.54
ClO_4^-	2.36	BeF_4^-	2.45	SbO_4^{3-}	2.60
PO_4^{3-}	2.38	AsO_4^{3-}	2.48	BiO_4^{3-}	2.68
OH^-	1.40	O_2^{2-}	1.80	CO_3^{2-}	1.85
NO_2^-	1.55	CN^-	1.82	NO_3^-	1.89

melting points may be seen in series such as:

$$MgO\,(2800\,°C), \quad CaO\,(2572\,°C) \quad \text{and} \quad BaO\,(1923\,°C)$$

Kapustinskii's equation

Kapustinskii (1956) noted an empirical increase in the value of the Madelung constant, A, as the coordination number of the ions in the structure increased, e.g. in the series ZnS, NaCl, CsCl (Table 2.5). Since, for a particular anion and cation, r_e also increases with coordination number (e.g. Fig. 2.3), Kapustinskii proposed a general equation for U in which variations in A and r_e are auto-compensated. He suggested using the rock salt value for A and octahedral ionic radii (Goldschmidt) in calculating r_e; substituting $r_e = r_c + r_a$, $\rho = 0.345$, $A = 1.745$ and values for N and e into equation (2.25) gives the Kapustinskii equation:

$$U = \frac{1200.5 V Z_+ Z_-}{r_c + r_a}\left(1 - \frac{0.345}{r_c + r_a}\right) \quad kJ\,mol^{-1} \tag{2.27}$$

where V is the number of ions per formula unit (two in NaCl, three in PbF_2, etc.). This formula may be used to calculate the lattice energy of any known or hypothetical ionic compound and in spite of the assumptions involved, the answers obtained are surprisingly accurate.

The Kapustinskii equation has been used to predict successfully the stable existence of several previously unknown compounds. In cases where U is known from Born–Haber cycle calculations (see later), it has been used to derive values for ionic radii. This has been particularly useful for complex anions, e.g. SO_4^{2-}, PO_4^{3-}, whose effective size in crystals is difficult to measure by other means. Radii determined in this way are known as *thermochemical radii* and some values are given in Table 2.7. It should be noted that radii for non-spherical ions such as CN^- represent gross simplifications and are really applicable only to other lattice energy calculations.

The Born–Haber cycle and thermochemical calculations

The lattice energy of a crystal is equivalent to its heat of formation from one mole of its ionic constituents in the gas phase:

$$Na^+(g) + Cl^-(g) \rightarrow NaCl(s), \quad \Delta H = U$$

It cannot be measured experimentally. However, the heat of formation of a crystal, ΔH_f, can be measured relative to the reagents in their standard states:

$$Na(s) + \tfrac{1}{2}Cl_2(g) \rightarrow NaCl(s), \quad \Delta H = \Delta H_f$$

ΔH_f may be related to U by constructing a thermochemical cycle known as a Born–Haber cycle, in which ΔH_f is given by the summation of energy terms in a hypothetical reaction pathway. For NaCl, the individual steps in the pathway, commencing with the elements in their standard states, are:

Sublimation of solid Na	$\Delta H = S$
Ionization of gaseous Na atoms	$\Delta H = IP$
Dissociation of Cl_2 molecules	$\Delta H = \tfrac{1}{2}D$
Formation of the Cl^- ion	$\Delta H = EA$
Coalescence of gaseous ions to give crystalline NaCl	$\Delta H = U$

Addition of these five terms is equivalent to forming crystalline NaCl from solid Na and Cl_2 molecules, as shown:

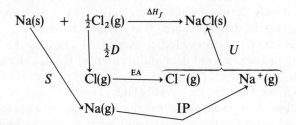

From Hess' Law,

$$\Delta H_f = S + \tfrac{1}{2}D + IP + EA + U \tag{2.28}$$

Applications. The Born–Haber cycle and equation (2.28) have various uses:

(a) Six enthalpy terms are present in equation (2.28). If all six can be determined independently for a particular compound, then the cycle gives a check on the internal consistency of the data. The values for NaCl are as follows:

S	$109\,kJ\,mol^{-1}$
IP	$493.7\,kJ\,mol^{-1}$
$\tfrac{1}{2}D$	$121\,kJ\,mol^{-1}$
EA	$-356\,kJ\,mol^{-1}$
U	$-764.4\,kJ\,mol^{-1}$
ΔH_f	$-410.9\,kJ\,mol^{-1}$

Summation of the first five terms gives a calculated H_f of $-396.7\,kJ\,mol^{-1}$, which compares reasonably well with the measured ΔH_f value of $-410.9\,kJ,\,mol^{-1}$.

(b) If only five of the energy terms are known, then the sixth may be evaluated using equation (2.28). An early application (~ 1918) was in the calculation of electron affinities, for which data were not then available.

Table 2.8   *Lattice energies* $(kJ\,mol^{-1})$ *of some Group I halides.*
(Data from Waddington, *Adv. Inorg. Chem. Radiochem.*, **1**, 157–221, 1959)

	U_{calc}	$U_{Born\ Haber}$	ΔU
AgF	920	953	33
AgCl	832	903	71
AgBr	815	895	80
AgI	777	882	105

(c) The possible stability of an unknown compound may be estimated. It is necessary to assume a structure for the compound in order to calculate U and while there are obviously errors involved, e.g. in choosing r_e, these are usually unimportant compared with the effect of some of the other energy terms involved in equation (2.28). Having estimated U, ΔH_f may then be calculated. If ΔH_f is large and positive then this explains why the compound is unknown—it is unstable relative to its elements. If ΔH_f(calc.) is negative, however, it may be worthwhile to try and prepare the compound under certain conditions. Examples are given in the next section.

(d) Differences between values of lattice energies obtained by the Born–Haber cycle using thermochemical data and theoretical values calculated from an ionic model of the crystal structure may be used as evidence for non-ionic bonding effects. Data for the silver halides (Table 2.8) and for thallium and copper halides (not given) show that the differences between the two lattice energies are least for the fluorides and greatest for the iodides. This is attributed to the presence of a strong covalent contribution to the bonding in the iodides which leads to an increase in the values of the thermochemical lattice energy. A correlation also exists between the insolubility of the Ag salts, especially AgI, in water and the presence of partial covalent bonding. Data for the corresponding alkali halides show that the differences between thermochemical and calculated lattice energies are small and indicate that the ionic bonding model may be applied satisfactorily to them.

While covalent bonding is present in, for example, AgCl and AgBr, as evidenced by the lattice energies, it is not strong enough to change the crystal structure from that of rock salt to one of lower coordination number. AgI does have a different structure, however; it is polymorphic and can exist in at least three structure types, all of which have low coordination numbers, usually four. Changes in structure type and coordination number due to increased covalent bonding are described later.

(e) Certain transition elements have *crystal field stabilization energies* due to their d electron configuration and this gives an increased lattice energy in their compounds. For example, the difference between experimental and calculated lattice energies in CoF_2 is $83\,kJ\,mol^{-1}$, which is in fair agreement with the CFSE value calculated for the high spin state of Co in CoF_2 of

$104 \, \text{kJ mol}^{-1}$. Ions which do not exhibit CFSE effects are those with configurations d^0 (e.g. Ca^{2+}), d^5 high spin (e.g. Mn^{2+}) and d^{10} (e.g. Zn^{2+}).

(f) The Born–Haber cycle has many other uses, e.g. in solution chemistry to determine energies of complexation and hydration of ions. These usually require a knowledge of the lattice energy of the appropriate solids, but since these applications do not provide any new information about solids, they are not discussed further.

Stabilities of real and hypothetical ionic compounds

Inert gas compounds

One may ask, whether it is worthwhile trying to synthesize, for example, ArCl. Apart from ΔH_f, the only unknown in equation (2.28) is U. Suppose that hypothetical ArCl had the rock salt structure and the radius of the Ar^+ ion is between that of Na^+ and K^+. An estimated lattice energy for ArCl is, then, $-745 \, \text{kJ mol}^{-1}$ ($NaCl = -764.4$; $KCl = -701.4$). Substitution in equation (2.28) gives, in kilojoules per mole:

$$
\begin{array}{cccccc}
S & \tfrac{1}{2}D & IP & EA & U & \Delta H_f(\text{calc.}) \\
0 & 121 & 1524 & -356 & -745 & +544
\end{array}
$$

from which it can be seen that ArCl has a large positive heat of formation and would be thermodynamically unstable, by a large amount, relative to the elements.

Such a calculation also tells us *why* ArCl is unstable and cannot be synthesized. Comparing the calculations for ArCl and NaCl, it is clear that the instability of ArCl is due to the very high ionization potential of argon (stability is strictly governed by free energies of formation, but ΔS is small and hence $\Delta G \simeq \Delta H$). The heats of formation calculated for several other hypothetical compounds are given in Table 2.9.

There is now a large number of inert gas compounds known, following on from the preparation of $XePtF_6$ by Bartlett in 1962. Consideration of lattice energies and enthalpies of formation led Bartlett to try and synthesize $XePtF_6$ by direct reaction of Xe and PtF_6 gases. He had previously prepared O_2PtF_6 as an ionic salt, $(O_2)^+(PtF_6)^-$, by reacting (by accident) O_2 with PtF_6. From a knowledge of the similarity in the first ionization potentials of molecular oxygen

Table 2.9 *Enthalpies of formation* $(kJ \, mol^{-1})$ *of some hypothetical* (*) *and real compounds*

HeF*	+ 1066	NeCl*	+ 1028	CsCl$_2^*$	+ 213	CuI$_2$	− 21
ArF*	+ 418	NaCl	− 411	CsF$_2^*$	− 125	CuBr$_2$	− 142
XeF*	+ 163	MgCl*	− 125	AgI$_2^*$	+ 280	CuCl$_2$	− 217
MgCl$_2$	− 639	AlCl*	− 188	AgCl$_2$	+ 96	CuF$_2$	− 890
NaCl$_2^*$	+ 2144	AlCl$_3$	− 694	AgF$_2$	− 205		

$(1176 \, \text{kJ mol}^{-1})$ and xenon $(1169 \, \text{kJ mol}^{-1})$, he reasoned, correctly, that the corresponding xenon compound should be stable.

Lower and higher valence compounds

Consider alkaline earth compounds. In these, the metal is always divalent. Since a great deal of extra energy is required to doubly ionize the metal atoms, it is reasonable to ask why monovalent compounds, such as 'MgCl', do not form. Data in Table 2.9 show that MgCl is indeed stable relative to the elements $(\Delta H(\text{calc.}) = -125 \, \text{kJ mol}^{-1})$ but that $MgCl_2$ is much more stable $(\Delta H = -639 \, \text{kJ mol}^{-1})$. This is shown by the following sequence:

$$2Mg + Cl_2 \xrightarrow{\makebox[3cm]{-250}} 2MgCl \xrightarrow{-389} Mg + MgCl_2$$
$$\underset{-639}{\xrightarrow{\hspace{6cm}}}$$

In any attempt to synthesize MgCl, attention should be directed towards keeping the reaction temperature low and/or isolating the MgCl product, in order to prevent it from reacting further or disproportionating. Similar trends are observed for other hypothetical compounds such as ZnCl, Zn_2O, AlCl and $AlCl_2$.

From a consideration of the factors that affect the stability of compounds, the following conclusions may be drawn about compounds with metals in unusual oxidation states:

(a) The formation and stability of compounds with lower than normal valence states appears to be favourable when (i) the second, and higher, ionization potentials of the metal are particularly high and (ii) the lattice energy of the corresponding compounds with the metal in its normal oxidation state is reduced.

(b) Conversely, in order to prepare compounds in which the metal has a higher than normal oxidation state and in which it is probably necessary to break into a closed electron shell, it is desirable to have (i) low values for the second (or higher) ionization potential of the metal atoms and (ii) large lattice energies of the resulting higher valence compounds.

As examples of these trends, calculations for the alkaline earth monohalides show that while all are unstable relative to the dihalides, the enthalpy of disproportionation is least in each case for the iodide $(U_{(MI_2)} < U_{(MBr_2)}$, etc.— effect a, ii). On the other hand, higher valence halogen compounds of the Group I elements are most likely to occur with caesium and the copper subgroup elements (effect b, i), in combination with fluorine (effect b, ii). Thus, from Table 2.9, all caesium dihalides, apart from CsF_2, have positive ΔH_f values and would be unstable. CsF_2 is stable in principle, with its negative ΔH_f value, but has not been prepared because its disproportionation to CsF has a large negative ΔH:

$$Cs + F_2 \xrightarrow{\makebox[3cm]{-125}} CsF_2 \xrightarrow{405} CsF + \tfrac{1}{2}F_2$$
$$\underset{-530}{\xrightarrow{\hspace{6cm}}}$$

For the silver dihalides, ΔH_f becomes less positive and finally negative across the series AgI_2 to AgF_2. This again shows the effect of r_e on U and hence on ΔH_f (effect b, ii). Unlike CsF_2, AgF_2 is a stable compound since AgF and AgF_2 have similar enthalpies of formation and therefore the disproportionation enthalpy of AgF_2 (to $AgF + \frac{1}{2}F_2$) is approximately zero.

The copper halides are particularly interesting. The divalent state of copper, in which the d^{10} shell of copper is broken into, is the most common state and again the dihalides show decreasing stability across the series CuF_2 to CuI_2 (Table 2.9): CuI_2 appears not to exist and its calculated ΔH_f is barely negative. On the other hand, in the monovalent state the situation is reversed and all the halides apart from CuF are known. CuF is calculated to be stable relative to the elements but not relative to CuF_2:

$$2Cu + F_2 \xrightarrow[\makebox[5cm]{-127}]{\makebox[2cm]{-36}} 2CuF \xrightarrow{-91} CuF_2 + Cu$$

The above examples show that several factors affect the formulae and stability of compounds: ionization potentials, lattice energies (via internuclear distances and the charges on ions) and the relative stability of elements in different oxidation states. Often a delicate balance between opposing factors controls the stability or instability of a compound and, as with the copper halides, detailed calculations are needed to assess the factors involved.

Partial covalent bonding

Covalent bonding, partial or complete, occurs when the outer electronic charge density on an anion is polarized towards and by a neighbouring cation. The net effect is that an electron pair which would be associated entirely with the anion in a purely ionic structure is displaced to occur between the anion and cation such that the electron pair is common to both. In cases of partial covalent bonding, some of the electron density is common to both atoms but the rest is still associated with the more electronegative atom.

Some clear-cut examples of the influence of bonding type on crystal structure are as follows:

(a) *SrO, BaO, HgO.* SrO and BaO both have the rock salt structure with octahedrally coordinated M^{2+} ions. Based on size considerations alone, and if it were ionic, HgO would also be expected to have the same structure. However, mercury is only *two* coordinate in HgO and the structure may be regarded as covalent. Linear O—Hg—O segments occur in the structure and may be rationalized on the basis of sp hybridization of mercury. The ground state of atomic mercury is

$$\text{(Xe core)} \quad 4f^{14} \quad 5d^{10} \quad 6s^2$$

The first excited state, corresponding to mercury (II), is

$$\text{(Xe core)} \quad 4f^{14} \quad 5d^{10} \quad 6s^1 \quad 6p^1$$

Hybridization of the 6s and one 6p orbital gives rise to two, linear sp hybrid orbitals, each of which forms a normal, electron pair, covalent bond by overlap with an orbital on oxygen. Hence mercury has a CN of two in HgO.

(b) AlF_3, $AlCl_3$, $AlBr_3$ and AlI_3. These compounds show a smooth transition from ionic to covalent bonding as the electronegativity difference between the two elements decreases. Thus, AlF_3 is a high melting, essentially ionic solid with a distorted octahedral coordination of Al^{3+} ions; its structure is related to that of $ReO_3 \cdot AlCl_3$ has a layered, polymeric structure in the solid state similar to the structure of $CrCl_3$, which is related to the $CdCl_2$ and CdI_2 structures. The bonds may be regarded as part ionic/part covalent. $AlBr_3$ and AlI_3 have molecular structures with dimeric units of formula Al_2X_6. Their structure and shape are shown in Fig. 1.20 and the bonding between aluminium and bromine or iodine is essentially covalent.

Halides of other elements, e.g. Be, Mg, Ga, In, also show variations in bond type and structure, depending on the halide. The trends are always the same; with the fluorides there is the largest difference in electronegativity between the two elements and these structures are the most ionic. With the other halides increasing covalent character occurs in the series, chloride–bromide–iodide. The occurrence of partial covalent bonding in an otherwise ionic structure may sometimes be detected by abnormally large values of the lattice energy (Table 2.8). In other cases, it is clear from the nature of the structure and the coordination numbers of the atoms that the bonding cannot be purely ionic.

Until comparatively recently, it has been difficult to quantify the degree of partial covalence in a particular structure, although one has intuitively felt that most so-called ionic structures must have a considerable degree of covalent character. Two new approaches which have had considerable success are the coordinated polymeric model of Sanderson and the ionicity plots of Mooser and Pearson.

Coordinated polymeric structures—Sanderson's model

A fresh approach to the theory and description of non-molecular crystal structures has been developed by Sanderson. Basically, he regards all bonds in non-molecular crystal structures as being polar covalent. The atoms contain partial charges whose values may be calculated readily using a new scale of electronegativities developed by Sanderson. From the partial charges, the relative contributions of ionic and covalent bonding to the total bond energy may be estimated. Since all non-molecular crystals contain only partially charged atoms, the ionic model represents an extreme form of bonding that is not attained in practice. Thus in, for example, KCl, the charges on the atoms are calculated to be ± 0.76 instead of ± 1.0 for a purely ionic structure.

The essential features of an atom which control its physical and chemical properties are (a) its electronic configuration and (b) the effective nuclear charge felt by the valence electrons. Recognition of the importance of the latter effect provides the starting point for Sanderson's theory.

Effective nuclear charge

The *effective nuclear charge* of an atom is the positive charge that would be felt by a foreign electron on arriving at the periphery of the atom. Atoms are, of course, electrically neutral overall but, nevertheless, the valence electrons of an atom are not very effective in shielding the outside world from the positive charge on the nucleus of the atom. Consequently, an incoming electron (e.g. an electron that belongs to a neighbouring atom and is coming to investigate the possibility of bond formation) feels a positive, attractive charge. Were this not the case and the surface of an atom were completely shielded from the nuclear charge, then atoms would have zero electron affinity and no bonds, ionic or covalent, would ever form.

The effective nuclear charge is greatest in elements which have a single vacancy in their valence shell, i.e. in the halogens. In the inert gases, the outermost electron shell is full and no foreign electrons can enter it. Consequently, incoming electrons would have to occupy vacant orbitals which are essentially 'outside' the atom and the effective nuclear charge experienced in such orbitals would be greatly reduced. Calculations of 'screening constants' were made by Slater who found that outermost electrons are much less efficient at screening nuclear charge than are electrons in inner shells. The screening constants of outermost electrons are calculated to be approximately one third; this means that for each unit increase in atomic number and positive nuclear charge across the series, e.g. sodium to chlorine, the additional positive charge is screened by only one third. Therefore, the effective nuclear charge increases in steps of two thirds and the valence electrons experience an increasingly strong attraction to the nucleus on going from sodium to chlorine. Similar effects occur throughout the periodic table: the effective nuclear charge is small for the alkali metals and increases to a maximum in the halogens.

Many atomic properties may be correlated with effective nuclear charge:

(a) Ionization potentials gradually increase from left to right across the periodic table.
(b) Electron affinities become increasingly negative in the same direction.
(c) Atomic radii progressively decrease from left to right.
(d) Electronegativities increase from left to right.

Let us consider two of these properties, atomic radii and electronegativities, in more detail.

Atomic radii

Atomic radii vary considerably for a particular atom depending on bond type and coordination number and many tabulations of radii are available; indeed, the subject of ionic radii is still controversial. Fortunately, however, the *non-polar covalent radii* of atoms can be measured accurately and represent a point of reference with which to compare other radii. Thus the atomic, non-polar covalent radius of carbon, given by half the C—C single bond length, is constant at 0.77 Å in materials as diverse as diamond and gaseous, paraffin hydrocarbons. Non-polar radii of atoms are listed in Table 2.10; from these radii, it is possible, using Sanderson's method, to estimate the effect that partial charges on atoms have on their radii. The general trends are that with increasing amounts of partial positive charge, the radii become smaller (i.e. as electrons are removed from the valence shell, but with the nuclear charge unchanged, so the remaining valence shell electrons feel a stronger attraction to the nucleus and the atom contracts). Conversely, with increasing negative charge on the atom, the radius becomes

Table 2.10 *Some electronegativity and size parameters of atoms. (After Sanderson, 1976)*

Element	S	$r_c(\text{Å})$	B (solid)	ΔS_c	$r_i(\text{Å})$
H	3.55	0.32		3.92	
Li	0.74	1.34	0.812	1.77	0.53
Be	1.99	0.91	0.330	2.93	0.58
B	2.93	0.82		2.56	
C	3.79	0.77		4.05	
N	4.49	0.74		4.41	
O	5.21	0.70	4.401	4.75	1.10
F	5.75	0.68	0.925	4.99	1.61
Na	0.70	1.54	0.763	1.74	0.78
Mg	1.56	1.38	0.349	2.60	1.03
Al	2.22	1.26		3.10	
Si	2.84	1.17		3.51	
P	3.43	1.10		3.85	
S	4.12	1.04	0.657	4.22	1.70
Cl	4.93	0.99	1.191	4.62	2.18
K	0.42	1.96	0.956	1.35	1.00
Ca	1.22	1.74	0.550	2·30	1.19
Zn	2.98			3.58	
Ga	3.28			3.77	
Ge	3.59	1.22		3.94	
As	3.90	1.19		4.11	
Se	4.21	1.16	0.665	4.27	1.83
Br	4.53	1.14	1.242	4.43	2.38
Rb	0.36	2.16	1.039	1.25	1.12
Sr	1.06	1.91	0.429	2.14	1.48
Ag	2.59	1.50	0.208		1.29
Cd	2.84	1.46	0.132	3.35	1.33
Sn	3.09	1.40		3.16, 3.66	
Sb	3.34	1.38		3.80	
Te	3.59	1.35	0.692	3.94	2.04
I	3.84	1.33	1.384	4.08	2.71
Cs	0.28	2.35	0.963	1.10	1.39
Ba	0.78	1.98	0.348	1.93	1.63
Hg	2.93			3.59	
Tl	3.02	1.48		2.85	
Pb	3.08	1.47		3.21, 3.69	
Bi	3.16	1.46		3.74	

larger. Sanderson developed a simple, empirical formula to quantify the variation of radius, r, with partial charge:

$$r = r_c - B\delta \tag{2.29}$$

where r_c is the non-polar covalent radius, B is a constant for a particular atom and δ is its partial charge. B values are also given in Table 2.10. The partial charges on atoms cannot be measured directly but may be estimated from Sanderson's electronegativity scale, as follows.

Electronegativity and partially charged atoms

The electronegativity of an atom is a measure of the net attractive force experienced by an outermost electron interacting with the nucleus. The concept of electronegativity was originated by Pauling as a parameter that would correlate with the observed polarity of bonds between unlike atoms. Atoms of high electronegativity attract electrons (in a covalent bond) more than do atoms of low electronegativity and, hence, they acquire a partial negative charge. The magnitude of the partial charge depends on the initial difference in electronegativity between the two atoms. The difficulty in working with electronegativities in the past has been that electronegativity has not had a precise definition and, therefore, it has not been possible to use an absolute scale of electronegativities and make accurate calculations. Pauling observed a correlation between the strengths of polar bonds and the degree of polarity in the bonds. He proposed that bond strengths are a combination of (a) a homopolar bond energy and (b) an 'extra ionic energy' due to bond polarity and, hence, electronegativity difference. He then used this correlation between polarity and extra ionic energy to establish his scale of electronegativities. These ideas have since been extended and modified by Sanderson to permit quantitative calculations of bond energies to be made for a wide variety of compounds. However, Sanderson used a different method to derive a scale of electronegativities. Since electronegativity is a measure of the attractive force between the effective nuclear charge and an outermost electron, it is related to the compactness of an atom. He used the relation:

$$S = \frac{D}{D_a} \tag{2.30}$$

to evaluate the electronegativity, S, in which D is the electron density of the atom (given by the ratio of atomic number: atomic volume) and D_a is the electron density that would be expected for the atom by linear interpolation of the D values for the inert gas elements. The electronegativity values so obtained are listed in Table 2.10, with some minor modifications made by Sanderson.

An important contribution to our understanding of bond formation is *the principle of electronegativity equalization* proposed by Sanderson. It may be stated: 'When two or more atoms initially different in electronegativity combine chemically, they become adjusted to the same intermediate electronegativity within the compound.' It is found in practice that the value of the intermediate electronegativity is given by the geometric mean of the individual electronegativities of the component atoms, e.g. for NaF:

$$S_b = \sqrt{S_{Na}S_F} = 2.006 \tag{2.31}$$

This means that in a bond between unlike atoms, the bonding electrons are preferentially and partially transferred from the less electronegative to the more electronegative atom. A partial negative charge on the more electronegative atom results whose magnitude is defined, by Sanderson, as follows: '*Partial charge* is defined as the ratio of the change in electronegativity undergone by an atom on bond formation to the change it would have undergone on becoming completely ionic with charge $+$ or -1.'

In order to make calculations of partial charges, a point of reference was necessary and for this it was assumed that the bonds in NaF are 75 per cent ionic; this assumption appears to have been well justified by subsequent developments. Further, it was necessary to assume that electronegativity changes linearly with charge. It can then be shown that the change in electronegativity, ΔS_c, of any atom on acquiring a unit positive or unit negative charge is given by

$$\Delta S_c = 2.08 \sqrt{S} \tag{2.32}$$

Partial charge, δ, may therefore be defined as

$$\delta = \frac{\Delta S}{\Delta S_c} \tag{2.33}$$

where $\Delta S = S - S_b$. Values of ΔS_c are also given in Table 2.10.

Returning now to the radii of partially charged atoms, the problem in assigning individual radii to atoms or ions has been the question of how to divide an experimentally observed internuclear distance into its component radii. Many methods have been tried and various tabulations of radii are available. All of them are additive in that they correctly predict bond distances. The method adopted by Pauling (and Sanderson) has been to divide the experimental internuclear distance in an isoelectronic 'ionic' crystal, such as NaF, according to the inverse ratio of the effective nuclear charges on the two 'ions'. The effective nuclear charges can be calculated from the screening constants. From a knowledge of the partial charges on the atoms and assuming that radii change systematically with partial charge (according to the relation $r = r_c - B\delta$), the ionic radii of Na^+ and F^- may be calculated; these then serve as a point of reference for calculating radii of other ions in materials that are not isoelectronic. In Table 2.10 are listed the radii of singly charged ions in the solid state calculated by the above method; also given are B and r_c data to enable the radii of partially charged atoms to be calculated and electronegativity data S, ΔS_c to enable the partial charges, δ, to be calculated.

Let us consider the use of these data for one example, BaI_2. From Table 2.10, $S_{Ba} = 0.78$ and $S_I = 3.84$; the intermediate electronegativity, S_b, is given by $S_b = \sqrt[3]{S_{Ba}S_I^2} = 2.26$. Therefore, for barium, $\Delta S = 2.26 - 0.78 = 1.48$ and for iodine, $\Delta S = 3.84 = 2.26 = 1.58$. The values of ΔS_c are 1.93 (Ba) and 4.08 (I). Hence $\delta_{Ba} = 1.48/1.93 = 0.78$ and $\delta_I = 1.58/4.08 = -0.39$, i.e. BaI_2 is ~ 39 per cent ionic, 61 per cent covalent (using the δ_I value). The radii of the partially charged atoms may now be calculated. For barium, $r_c = 1.98$, $B = 0.348$ and δ is calculated to be 0.78; hence $r_{Ba} = r_c - B\delta = 1.71$ Å. For iodine, $r_c = 1.33$, $B = 1.384$ and $\delta = -0.39$; hence $r_I = 1.87$ Å. Therefore, the barium–iodine distance is calculated to be $1.87 + 1.71 = 3.58$ Å, which compares very well with the experimental value of 3.59 Å.

Using the above methods, Sanderson has evaluated the partial charges and atomic radii in a large number of solid compounds. For example, data are given in Table 2.11 for a number of mono- and divalent chlorides; the charge on the chloride atom varies from -0.21 in $CdCl_2$ to -0.81 in CsCl and at the same time the calculated radius of the chlorine atom varies from 1.24 to 1.95 Å. These compare with a non-polar covalent radius of 0.99 Å and an ionic radius of 2.18 Å (Table 2.10). While these radii and partial charges may not be quantitatively correct, because some of the assumptions involved in their calculation are rather empirical, they nevertheless appear to be realistic. Most of the compounds in Table 2.11 are normally regarded as ionic, and if the data are in any way correct, they clearly show that it is unrealistic and misleading to assign a radius to the chloride ion which is constant for all solid chlorides.

A similar but more extensive list of the partial charges on oxygen in a variety of oxides is given in Table 2.12, where values cover almost the entire range between 0 and -1. Although oxides are traditionally regarded as containing the oxide ion, O^{2-}, the calculations show that the actual charge carried by an oxygen never exceeds -1 and is usually much less than -1.

From the principle of electronegativity equalization, electrons in a hypothetical covalent bond are partially transferred to the more electronegative atom. This removal of electrons from the electropositive atom leads to an increase in its

Table 2.11 *Partial charge and radius of the chlorine atom in some solid chlorides*

Compound	$-\delta_{Cl}$	r_{Cl}(Å)
$CdCl_2$	0.21	1.24
$BeCl_2$	0.28	1.26
CuCl	0.29	1.34
AgCl	0.30	1.35
$MgCl_2$	0.34	1.39
$CaCl_2$	0.40	1.47
$SrCl_2$	0.43	1.50
$BaCl_2$	0.49	1.57
LiCl	0.65	1.76
NaCl	0.67	1.79
KCl	0.76	1.90
RbCl	0.78	1.92
CsCl	0.81	1.95

Table 2.12 *Partial charge on oxygen in some solid oxides*

Compound	$-\delta_0$	Compound	$-\delta_0$	Compound	$-\delta_0$	Compound	$-\delta_0$
Cu_2O	0.41	HgO	0.27	Ga_2O_3	0.19	CO_2	0.11
Ag_2O	0.41	ZnO	0.29	Tl_2O_3	0.21	GeO_2	0.13
Li_2O	0.80	CdO	0.32	In_2O_3	0.23	SnO_2	0.17
Na_2O	0.81	CuO	0.32	B_2O_3	0.24	PbO_2	0.18
K_2O	0.89	BeO	0.36	Al_2O_3	0.31	SiO_2	0.23
Rb_2O	0.92	PbO	0.36	Fe_2O_3	0.33	MnO_2	0.29
Cs_2O	0.94	SnO	0.37	Cr_2O_3	0.37	TiO_2	0.39
		FeO	0.40	Sc_2O_3	0.47	ZrO_2	0.44
		CoO	0.40	Y_2O_3	0.52	HfO_2	0.45
		NiO	0.40	La_2O_3	0.56		
		MnO	0.41				
		MgO	0.50				
		CaO	0.56				
		SrO	0.60				
		BaO	0.68				

effective nuclear charge, decrease in its size and hence an increase in its effective electronegativity. Likewise, as the electronegative atom acquires electrons, so its ability to attract still more electrons diminishes and its electronegativity decreases. In this way the electronegativities of the two atoms adjust themselves until they are equal. This principle of electronegativity equalization may be applied equally to diatomic gas molecules, in which only one bond is involved, or to three-dimensional solid structures in which each atom is surrounded by and bonded to several others. It illustrates how covalent bonds that are initially non-polar may become polar due to electronegativity equalization.

An alternative approach is to start with purely ionic bonds and consider how they may acquire some covalent character. In an ionic structure, M^+X^-, the cations are surrounded by anions (usually 4, 6 or 8). However, the cations have empty valence shells and are potential electron pair acceptors; likewise, anions with their filled valence shells are potential electron pair donors. The cations and anions therefore interact in the same way as do Lewis acids and bases: the anions, with their lone pairs of electrons, coordinate to the surrounding cations. The strength of this interaction, and hence the degree of covalent bonding which results, is again related to the electronegativities of the two atoms. Thus electronegative cations such as Al^{3+} are much stronger electron pair acceptors (and Lewis acids) than are electropositive cations such as K^+. The *coordinated polymeric model* of structures, proposed by Sanderson, is based on this idea of acid–base interactions between ions. It therefore forms a bridge between the ionic and covalent extremes of bond type.

Mooser–Pearson plots and ionicities

While the radius ratio rules appear to be rather unsatisfactory for predicting and explaining the structure adopted by, for instance, particular AB compounds,

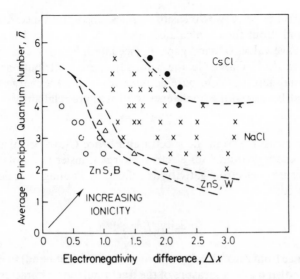

Fig. 2.6 Mooser–Pearson plot for AB compounds containing A group cations. (From Mooser and Pearson, *Acta Cryst.*, **12**, 1015, 1959). Arrow indicates direction of increasing bond ionicity

an alternative approach by Mooser and Pearson has had considerable success. This approach focuses on the directionality or covalent character of bonds. The two factors which are regarded as influencing covalent bond character in crystals are (a) the average principal quantum number, $\bar{n}$, of the atoms involved and (b) their difference in electronegativity, Δx. In constructing Mooser–Pearson plots, these two parameters are plotted against each other, as shown for AB compounds in Fig. 2.6. The most striking feature of the plot is the almost perfect separation of compounds into four groups corresponding to the structures: zinc blende (ZnS, B), wurtzite (ZnS, W), rock salt and CsCl. Similarly, successful diagrams have been presented for other formula types—AB_2, AB_3, etc. Of the four simple AB structure types, one's intuitive feeling is that zinc blende (or sphalerite) is the most covalent and either rock salt or CsCl is the most ionic. Mooser–Pearson plots present this intuitive feeling in diagrammatic form.

The term *ionicity* has been used to indicate the degree of ionic character in bonds; in Fig. 2.6, ionicity increases from the bottom left to the top right of the diagram, as shown by the arrow. Thus it appears that ionicity is not governed by electronegativity alone but also depends on the principal valence shell of the atoms and hence on atomic size. There is a general trend for highly directional covalent bonds to be associated with lighter elements, i.e. at the bottom of Fig. 2.6, and with small values of Δx, i.e. at the left-hand side of Fig. 2.6.

The fairly sharp crossover between the different 'structure fields' in Fig. 2.6 suggests that for each structure type there are critical ionicities which place a limit on the compounds that can have that particular structure. Theoretical support for Mooser–Pearson plots has come from the work of Phillips (1970) and van

Vechten and Phillips (1970) who measured optical absorption spectra of some AB compounds and, from these, calculated electronegativities and ionicities. The spectral data gave values of band gaps, E_g (see later). For isoelectronic series of compounds, e.g. ZnSe, GaAs, Ge, the band gaps have contributions from: (a) a homopolar band gap, E_h, as in pure germanium, and (b) a charge transfer, C, between A and B, termed the 'ionic energy'. These are related by

$$E_g^2 = E_h^2 + C^2 \qquad (2.34)$$

E_g and E_h are measured from the spectra and hence C can be calculated. C is related to the energy required for electron charge transfer in a polar bond and hence is a measure of electronegativity, as defined by Pauling. A scale of ionicity has been devised:

$$\text{Ionicity, } f_i = \frac{C^2}{E_g^2} \qquad (2.35)$$

Values of f_i range from zero ($C = 0$ in a homopolar covalent bond) to one ($C = E_g$ in an ionic bond) and give a measure of the fractional ionic character of a bond. Phillips analysed the spectroscopic data for 68 AB compounds with either octahedral or tetrahedral structures and found that the compounds fall into two groups separated by a critical ionicity, f_i, of 0.785.

The link between Mooser–Pearson plots and Phillips–Van Vechten ionicities is that

$$\Delta x (\text{Mooser–Pearson}) \simeq C (\text{Phillips})$$

and

$$\bar{n} (\text{Mooser–Pearson}) \simeq E_h \ (\text{Phillips})$$

The explanation of the latter is that as the principal quantum number of an element increases, the outer orbitals become larger and more diffuse and the energy differences between outer orbitals (s, p, d and/or f) decrease; the band gap, E_h, decreases until metallic behaviour occurs at $E_h = 0$. Use of $\bar{n}$ gives the average behaviour of anion and cation. The Phillips–Van Vechten analysis has so far been restricted to AB compounds but this has provided a theoretical justification for the more widespread use and application of the readily constructed Mooser–Pearson plots.

Bond valence and bond length

The structures of most *molecular* materials—organic and inorganic—may be described satisfactorily using valence bond theory in which single, double, triple and occasionally partial bonds occur between atoms. Difficulties rapidly arise, however, when valence bond theory is applied to crystalline, *non-molecular* inorganic materials, even though the bonding in them may be predominantly covalent. Ths is because there are usually insufficient bonding electrons available for each bond to be treated as an electron pair single bond. Instead, most bonds must be regarded as partial bonds.

An empirical but nevertheless useful approach to describing such bonds has been developed by Pauling, Brown, Shannon, Donnay and others, and involves the evaluation of *bond orders* or *bond valences* in a structure. Bond valences are defined in a similar way to electrostatic bond strengths in Pauling's electrostatic valence rule for ionic structures. As such, bond valences represent an extension of Pauling's rule to structures that are not necessarily ionic. Bond valences are defined empirically, using information on atom oxidation states and experimental bond lengths; no reference is made, at least not initially, to the nature of the bonding, whether it be covalent, ionic or some blend of the two.

Pauling's electrostatic valence rule requires that the sum of the electrostatic bond strengths between an anion and its neighbouring cations should be equal in magnitude to the formal charge on the anion (equation (2.2) and (2.3)). This rule may be modified to include structures which are not necessarily ionic by replacing (a) the electrostatic bond strength by the bond valence and (b) the formal charge on the anion by the valence of that atom (valence being defined as the number of electrons that take part in bonding). This leads to the *valence sum rule* which relates the valence, V_i, of atom i to the bond valence, b_{ij}, between atom i and neighbouring atom j; i.e.

$$V_i = \sum_j b_{ij} \qquad (2.36)$$

Thus, the valence of an atom must be equal to the sum of the bond valences for all the bonds that it forms. For cases in which b_{ij} is an integer, this rule becomes the familiar rule that is used for evaluating the number of bonds around an atom in molecular structures, i.e. the valence of an atom is equal to the number of bonds that it forms (counting double bonds as two bonds, etc.). In non-molecular, inorganic structures, however, integral bond valences are the exception rather than the rule.

The electrostatic bond strength in Pauling's rule is given simply by the ratio of cation charge: cation coordination number. Thus for structures in which the cation coordination is irregular or the cation–anion bonds are not all of the same length, only an average electrostatic bond strength is obtained. One advantage of the bond valence approach is that each bond is treated as an individual and hence irregularities or distortions in coordination environments can be taken into account.

For a given pair of elements, an inverse correlation between bond valence and bond length exists, as shown in Fig. 2.7 for bonds between oxygen and atoms of the second row in their group valence states, i.e. Na^I, Mg^{II}, Al^{III}, Si^{IV}, P^V and S^{VI}. While each atom and oxidation (or valence) state has its own bond valence–bond length curve, it is a considerable simplification that 'universal curves' such as Fig. 2.7 may be used for isoelectronic series of ions. Various analytical expressions have been used to fit curves such as Fig. 2.7, including

$$b_{ij} = \left(\frac{R_0}{R}\right)^N \qquad (2.37)$$

where R is the bond length and R_0, N are constant (R_0 is the value of the bond

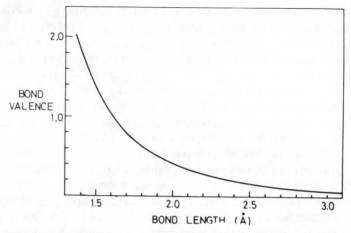

Fig. 2.7 Bond valence–bond length universal correlation curve for bonds between oxygen and second row atoms: Na, Mg, Al, Si, P and S. (From Brown, *Chem. Soc. Revs.*, **7** (3), 359, 1978)

length for unit bond valence); for the elements represented by Fig. 2.7, $R_0 = 1.622$ and $N = 4.290$.

From Fig. 2.7, bond length clearly increases with decreasing bond valence and since, for a given atom, bond valence inevitably decreases with increasing coordination number, a correlation also exists between increasing coordination number and increasing bond length. This correlation has already been presented, in a slightly different form, in Fig. 2.3, in which the ionic radii of cations increase with increasing coordination number. Since the data of Fig. 2.3 assume a constant radius of the fluoride or oxide ion, the ordinate in Fig. 2.3 could be changed from the ionic radius to the metal–oxygen bond length.

Curves such as Fig. 2.7 are important for several reasons, the most important of which is that they lead to an increased rationalization and understanding of crystal structures. They have several applications that are specifically associated with the determination of crystal structures; for example:

(a) As a check on the correctness of a proposed structure, the valence sum rule should be obeyed by all atoms of the structure to within a few per cent.

(b) To locate hydrogen atoms. Hydrogen atoms are often 'invisible' in X-ray structure determinations because of the very low scattering power of hydrogen. It may be possible to locate them by evaluating the bond valence sums around each atom and noting which atoms (e.g. oxygen atoms in hydrates) show a large discrepancy between the atom valence and the bond valence sum. The hydrogen is then likely to be bonded to such atoms.

(c) To distinguish between Al^{3+} and Si^{4+} positions in aluminosilicate structures. By X-ray diffraction, Al^{3+} and Si^{4+} cannot be distinguished because of their very similar scattering power, but in sites of similar coordination number they give different bond valences, e.g. in regular MO_4 tetrahedra, Si—O bonds have a bond valence of one but Al—O bonds have a bond valence of 0.75. Site occupancies may therefore be determined using the valence sum rule and/or by comparing the M—O bond lengths with values expected for Si—O and Al—O.

Non-bonding electron effects

In this section, the influence on structure of two types of non-bonding electrons is considered: the d electrons in transition metal compounds and the s^2 pair of electrons in compounds of the heavy p-block elements in low oxidation states. These two types of electron do not take part in bonding as such, but nevertheless exert a considerable influence on the coordination number and environment of the metal atom.

d-electron effects

In transition metal compounds, the majority of the d electrons on the metal atom do not usually take part in bond formation but do influence the coordination environment of the metal atom and are responsible for properties such as magnetism. For present purposes, basic crystal field theory (CFT) is adequate to describe qualitatively the effects that occur. It is assumed that the reader is acquainted with CFT and only a summary will be given here.

Crystal field splitting of energy levels

In an octahedral environment, the five d orbitals on a transition metal atom are no longer degenerate but split into two groups, the t_{2g} group of lower energy and

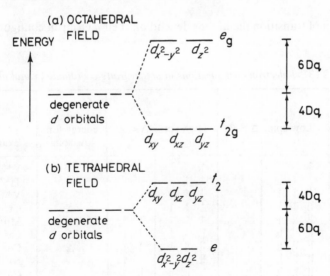

Fig. 2.8 Splitting of d energy levels in (a) an octahedral and (b) a tetrahedral field

low spin, $\Delta > P$ high spin, $\Delta < P$

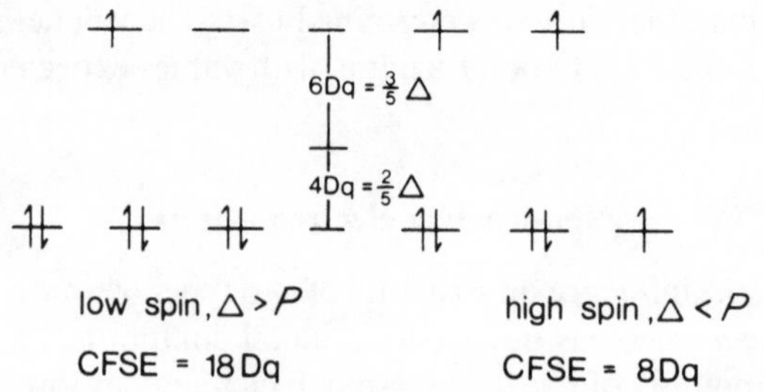

CFSE = 18 Dq   CFSE = 8 Dq

Fig. 2.9 Low spin and high spin states for a d^7 transition metal ion in an octahedral coordination environment

the e_g group of higher energy (Fig. 2.8a). If possible, electrons occupy orbitals singly, according to Hund's rule of maximum multiplicity. For d^4 to d^7 atoms or ions, two possible configurations occur, giving low spin (LS) and high spin (HS) states; these are shown for a d^7 ion in Fig. 2.9. In these, the increased energy, Δ, required to place an electron in an e_g orbital, and hence maximize the multiplicity, has to be balanced against the repulsive energy or pairing energy, P, which arises when two electrons occupy the same t_{2g} orbital. The magnitude of Δ depends on the ligand or anion to which the metal is bonded: for weak field anions, Δ is small and the HS configuration occurs, and vice versa for strong field ligands. Δ also depends on the metal and, in particular, to which row it belongs: generally $\Delta(5d) > \Delta(4d) > \Delta(3d)$. Consequently HS behaviour is rarely observed in the $4d$ and $5d$ series. Δ values may be determined experimentally from electronic spectra. The possible spin configurations for the different numbers of d electrons are given in Table 2.13.

The radii of transition metal ions depend on their d electron configuration, as

Table 2.13   *d-electron configurations in octahedrally coordinated metal atoms*

Number of electrons	Low spin, $\Delta > P$ t_{2g}	e_g	High spin, $\Delta < P$ t_{2g}	e_g	Gain in orbital energy for low spin	Example
1	↑		↑			V^{4+}
2	↑ ↑		↑ ↑			Ti^{2+}, V^{3+}
3	↑ ↑ ↑		↑ ↑ ↑			V^{2+}, Cr^{3+}
4	↑↓ ↑ ↑		↑ ↑ ↑	↑	Δ	Cr^{2+}, Mn^{3+}
5	↑↓ ↑↓ ↑		↑ ↑ ↑	↑ ↑	2Δ	Mn^{2+}, Fe^{3+}
6	↑↓ ↑↓ ↑↓		↑↓ ↑ ↑	↑ ↑	2Δ	Fe^{2+}, Co^{3+}
7	↑↓ ↑↓ ↑↓	↑	↑↓ ↑↓ ↑	↑ ↑	Δ	Co^{2+}
8	↑↓ ↑↓ ↑↓	↑ ↑	↑↓ ↑↓ ↑↓	↑ ↑		Ni^{2+}
9	↑↓ ↑↓ ↑↓	↑↓ ↑	↑↓ ↑↓ ↑↓	↑↓ ↑		Cu^{2+}
10	↑↓ ↑↓ ↑↓	↑↓ ↑↓	↑↓ ↑↓ ↑↓	↑↓ ↑↓		Zn^{2+}

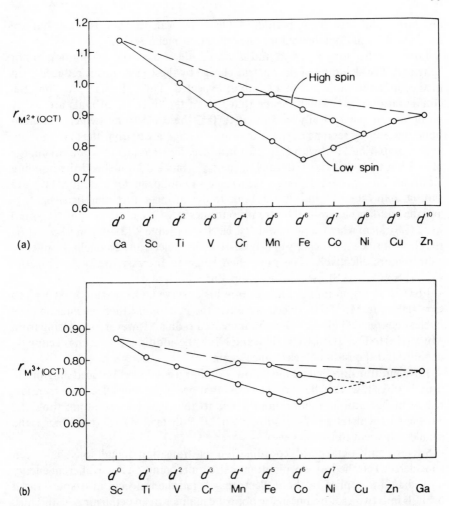

Fig. 2.10 (a) Radii of octahedrally coordinated divalent transition metal ions. (b) Radii of octahedrally coordinated trivalent transition metal ions. Data from Shannon and Prewitt, relative to $r_{F^-} = 1.19\,\text{Å}$

shown in Fig. 2.10(a) for the octahedrally coordinated divalent ions. With increasing atomic number, several trends occur. First, there is a gradual, overall decrease in radius as the d shell is filled, as shown by the dashed line that passes through Ca^{2+}, Mn^{2+}(HS) and Zn^{2+}. For these three ions, the distribution of d electron density around the M^{2+} ion is spherically symmetrical because the d orbitals are either empty (Ca), singly occupied (Mn) or doubly occupied (Zn). The gradual decrease in radius with increasing atomic number is associated with poor shielding of the nuclear charge by the d electrons; hence a greater effective nuclear charge is experienced by the outer, bonding electrons which results in a steady contraction in radius with increasing atomic number. Similar effects occur across any horizontal row of the periodic table as the valence shell is filled, but are particularly well documented for the transition metal series.

For the other ions d^1 to d^4 and d^6 to d^9, the d electron distribution is not spherical. The shielding of the nuclear charge by these electrons is reduced even further and the radii are smaller than expected. Thus, the Ti^{2+} ion has the configuration $(t_{2g})^2$, which means that two of the three t_{2g} orbitals are singly occupied. In octahedrally coordinated Ti^2, these electrons (which are non-bonding) occupy regions of space that are directed away from the (Ti^{2+}-anion) axes. Comparing Ti^{2+} with Ca^{2+}, for instance, Ti^{2+} has an extra nuclear charge of $+2$ but the two extra electrons in the t_{2g} orbitals do not shield the bonding electrons from this extra charge. Hence, Ti—O bonds in, for example, TiO, are shorter than Ca—O bonds in CaO due to the stronger attraction between Ti^{2+} and the bonding electrons. This trend continues in V^{2+}, Cr^{2+} (LS), Mn^{2+} (LS) and Fe^{2+} (LS), all of which contain only t_{2g} electrons (Table 2.13). Beyond Fe^{2+} (LS), the electrons begin to occupy e_g orbitals and these electrons do shield the nuclear charge more effectively. The radii then begin to increase again in the series Fe^{2+} (LS), Co^{2+} (LS), Ni^{2+}, Cu^{2+} and Zn^{2+}.

For the high spin ions a different trend is observed. On passing from V^{2+} to Cr^{2+}(HS) and Mn^{2+}(HS), electrons enter the e_g orbitals, thereby shielding the nuclear charge and giving rise to an increased radius. However, on passing from Mn^{2+}(HS) to Fe^{2+}(HS), Co^{2+}(HS) and Ni^{2+}, the additional electrons occupy t_{2g} orbitals and the radii decrease once again.

Trivalent transition metal ions show a similar trend but of reduced magnitude (Fig. 2.10b). However, the various effects that occur are now effectively transferred to higher atomic number and hence to the right by one atom; thus, the trivalent ion with the smallest radius is Co^{3+} (LS) instead of Fe^{2+} (LS) which is the smallest divalent ion.

So far, only octahedral coordination of transition metal ions has been considered. Tetrahedral coordination is also quite common but a different energy level diagram applies to the d electrons. A tetrahedral field also splits the d orbitals into two groups, but in the opposite manner to an octahedral field. Thus, three orbitals have higher energy d_{xy}, d_{xz} and d_{yz} whereas the other two, $d_{x^2-y^2}$ and d_{z^2} have lower energy (Fig. 2.8b).

It was mentioned earlier that crystal field splitting of d orbitals in transition metal ions may result in *crystal field stabilization energies* (CFSE) and increased lattice energies of ionic compounds. For example, CoF_2 has the rutile structure with octahedrally coordinated $Co^{2+}(d^7)$ ions in which Co^{2+} adopts the high spin configuration (Fig. 2.9). The energy difference, Δ, between t_{2g} and e_g orbitals is set equal to 10 Dq and the t_{2g} orbitals are stabilized by an amount 4 Dq whereas the e_g orbitals are destabilized by an amount 6 Dq. Relative to the situation of five degenerate orbitals without crystal field splitting (Fig. 2.8), the crystal field stabilization energy of Co^{2+} in HS and LS states may be calculated. For the LS state, the CFSE is 6×4 Dq $- 1 \times 6$ Dq $= 18$ Dq. For the HS state, the CFSE is 5×4 Dq $- 2 \times 6$ Dq $= 8$ Dq.

The occurrence of CFSE leads to an increased lattice energy. The value of the

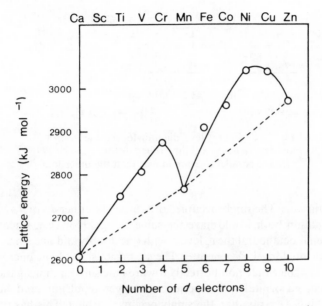

Fig. 2.11 Lattice energies of divalent transition metal fluorides determined from Born–Haber cycle calculations. (From Waddington, *Adv. Inorg. Chem. Radiochem.*, 1, 157–221, 1959)

CFSE calculated for Co^{2+}(HS) in rutile is $104 \, kJ \, mol^{-1}$. This compares fairly well with the value of $83 \, kJ \, mol^{-1}$ given by the difference between the lattice energy determined from a Born–Haber cycle ($2959 \, Kj \, mol^{-1}$) and that calculated using the Born–Mayer equation ($2876 \, kJ \, mol^{-1}$).

The lattice energies of the divalent fluorides of the first row transition elements, determined from a Born–Haber cycle, are shown in Fig. 2.11. Similar trends are observed for other halides. Ions that do not exhibit CFSE are d^0(Ca), d^5(HS)(Mn) and d^{10}(Zn), and their lattice energies fall on the lower, dashed curve. Most ions do show some degree of CFSE, however, and their lattice energies fall on the upper, solid curve. For the fluorides (Fig. 2.11), the agreement between the calculated CFSE and the difference in lattice energy ΔU (i.e. U(Born–Haber)–U(Born–Mayer)) is reasonable and indicates that the bonding may be treated as ionic. For the other halides, however, $\Delta U \gg$ CFSE and indicates that other effects, perhaps covalent bonding, must be present.

Jahn–Teller distortions

In many transition metal compounds, the metal coordination is distorted octahedral and the distortions are such that the two axial bonds are either shorter than or longer than the other four bonds. The Jahn–Teller effect is responsible for these distortions in d^9, d^7(LS) and d^4(HS) ions. Consider the d^9 ion Cu^{2+} whose configuration is $(t_{2g})^6(e_g)^3$. One of the e_g orbitals contains two electrons and the

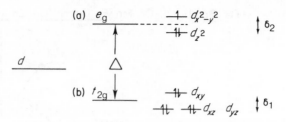

Fig. 2.12 Energy level diagram for the d-levels in a d^9 ion experiencing a Jahn–Teller distortion. The two bonds parallel to z are longer than the other four

other contains one. The singly occupied orbital can be either d_{z^2} or $d_{x^2-y^2}$ and in a free ion situation both would have the same energy. However, since the metal coordination is octahedral the e_g levels, with one doubly and one singly occupied orbitals, are no longer degenerate. The e_g orbitals are high energy orbitals (relative to t_{2g}) since they point directly towards the surrounding ligands and the doubly occupied orbital will experience stronger repulsions and hence have somewhat higher energy than the singly occupied orbital. This has the effect of lengthening the metal–ligand bonds in the directions of the doubly occupied orbital, e.g. if the d_{z^2} orbital is doubly occupied, the two metal–ligand bonds along the z axis will be longer than the other four metal–ligand bonds. The energy level diagram for this latter situation is shown in Fig. 2.12(a). Lengthening of the metal–ligand bond along the z axis leads to a *lowering* of energy of the d_{z^2} orbital. The distorted structure is stabilized by an amount $\frac{1}{2}\delta_2$ relative to the regular octahedral arrangement and, hence, the distorted structure becomes the observed, ground state.

High spin d^4 and low spin d^7 ions also have odd numbers of e_g electrons and show Jahn–Teller distortions. It is not clear which type of distortion is preferred (i.e. two short and four long bonds, or vice versa) and the actual shapes in a particular structure must be determined experimentally. The degeneracy of the t_{2g} levels may also be removed by the Jahn–Teller effect, but the magnitude of the splitting, δ_1 in Fig. 2.12(b), is small and the effect is relatively unimportant.

The normal coordination environment of the Cu^{2+} ion is distorted octahedral with four short and two long bonds. The distortion varies from compound to compound. In, for example, CuF_2 (distorted rutile structure), the distortion is fairly small (four fluorine atoms at $1.93\,\text{Å}$, two at $2.27\,\text{Å}$), but is larger in $CuCl_2$ (four chlorine atoms at $2.30\,\text{Å}$, two at $2.95\,\text{Å}$) and extreme in tenorite, CuO, which is effectively square planar (four oxygen atoms at $1.95\,\text{Å}$, two at $2.87\,\text{Å}$).

The importance of Jahn–Teller distortions in Cu^{2+} and Cr^{2+} (d^4 ion) compounds is seen by comparing the structures of the oxides and fluorides of the first-row divalent, transition metal ions. For the oxide series, MO (M^{2+} is Ti, V, Cr, Mn, Fe, Co, Ni and Cu), all have the rock salt structure with regular octahedral coordination apart from (a) CuO which contains grossly distorted (CuO_6) octahedra and possibly (b) CrO, whose structure is not known. For the fluoride series, MF_2, all have the regular rutile structure apart from CrF_2 and CuF_2 which have distorted rutile structures.

Other examples of distorted octahedral coordination due to the Jahn–Teller effect are found in compounds of Mn^{3+} (HS) and Ni^{3+} (LS).

Square planar coordination

The d^8 ions—Ni^{2+}, Pd^{2+}, Pt^{2+}—commonly have square planar or rectangular planar coordination in their compounds. In order to understand this, consider the d energy level diagram for such ions in (a) octahedral and (b) distorted octahedral fields:

(a) The normal configuration of a d^8 ion in an octahedral field is $(t_{2g})^6(e_g)^2$. The two e_g electrons singly occupy the d_{z^2} and $d_{x^2-y^2}$ orbitals, which are degenerate, and the resulting compounds, with unpaired electrons, are paramagnetic.

(b) Consider, now, the effect of distorting the octahedron and lengthening the two metal–ligand bonds along the z axis. The e_g orbitals lose their degeneracy, and the d_{z^2} orbital becomes stabilized by an amount $\frac{1}{2}\delta_2$ (Fig. 2.12). For small elongations along the z axis, the pairing energy required to doubly occupy the d_{z^2} orbital is larger than the energy difference between d_{z^2} and $d_{x^2-y^2}$, i.e. $P > \delta_2$. There is no gain in stability if the d_{z^2} orbital is doubly occupied, therefore, and no reason why small distortions from octahedral coordination should be stable. With increasing elongation along the z axis, however, a stage is reached where $P < \delta_2$ and in which the doubly occupied d_{z^2} orbital becomes stabilized and is the preferred ground state for a d^8 ion. The distortion from octahedral coordination is now sufficiently large that the coordination is regarded as square planar; in many cases, e.g. PdO, there are, in fact, no axial ligands along z and, hence, the transformation from octahedral to square planar coordination is complete. Because they have no unpaired electrons, square planar compounds are diamagnetic.

Square planar coordination is more common with $4d$ and $5d$ transition elements than with $3d$ elements because the $4d$ and, especially, $5d$ orbitals are more diffuse and extend to greater radial distances from the nucleus. Consequently, the magnitude of the crystal (or ligand) field splitting (Δ, δ) caused by a particular ligand, e.g. O^{2-}, increases in the series $3d < 4d < 5d$. Thus, NiO has the rock salt structure with regular octahedral coordination of Ni^{2+} ($3d$ ion) whereas PdO and PtO both have square planar coordination for the metal atoms ($4d$ and $5d$). The only known compound of palladium with octahedral Pd^{2+} is PdF_2 (rutile structure) and no octahedral Pt^{2+} or Au^{3+} compounds are known.

Tetrahedral coordination

As stated earlier, a tetrahedral field causes splitting of the d energy levels, but in the opposite sense to an octahedral field (Fig. 2.8b). Further, the magnitude of the

104

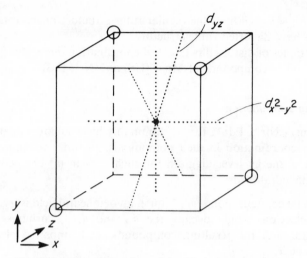

Fig. 2.13 The orientation of d orbitals in a tetrahedral field

splitting, Δ, is generally less in a tetrahedral field since none of the d orbitals point directly towards the four ligands. Rather, the d_{xy}, d_{xz} and d_{yz} orbitals are somewhat closer to the ligands than are the other two orbitals, Fig. 2.13 (only two orbital directions, d_{yz} and $d_{x^2-y^2}$ are shown). Jahn–Teller distortions again occur, especially when the upper t_2 orbitals contain 1, 2, 4 or 5 electrons (e.g. d^3(HS), d^4(HS), d^5 and d^9). Details will not be given since various types of distortion are possible (e.g. tetragonal or trigonal distortions) and these have not been as well studied as have the octahedral distortions. A common type of distortion is a flattening or elongation of the tetrahedron in the direction of one of the twofold axes of the tetrahedron (e.g. along z in Fig. 2.13). An example is the flattened $CuCl_4$ tetrahedron in Cs_2CuCl_4.

Tetrahedral versus octahedral coordination

Most transition metal ions prefer octahedral coordination or a distorted variant of octahedral and an important factor is their large CFSE in octahedral sites. This can be estimated as follows. In octahedral coordination, each t_{2g} electron experiences a stabilization of $(4/10)\Delta^{oct}$ and each e_g electron a destabilization of $(6/10)\Delta^{oct}$. Thus Cr^{3+}, $d^3(t_{2g}^3)$ has a CFSE of $1.2\Delta^{oct}$ whereas Cu^{2+} $d^9(t_{2g})^6(e_g)^3$, has a CFSE of $0.6\Delta^{oct}$. In tetrahedral coordination, each e electron has a stabilization of $(6/10)\Delta^{tet}$ and each t_2 electron has a destabilization of $(4/10)\Delta^{tet}$ (Fig. 2.8). Thus, Cr^{3+} would have a CFSE of $0.8\Delta^{tet}$ and Cu^{2+} would have a CFSE of $0.4\Delta^{tet}$. With the guideline that

$$\Delta^{tet} \simeq 0.4\Delta^{oct}$$

the values of Δ^{oct} and Δ^{tet} for ions may be used to predict site preferences. More

Table 2.14 *Crystal field stabilization energies ($kJ\,mol^{-1}$) estimated for transition metal oxides.* (Data from Dunitz and Orgel, *Adv. Inorg Radiochem.*, **2**, 1–60, 1960)

Ion		Octahedral stabilization	Tetrahedral stabilization	Excess octahedral stabilization
Ti^{3+}	d^1	87.4	58.5	28.9
V^{3+}	d^2	160.1	106.6	53.5
Cr^{3+}	d^3	224.5	66.9	157.6
Mn^{3+}	d^4	135.4	40.1	95.3
Fe^{3+}	d^5	0	0	0
Mn^{2+}	d^5	0	0	0
Fe^{2+}	d^6	49.7	33.0	16.7
Co^{2+}	d^7	92.8	61.9	30.9
Ni^{2+}	d^8	122.1	35.9	86.2
Cu^{2+}	d^9	90.3	26.8	63.5

accurate values may be obtained spectroscopically and are given in Table 2.14 for some oxides of transition metal ions. It can be seen that high spin d^5 ions, as well as d^0 and d^{10} ions, have no particular preference for octahedral or tetrahedral sites insofar as crystal field effects are concerned. Ions such as Cr^{3+}, Ni^{2+} and Mn^{3+} show the strongest preference for octahedral coordination: thus tetrahedral coordination is rare for Ni^{2+}.

The coordination preferences of ions are shown by the type of spinel structure that they adopt. Spinels have the formula AB_2O_4 (Chapter 1) and may be:

(a) normal—A tetrahedral, B octahedral;
(b) inverse—A octahedral, B tetrahedral and octahedral;
(c) some intermediate between normal and inverse.

The parameter, γ, is the fraction of A ions on octahedral sites. For normal spinel, $\gamma = 0$; for inverse, $\gamma = 1$; for a random arrangement of A and B ions, $\gamma = 0.67$. Lattice energy calculations show that, in the absence of CFSE effects, spinels of the type 2,3 (i.e. $A = M^{2+}$, $B = M^{3+}$, e.g. $MgAl_2O_4$) tend to be normal whereas spinels of type 4, 2 (i.e. $A = M^{4+}$, $B = M^{2+}$, e.g. $TiMg_2O_4$) tend to be inverse. However, these preferences may be changed by the intervention of CFSE effects, as shown by the γ parameters of some 2, 3 spinels in Table 2.15. Examples are:

(a) All chromate spinels contain octahedral Cr^{3+} and are normal. This is consistent with the very large CFSE or Cr^{3+} and ions such as Ni^{2+} are forced into tetrahedral sites in $NiCr_2O_4$.
(b) Most 2, 3 Mg^{2+} spinels are normal apart from $MgFe_2O_4$ which is essentially inverse. This reflects the lack of any CFSE for Fe^{3+}.
(c) $Co_3O_4(\equiv CoO \cdot Co_2O_3)$ is normal because low spin Co^{3+} gains more CFSE by going into the octahedral site than Co^{2+} loses by occupying the tetrahedral site. Mn_3O_4 is also normal. Magnetite, Fe_3O_4, however, is inverse because whereas Fe^{3+} has no CFSE in either tetrahedral or octahedral coordination, Fe^{2+} has a preference for octahedral sites.

Table 2.15 *The parameters of some spinels.* (Data from Greenwood, *Ionic Crystals, Lattice Defects and Nonstoichiometry*, Butterworths, 1968, and Dunitz and Orgel, *Adv. Inorg. Radiochem*, **2**, 1, 1960)

M^{3+}	M^{2+}	Mg^{2+}	Mn^{2+}	Fe^{2+}	Co^{2+}	Ni^{2+}	Cu^{2+}	Zn^{2+}
Al^{3+}		0	0.3	0	0	0.75	0.4	0
Cr^{3+}		0	0	0	0	0	0	0
Fe^{3+}		0.9	0.2	1	1	1	1	0
Mn^{3+}		0	0	0.67	0	1	0	0
Co^{3+}		—	—	—	0	—	—	0

Spinels usually have cubic symmetry but some show tetragonal distortions in which one of the cell edges is of a different length to the other two. The Jahn–Teller effect gives rise to such distortions in Cu^{2+} containing spinels, $CuFe_2O_4$ (the tetragonal unit cell parameter ratio, $c/a = 1.06$) and $CuCr_2O_4$ ($c/a = 0.9$). $CuFe_2O_4$ is an inverse spinel with octahedral Cu^{2+} ions and the Jahn–Teller effect distorts the CuO_6 octahedra so that the two Cu—O bonds along z are longer than the four Cu—O bonds in the xy plane. On the other hand, $CuCr_2O_4$ is normal and the CuO_4 tetrahedra are flattened in the z direction, again due to the Jahn–Teller effect, thereby causing a shortened c axis.

Inert pair effect

The heavy, post-transition elements, especially Tl, Sn, Pb, Sb and Bi, exhibit, in some compounds, a valence that is two less than the group valence (e.g. the divalent state in Group IV elements, Sn and Pb). This is the so-called 'inert pair effect' and manifests itself structurally by a distortion of the metal ion coordination environment. Thus, Pb^{2+} has the configuration: (Xe core) $4f^{14}5d^{10}6s^2$, and the $6s^2$ pair is 'stereochemically active' in that these electrons are not in a spherically symmetrical orbital but stick out to one side of the Pb^{2+} ion (perhaps in some kind of s–p hybridized orbital).

Various kinds of distorted coordination polyhedra occur. Sometimes, the lone pair comes between the metal ion and some of its immediate anionic neighbours and causes a variation of bond length about the metal ion, e.g. red PbO has a structure that is a tetragonal distortion of the CsCl structure (Fig. 2.14). Four oxygens are situated at 2.3 Å from the Pb^{2+} ion, which is a reasonable Pb—O bond length, but the other four oxygens are at 4.3 Å. Although the lone pair is not directly visible, its presence is apparent from the distorted nature of the cubic coordination of Pb^{2+}.

A related distortion occurs in SnS which has a distorted rock salt structure. In this case, the SnS_6 octahedra are distorted along a [111] direction such that three sulphur atoms on one side of the tin are at ~ 2.64 Å but the other three are repelled by the lone pair to a distance of ~ 3.31 Å.

Another common kind of distortion occurs when the lone pair simply takes the place of an anion and its associated pair of bonding electrons. Five-coordinated

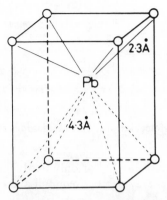

Fig. 2.14 The structure of red PbO, showing the presence of the inert pair effect by the variation in Pb—O bond distances

structures result in, for example, TlI, in which one corner (anion) of the octahedron 'TlI$_6$' is missing.

Metallic bonding and band theory

Metallic structures and bonding are characterized by delocalized valence electrons and these are responsible for properties such as the high electrical conductivity of metals. This contrasts with ionic and covalent bonding in which the valence electrons are localized on particular atoms or ions and, in general, are not free to migrate through the structure.

The bonding theory used to account for delocalized electons is the band theory of solids. We shall not be concerned here with the mathematics of band theory but shall see qualitatively how it can be applied to certain groups of solids, especially metals and some semiconductors.

In a metal such as aluminium, the inner core electrons—1s, 2s and 2p—are localized in discrete atomic orbitals on the individual aluminium atoms. However, the 3s and 3p electrons that form the valence shell occupy energy levels that are delocalized over the whole of the metal crystal. These levels are like giant molecular orbitals, each of which can contain only two electrons. In practice, in a metal there must be an enormous number of such levels and they are separated from each other by very small energy differences. Thus in a crystal of aluminium that contains N atoms, each atom contributes one 3s orbital and the result is a band that contains N closely spaced energy levels. This band is called the 3s *valence band*. The 3p levels in aluminium are similarly present as a delocalized 3p band of energy levels.

The band structure of other materials may be regarded in a similar way. The differences between metals, semiconductors and insulators depend on:

(a) the band structure of each,

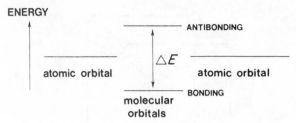

Fig. 2.15   Molecular orbitals in a diatomic molecule

(b) whether the valence bands are full or only partly full,
(c) the magnitude of any energy gap between full and empty bands.

The band theory of solids is well supported by X-ray spectroscopic data and by two independent theoretical approaches. The 'chemical approach' to band theory is to take molecular orbital theory, as it is usually applied to small, finite-sized molecules and to extend the treatment to infinite, three-dimensional structures. In the molecular orbital theory of diatomic molecules, an atomic orbital from atom 1 overlaps with an atomic orbital on atom 2, resulting in the formation of two molecular orbitals that are delocalized over both atoms. One of the molecular orbitals is 'bonding' and has lower energy than that of the atomic orbitals. The other is 'antibonding' and is of higher energy (Fig. 2.15).

Extension of this approach to larger molecules leads to an increase in the number of molecular orbitals. For each atomic orbital that is put into the system, one molecular orbital is created. As the number of molecular orbitals increases, the average energy gap between adjacent molecular orbitals must decrease (Fig. 2.16). The gap between bonding and antibonding orbitals also may decrease until the situation is reached in which there is essentially a continuum of energy levels.

Metals may be regarded as infinitely large 'molecules' in which an enormous number of energy levels or 'molecular' orbitals is present, $\sim 6 \times 10^{23}$ for one mole of metal. It is, however, no longer appropriate to refer to each level as a 'molecular' orbital since each is delocalized over all the atoms in the metal crystal. Instead, they are usually referred to as energy levels or energy states.

The band structure of metallic sodium, calculated using the 'tight binding

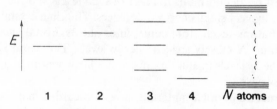

Fig. 2.16   Splitting of energy levels on molecular
orbital theory

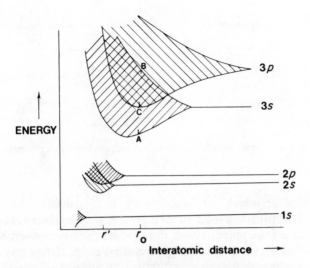

Fig. 2.17 Effect of interatomic spacing on atomic energy levels and bands for sodium, calculated using tight binding theory. Shaded areas represent bands of energy levels, formed by significant overlap of atomic orbitals on adjacent atoms

approximation', is shown in Fig. 2.17. It can be seen that the width of a particular energy band is related to the interatomic separation and hence to the degree of overlap between orbitals on adjacent atoms. Thus, at the experimentally determined value for the interatomic separation, r_0 , the 3s and 3p orbitals on adjacent atoms are calculated to overlap significantly to form broad 3s and 3p bands (shaded). The upper levels of the 3s band with energies in the range C to B, have similar energies to the lower levels of the 3p band. Hence, there is no discontinuity in energy between 3s and 3p bands. Such overlap of bands is important in explaining the metallic properties of elements such as the alkaline earths.

At the interatomic distance, r_0, the 1s, 2s and 2p orbitals on adjacent sodium atoms do not overlap. Instead of forming bands they remain as discrete atomic orbitals associated with the individual atoms. They are represented in Fig. 2.17 as thin lines. If it were possible to reduce the internuclear separation from r_0 to r' by compression of sodium under pressure, then the 2s and 2p orbitals would also overlap to form bands of energy levels (shaded). The 1s levels would, however, still be present as discrete levels at distance r'. It has been suggested that similar effects would occur in other elements subjected to high pressure. For instance, it has been calculated that hydrogen would become metallic at a pressure $\gtrsim 10^6$ atm.

Sodium has the electronic configuration $1s^2\ 2s^2\ 2p^6\ 3s^1$. It therefore has one valence electron per atom. Since the 3s, 3p bands overlap (Fig. 2.17), the valence electrons are not confined to the 3s band but are distributed over the lower levels of both the 3s and 3p bands.

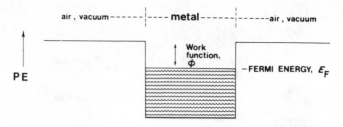

air , vacuum - - - - - - | - - - - **metal** - - - - - | - - - - - air , vacuum

Work
function,
ϕ

−FERMI ENERGY, E_F

PE

Fig. 2.18   Free electron theory of a metal; electrons in a potential
well

The 'physical approach' to band theory is to consider the energy and wavelength of electrons in a solid. In the early *free electron theory* of Sommerfeld, a metal is regarded as a potential well, inside which the more loosely held valence electrons are free to move. The energy levels that the electrons may occupy are quantized (analogy with the quantum mechanical problem of a particle in a box) and the levels are filled from the bottom of the well with two electrons per level. The highest filled level at absolute zero is known as the Fermi level. The corresponding energy is the Fermi energy, E_F (Fig. 2.18). The work function, ϕ, is the energy required to remove the uppermost valence electrons from the potential well. It is analogous to the ionization potential of an isolated atom.

A useful diagram is the *density of states*, $N(E)$, diagram which is a plot of the number of energy levels, $N(E)$, as a function of energy, E (Fig. 2.19). The number of available energy levels increases steadily with increasing energy in the Sommerfeld theory. Although the energy levels are quantized, there are so many of them and the energy difference between adjacent levels is so small that, effectively, a continuum occurs. At temperatures above absolute zero, some electrons in levels near to E_F have sufficient thermal energy to be promoted to empty levels above E_F. Hence at real temperatures, some states above E_F are occupied and others below E_F must be vacant. The average occupancy of the energy levels at some temperature T above zero is shown as shading in Fig. 2.19.

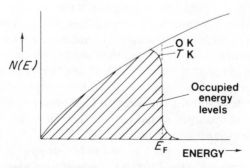

$N(E)$

0 K
T K

Occupied
energy
levels

E_F ENERGY →

Fig. 2.19   Density of states plot on the free
electron theory

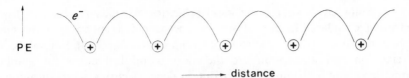

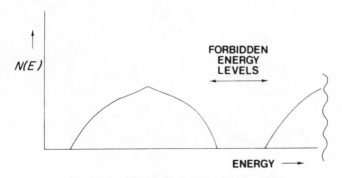

Fig. 2.20 Potential energy of electrons as a function of distance through a solid

The high electrical conductivity of metals is due to the drift of those electrons that are in half-occupied states close to E_F. Electrons that are in doubly occupied states lower down in the valence band cannot undergo any net migration in a particular direction whereas those in singly occupied levels are free to move. The promotion of an electron from a full level below E_F to an empty one above E_F therefore gives rise, effectively, to two mobile electrons.

The free electron theory is an oversimplification of the electronic structure of metals but it is a very useful starting model. In more refined theories, the potential inside the crystal or well is regarded as periodic (Fig. 2.20) and not constant as in Sommerfeld's theory. The positively charged atomic nuclei are arranged in a regularly repeating manner. The potential energy of the electrons passes through a minimum at the positions of the nuclei, due to coulombic attraction, and passes through a maximum midway between adjacent nuclei. Solution of the Schrödinger equation for a periodic potential function such as shown in Fig. 2.20 was carried out by Bloch using Fourier analysis. An important conclusion was that an uninterrupted continuum of energy levels does not occur but that, instead, only certain bands or ranges of energies are permitted for the electrons. (The forbidden energies correspond to electron wavelengths that satisfy Bragg's law for diffraction in a particular direction in the crystal). Consequently, the density of states diagrams given by Bloch's results show discontinuities, as in Fig. 2.21.

Similar conclusions about the existence of energy bands in solids are obtained from both the molecular orbital and periodic potential approaches. From either theory, one obtains a model with bands of energy levels for the valence electrons. In some materials, overlap of different bands occurs. In other materials, a forbidden gap exists between energy bands.

Fig. 2.21 Density of states on band theory

Experimental evidence for the band structure of solids is obtained from spectroscopic studies. Electronic transitions between different energy levels may be observed using various spectroscopic techniques (Chapter 4). For solids, X-ray emission and absorption are the most useful techniques for gaining information about both inner core and outer valence electrons. A certain amount of information about outer valence electrons also comes from visible and UV spectroscopy. X-ray emission spectra of solids usually contain peaks or bands of various width. Transitions between inner or core levels appear as sharp peaks and indicate that these levels are discrete atomic orbitals. Transitions involving valence shell electrons may give broad spectral peaks, however, especially for metals, and indicate that such valence electrons have a broad distribution of energies and are therefore located in energy bands.

Band structure of metals

Metals are characterized by a band structure in which the highest occupied band, the valence band, is only part full (Fig. 2.22). The occupied levels are shown schematically by the shading; some levels just below the Fermi level are vacant whereas some above E_F are occupied. Electrons in singly occupied states

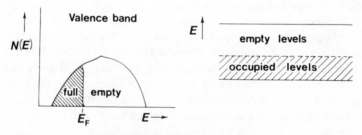

Fig. 2.22 Band structure of a metal

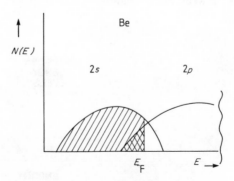

Fig. 2.23 Overlapping band structure of beryllium metal

close to E_F are able to move and are responsible for the high conductivity of metals.

In some metals, such as sodium, energy bands overlap (Fig. 2.17). Consequently, both 3s and 3p bands in sodium contain electrons. Overlap of bands is responsible for the metallic properties of, for example, the alkaline earth metals. The band structure of beryllium is shown schematically in Fig. 2.23. It has overlapping 2s, 2p bands, both of which are only partly full. If the 2s and 2p bands did not overlap, then the 2s band would be full, the 2p band empty and beryllium would not be metallic. This is the situation that holds in insulators and semiconductors.

Band structure of insulators

The valence band in insulators is full. It is separated by a large, forbidden gap from the next energy band, which is empty (Fig. 2.24). Diamond is an excellent insulator with a band gap of ~ 6 eV. Very few electrons from the valence band have sufficient thermal energy to be promoted into the empty band above. Hence the conductivity is negligibly small. The origin of the band gap in diamond is similar to that for silicon, discussed next.

Band structure of semiconductors: silicon

Semiconductors have a similar band structure to insulators but the band gap is not very large; usually it is in the range 0.5 to 3.0 eV. At least a few electrons have sufficient thermal energy to be promoted into the empty band.

Two types of conduction mechanism may be distinguished in semiconductors (Fig. 2.25). Any electrons that are promoted into an upper, empty band, called the *conduction band*, are regarded as negative charge carriers and would move towards a positive electrode under an applied potential. The vacant electron

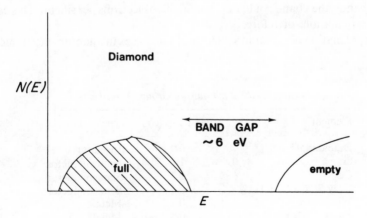

Fig. 2.24 Band structure of an insulator, carbon (diamond)

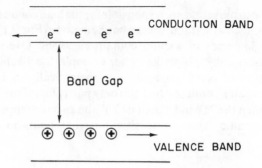

Fig. 2.25 Positive and negative charge carriers

levels that are left behind in the valence band may be regarded as *positive holes*. Positive holes move when an electron enters them, leaving its own position vacant as a fresh positive hole. Effectively, positive holes move in the opposite direction to electrons.

Intrinsic semiconductors are pure materials whose band structure may be as shown in Fig. 2.25. For these, the number of electrons, n, in the conduction band is governed entirely by (i) the magnitude of the band gap and (ii) temperature. Pure silicon is an intrinsic semiconductor. Its band gap, and that of other Group IV elements, is given in Table 2.16.

The band structure of silicon is, infact, quite different to that which might be expected by comparison with sodium and magnesium. In these elements, the 3*s*, 3*p* levels overlap to give two broad bands, both of which are part full. If the trend continued, it might be expected that two similar bands would be present in silicon. In this case, the bands would, on average, be half full and silicon would be metallic. Clearly this is not the case and instead silicon contains two bands separated by a forbidden gap. Further, the lower band contains four electrons per silicon atom and is full. If the forbidden gap merely corresponded to separation of *s* and *p* bands, the *s* band could only contain two electrons per silicon. This cannot be the explanation, therefore.

The explanation of the band structure of silicon lies in quantum mechanics and

Table 2.16 *Band gaps of Group IV elements*

Element	Band gap (eV)	Type of material
Diamond, C	6.0	Insulator
Si	1.1	Semiconductor
Ge	0.7	Semiconductor
Grey Sn ($> 13\,°C$)	0.1	Semiconductor
White Sn ($< 13\,°C$)	0	Metal
Pb	0	Metal

the fact that the crystal structure of silicon is quite different to that of sodium: the structure of sodium is body centred cubic with a coordination number of eight whereas that of silicon is derived from face centred cubic with a coordination number of four. However, a simplified and non-mathematical explanation of the band structure of silicon (and germanium, diamond, etc.) may also be given. It starts from the observation that each silicon forms four equal bonds arranged tetrahedrally; these bonds or orbitals may be regarded as sp^3 hybridized. Each hybrid orbital overlaps a similar orbital on an adjacent silicon to form a pair of molecular orbitals, one bonding, σ, and the other antibonding σ^*. Each can contain two electrons, one from each silicon atom. The only step that remains is to allow the individual σ molecular orbitals to overlap to form a σ band; this becomes the valence band. The σ^* orbitals overlap similarly and become the conduction band. The σ band is full since it contains four electrons per silicon atom. The σ^* band is empty.

Further discussion of semiconductors and their applications is deferred to Chapter 7.

Band structure of inorganic solids

So far, we have concentrated on materials that are conductors of electricity. However, many inorganic solids can also be treated profitably using band theory, whether or not they are regarded as electrical conductors. Band theory provides an additional insight into the structures, bonding and properties of inorganic solids that complements the insight obtained with the ionic/covalent models. Most inorganic materials have structures that are more complex than those of the metals and semiconducting elements. They have also received less theoretical attention of the type involving band structure calculations. Consequently, their band structures are usually known only approximately.

Above, we have considered Group IV elements, especially silicon; closely related are the III–V compounds, such as GaP. These latter compounds are isoelectronic with Group IV elements, at least as far as the numbers of electrons in the valence shells are concerned and have similar semiconducting properties. Let us now take this one stage further and consider more extreme cases, with I–VII compounds such as NaCl and II–VI compounds such as MgO.

The bonding in these materials is predominantly ionic. They are white, insulating solids with negligibly small electronic conductivity. Addition of dopants tends to produce ionic rather than electronic conductivity. On the assumption that NaCl is 100 per cent ionic, the ions have the configurations:

$$Na^+ : 1s^2 2s^2 2p^6$$
$$Cl^- : 1s^2 2s^2 2p^6 3s^2 3p^6$$

Hence, the 3s, 3p valence shell of Cl^- is full and that of Na^+ is empty. Adjacent Cl^- ions are approximately in contact in NaCl and the 3p orbitals may overlap somewhat to form a narrow 3p valence band. This band is composed of anion orbitals only. The 3s, 3p orbitals on Na^+ ions may also overlap to form a band,

Table 2.17   *Band gaps (eV) of some inorganic solids*

I–VII compounds		II–VI compounds		III–V compounds	
LiF	11	ZnO	3.4	AlP	3.0
LiCl	9.5	ZnS	3.8	AlAs	2.3
NaF	11.5	ZnSe	2.8	AlSb	1.5
NaCl	8.5	ZnTe	2.4	GaP	2.3
NaBr	7.5	CdO	2.3	GaAs	1.4
KF	11	CdS	2.45	GaSb	0.7
KCl	8.5	CdSe	1.8	InP	1.3
KBr	7.5	CdTe	1.45	InAs	0.3
KI	5.8	PbS	0.37	InSb	0.2
		PbSe	0.27	β-SiC	2.2
		PbTe	0.33	α-SiC	3.1

Some of these values, especially for the alkali halides, are only approximate.

the conduction band. This band is formed from cation orbitals only. It is completely empty of electrons under normal conditions since the band gap is large,' $\sim 7\,eV$. The band structure of NaCl is very like that of an insulator (Fig. 2.24), therefore, but with the additional detail that the valence band is composed of anion orbitals and the conduction band of cation orbitals. Any promotion of electrons from the valence band to the conduction band may therefore also be regarded as back transfer of charge from Cl^- to Na^+.

This conclusion leads us to expect some kind of correlation between the magnitude of the band gap and the difference in electronegativity between anion and cation. A large electronegativity difference favours ionic bonding. In such cases back transfer of charge from anion to cation is expected to be difficult and this correlates with the general observation that ionic solids have large band gaps. The band gaps of a variety of inorganic solids are given in Table 2.17. A quantitative relation between band gap and ionicity has been proposed by Phillips and van Vechten (equation (2.34)). The band gap is regarded as being made of two components: part is due to the 'homopolar band gap' which is the band gap that would be observed in the absence of any difference in electronegativity between the constituent elements; the other part is associated with the degree of ionic character in the bonds.

In compounds of transition metals, an additional factor that is of great significance is the presence of partly filled d orbitals on the metal ions. In some cases, these overlap to give a d band or bands and the material may have high conductivity. In other cases, d orbital overlap is very limited and the orbitals are effectively localized on the individual atoms. An example of the latter is stoichiometric NiO. Its pale green colour is due to internal d–d transitions within the individual Ni^{2+} ions. It has a very low conductivity, $\sim 10^{-14}\,ohm^{-1}\,cm^{-1}$ at 25 °C, and there is no evidence for any significant overlap of the d orbitals to form

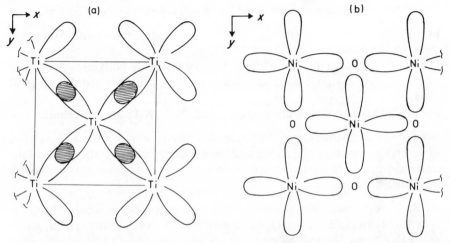

Fig. 2.26 (a) Section through the TiO structure, parallel to a unit cell face and showing Ti^{2+} positions only. Overlap of d_{xy} orbitals on adjacent Ti^{2+} ions, together with similar overlap of d_{xz} and d_{yz} orbitals, leads to a t_{2g} band. (b) Structure of NiO, showing $d_{x^2-y^2}$ orbitals pointing directly at oxide ions and, therefore, unable to overlap and form an e_g band

a partly filled d band. Examples at the other extreme are TiO and VO. These have the rock salt structure, as has NiO, but by contrast, d orbitals of the type d_{xy}, d_{xz}, d_{yz} on the M^{2+} ions overlap strongly (Fig. 2.26a) to form a broad t_{2g} band. This band is only partly filled by electrons. Consequently, TiO and VO have almost metallic conductivity, $\sim 10^3$ ohm^{-1} cm^{-1} at 25 °C.

An additional difference between TiO and NiO is that the t_{2g} band, capable of containing six electrons per metal atom, must be full in NiO. The two extra d electrons in Ni^{2+} are in e_g levels, d_{z^2}, $d_{x^2-y^2}$. These e_g orbitals point directly at the oxide ions (Fig. 2.26b). Because of the intervening oxide ions, the e_g orbitals on adjacent Ni^{2+} ions cannot overlap to form a band. Hence e_g orbitals remain localized on the individual Ni^{2+} ions.

Some general guidelines as to whether good overlap of d orbitals is likely to occur have been given by Phillips and Williams. In these guidelines, d band formation is likely to occur if:

(a) The formal charge on the cations is small.
(b) The cation occurs early in the transition series.
(c) The cation is in the second or third transition series.
(d) The anion is reasonably electropositive.

The reasoning behind these guidelines is fairly straightforward and is based on similar arguments that are used in ligand field theory. Effects (a) to (c) are concerned with keeping the d orbitals spread out as far as possible and reducing the amount of positive charge that they 'feel' from their parent transition metal ion nucleus. Effect (d) is associated with a reduction in ionicity and band gap, as discussed earlier in this section.

A variety of examples can be found to illustrate each of the guidelines:

For (a), TiO is metallic whereas TiO_2 is an insulator. Cu_2O and MoO_2 are semiconductors whereas CuO and MoO_3 are insulators.

For (b), TiO, VO are metallic whereas NiO and CuO are poor semiconductors.

For (c), Cr_2O_3 is a poor conductor whereas lower oxides of Mo, W are good conductors.

For (d), NiO is a poor conductor whereas NiS, NiSe, NiTe are good conductors.

The d electron structure of solid transition metal compounds is also sensitive to the crystal structure of the solid and to any variation in oxidation state of the transition metal. Some interesting examples are provided by complex oxides with the spinel structure:

(a) Both Fe_3O_4 and Mn_3O_4 have a spinel structure but whereas Mn_3O_4 is virtually an insulator, Fe_3O_4 has almost metallic conductivity. The structure of Fe_3O_4 may be written as:

$$[Fe^{3+}]^{tet}[Fe^{2+}, Fe^{3+}]^{oct}O_4; \quad \text{inverse spinel}$$

whereas the structure of Mn_3O_4 is:

$$[Mn^{2+}]^{tet}[Mn_2^{3+}]^{oct}O_4; \quad \text{normal spinel}$$

Since Fe_3O_4 is an inverse spinel, it contains Fe^{2+} and Fe^{3+} ions distributed over the octahedral sites. These octahedral sites are close together since they belong to edge-sharing octahedra. Consequently, positive holes can migrate easily from Fe^{2+} to Fe^{3+} ions and hence Fe_3O_4 is a good conductor.

In Mn_3O_4, the spinel structure is normal which means that the closely spaced octahedral sites contain only Mn^{3+} ions. The tetrahedral sites, containing Mn^{2+} ions, share corners only with the octahedral sites. The $Mn^{2+}-Mn^{3+}$ distance is greater, therefore, and electron exchange cannot take place easily.

(b) A related example, which is really an example of guideline (b) above, is provided by the lithium spinels, $LiMn_2O_4$ and LiV_2O_4. The structural formulae of these spinels are similar:

$$[Li^+]^{tet}[Mn^{3+}Mn^{4+}]^{oct}O_4$$

$$[Li^+]^{tet}[V^{3+}V^{4+}]^{oct}O_4$$

A mixture of $+3$ and $+4$ ions is present in the octahedral sites of both but since d orbital overlap is greater for vanadium than for manganese, this is reflected in the electrical properties: $LiMn_2O_4$ is a hopping semiconductor whereas LiV_2O_4 has metallic conductivity (see the sections on semiconductivity and metallic conductivity in Chapter 7 for a discussion of these properties).

Bands or bonds: a final comment

The three extreme types of bonding in solid materials are ionic, covalent and metallic. Within each category many good examples can be found. Most inorganic solids do not belong exclusively to one particular category, however. For solids with intermediate bond type, the question arises as to the most appropriate way of describing their bonding.

The band model for bonding is clearly appropriate in systems where there are freely mobile electrons, as in metals and some semiconductors. Experimental measurements of mobility show that these electrons are highly mobile and are not associated with individual atoms. In certain other semiconductors, such as doped nickel oxide (Chapters 6, 7), the band model would not appear to be suitable for explaining the electrical behaviour. The evidence indicates that these materials are best regarded as hopping semiconductors in which the electrons do not have high mobility. Instead, it appears to be more appropriate to regard the d electrons as occupying discrete orbitals on the nickel ions. It is important to remember, however, that the question of conduction in nickel oxide refers to only one or two sets of energy levels. Nickel oxide, like other materials, has many sets of energy levels. The lower lying levels are fully occupied by electrons. These are discrete levels associated with the individual anions and cations. At higher energy there are various excited levels that are usually completely empty of electrons. However, these levels may well overlap to form energy bands. In asking whether or not a bond or band model is the most suitable, one has to be clear about the particular property or set of energy levels to which the question refers. Thus many ionically bonded solids may, under UV irradiation, show electronic conductivity that is best described in terms of band theory.

Chapter 3

Crystallography and Diffraction Techniques

General comments

The simplest and most obvious first question to ask about an inorganic substance is 'What is it?'. The methods that are used to answer this come into two main categories, depending on whether the substance is molecular or non-molecular. If the substance is molecular, whether it be solid, liquid or gaseous, identification is usually carried out by some combination of spectroscopic methods and chemical analysis. If the substance is non-molecular and crystalline, identification is usually carried out by X-ray powder diffraction supplemented, where necessary, by chemical analysis. Each crystalline solid has its own characteristic X-ray powder pattern which may be used as a 'fingerprint' for its identification. The powder patterns of most known inorganic solids are included in an updated version of the *Powder Diffraction File* (described in the section on the powder method); by using an appropriate search procedure, unknowns can usually be identified rapidly and unambiguously.

Once the substance has been identified, the next stage is to determine its structure, if this is not known already. For molecular materials, details of the molecular geometry may be obtained from further spectroscopic measurements. Alternatively, if the substance is crystalline, X-ray crystallography may be used, in which case information is also obtained on the way in which the molecules pack together in the crystalline state. For molecular substances, this usually completes the story as far as identification and structure determination are concerned; attention may then be focused on other matters such as properties or chemical reactivity.

For non-molecular substances, however, the word 'structure' takes on a whole new meaning. In order for a solid to be well characterized, one needs to know about:

(a) the form of the solid, whether it is single crystal or polycrystalline and, if the latter, what is the number, size, shape and distribution of the crystalline particles;
(b) the crystal structure;
(c) the crystal defects that are present—their nature, number and distribution;

(d) the impurities that are present and whether they are distributed at random or are concentrated into small regions;

(e) the surface structure, including any compositional inhomogeneities or absorbed surface layers.

No single technique is capable of providing a complete characterization of a solid. Rather, a variety of techniques are used in combination.

There are three main categories of physical technique which may be used to characterize solids; these are diffraction, microscopic and spectroscopic techniques. In addition, other techniques such as thermal analysis, magnetic measurements and physical property measurements give valuable information in certain cases. This chapter deals with diffraction techniques; several of the other techniques are considered in Chapter 4.

X-ray diffraction has been in use since the early part of this century in two main areas, for the fingerprint characterization of crystalline materials and for determination of their structure. It is one of the principal techniques of solid state chemistry and accordingly, is given most space in this chapter. A brief description is also given of electron diffraction and neutron diffraction, two rather specialized techniques which have important applications.

X-ray diffraction

Generation of X-rays

X-rays are electromagnetic radiation of wavelength ~ 1 Å (10^{-10} m). They occur in that part of the electromagnetic spectrum between γ-rays and the ultraviolet. X-rays are produced when high-energy charged particles, e.g. electrons accelerated through 30,000 V, collide with matter. The X-ray spectra that are produced usually have two components, a broad spectrum of wavelengths known as 'white radiation' and a number of fixed, or monochomatic wavelengths. White radiation arises when the electrons are slowed down or stopped by the collision and some of their lost energy is converted into electromagnetic radiation. White radiation has wavelengths ranging upwards from a certain lower limiting value. This lower wavelength limit corresponds to the X-rays of highest energy and occurs when all the kinetic energy of the incident particles is converted into X-rays. It may be calculated from the formula, $\lambda_{min}(\text{Å}) = 12400/V$, where V is the accelerating voltage.

The X-rays which are used in almost all diffraction experiments are produced by a different process that leads to *monochromatic X-rays*. A beam of electrons, again accelerated through, say, 30 kV is allowed to strike a metal *target*, often copper. The incident electrons have sufficient energy to ionize some of the copper $1s$ (K shell) electrons (Fig. 3.1a). An electron in an outer orbital ($2p$ or $3p$) immediately drops down to occupy the vacant $1s$ level and the energy released in the transition appears as X-radiation. The transition energies have fixed values and so a spectrum of characteristic X-rays results (Fig. 3.1b). For copper, the

122

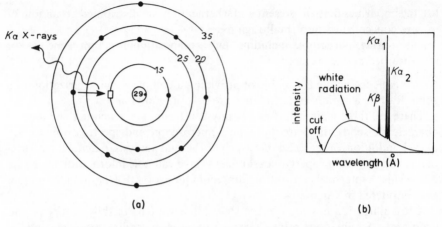

Fig. 3.1 (a) Generation of Cu $K\alpha$ X-rays. A $1s$ electron is ionized; a $2p$ electron falls into the empty $1s$ level ($\square$) and the excess energy is released as X-rays. (b) X-ray emission spectrum of copper metal

$2p \rightarrow 1s$ transition, called $K\alpha$, has a wavelength of 1.5418 Å and the $3p \rightarrow 1s$ transition, $K\beta$, 1.3922 Å. The $K\alpha$ transition occurs much more frequently than the $K\beta$ transition and it is the more intense $K\alpha$ radiation which results that is used in diffraction experiments. In fact, the $K\alpha$ transition is a doublet, $K\alpha_1 = 1.54051$ Å and $K\alpha_2 = 1.54433$ Å, because the transition has a slightly different energy for the two possible spin states of the $2p$ electron which makes the transition, relative to the spin of the vacant $1s$ orbital. In some X-ray experiments, diffraction by the $K\alpha_1$ and $K\alpha_2$ radiations is not resolved and a single line or spot is observed instead of a doublet (e.g. in powder diffractometry at low angle). In other experiments, separate diffraction peaks may be observed; if desired, this can be overcome by removing the weaker $K\alpha_2$ beam from the incident radiation.

The wavelengths of the $K\alpha$ lines of the target metals commonly used for X-ray generation are given in Table 3.1. The wavelengths are related to the atomic number Z, of the metal, by *Moseley's law*:

$$\lambda^{-1/2} = C(Z - \sigma) \tag{3.1}$$

Table 3.1   *X-ray wavelengths of commonly used target materials*

Target	$K\alpha_1$	$K\alpha_2$	$K\bar{\alpha}$*	Filter
Cr	2.2896	2.2935	2.2909	V
Fe	1.9360	1.9399	1.9373	Mn
Cu	1.5405	1.5443	1.5418	Ni
Mo	0.7093	0.7135	0.7107	Nb
Ag	0.5594	0.5638	0.5608	Pd

*$\bar{\alpha}$ is the intensity-weighted average of α_1 and α_2.

where C and σ are constants. Hence the wavelength of the $K\alpha$ line decreases with increasing atomic number.

The X-ray emission spectrum of an element such as copper, Fig. 3.1b, has two main features. The intense, monochromatic peaks, caused by electronic transitions within the atoms, have wavelengths that are characteristic of the element, i.e. copper. These monochromatic peaks are superposed on a background of 'white' radiation, mentioned earlier, which is produced by the general interaction of high-velocity electrons with matter. In order to generate the characteristic monochromatic radiation, the voltage used to accelerate the electrons needs to be sufficiently high ($\gtrsim 10\,\mathrm{kV}$) so that ionization of the copper $1s$ electrons may occur.

In the generation of X-rays (Fig. 3.2), the electron beam, provided by a heated tungsten filament, is accelerated towards an anode by a potential difference of $\sim 30\,\mathrm{kV}$. The electrons strike the target, a piece of copper fixed to the anode, and a spectrum of X-rays, such as shown in Fig. 3.1b, is emitted. The chamber, known as the *X-ray tube*, is evacuated in order to prevent oxidation of the tungsten filament. The X-rays leave the tube through 'windows' made of beryllium. The absorption of X-rays on passing through materials depends on the atomic weight of the elements present in the material. Beryllium with an atomic number of 4 is, therefore, one of the most suitable window materials. For the same reason, lead is a very effective material for shielding X-ray equipment and absorbing stray radiation. While an X-ray tube is in operation, continuous cooling of the anode is necessary. Only a small fraction of the energy of the incident electron beam is converted into X-rays. Most of the energy is converted into heat and the anode would soon melt if it were not cooled.

For most diffraction experiments, a monochromatic beam of X-rays is desired and not a continuous spectrum. In the spectrum of X-rays emitted by copper (or any metal), the $K\alpha$ line(s) is the most intense and it is desired to filter out all the other wavelengths, leaving the $K\alpha$ line for diffraction experiments. For copper radiation, a sheet of nickel foil is very effective in carrying out this separation. The energy required to ionize $1s$ electrons of nickel corresponds to a wavelength of 1.488 Å, which lies between the values for the $K\alpha$ and $K\beta$ lines of the copper emission spectrum. Cu $K\beta$ radiation, therefore, has sufficient energy to ionize $1s$

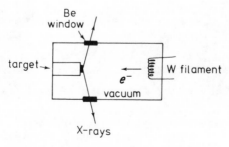

Fig. 3.2 Schematic design of a filament X-ray tube

124

electrons of nickel whereas Cu $K\alpha$ radiation does not. Nickel foil is effective in absorbing the Cu $K\beta$ radiation and most of the white radiation, leaving a monochromatic, reasonably clean beam of $K\alpha$ radiation. A lighter element, such as iron, would absorb Cu $K\alpha$ radiation as well as $K\beta$, because its *absorption edge* is displaced to higher wavelengths. On the other hand, a heavier element, such as zinc, would transmit both $K\alpha$ and $K\beta$ radiations while still absorbing much of the higher-energy white radiation. The atomic number of the element in the filter generally is one or two less than that of the target material (Table 3.1). An alternative method of obtaining monochromatic X-rays uses a single crystal monochromator and is discussed later.

An optical grating and diffraction of light

As an aid to understanding the diffraction of X-rays by crystals, let us consider the diffraction of light by an optical grating. This gives a 1-dimensional analogue of the three-dimensional process that occurs in crystals.

An optical grating consists of a piece of glass on which have been ruled a large number of accurately parallel and closely spaced lines. The separation of the lines should be a little larger than the wavelength of light, say 10,000 Å. The grating is shown in projection as a row of points in Fig. 3.3a. Consider what happens to a beam of light which hits the grating perpendicular to the plane of the grating. A piece of glass without the lines would simply transmit the light, but in the grating the lines act as secondary point (or, rather, line) sources of light and re-radiate light in all directions. Interference then occurs between the waves originating from each line source. In certain directions, adjacent beams are in phase with each other and *constructive interference* occurs to give a resultant diffracted beam in that direction. Two such directions are shown in (b). In direction 1, parallel to the incident beam, the diffracted beams are obviously in phase. In direction 2, the

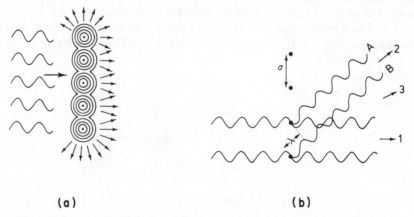

(a)   **(b)**

Fig. 3.3 (a) Lines on an optical grating act as secondary sources of light. (b) Constructive interference in directions 1 and 2

beams are also in phase although beam B is now exactly one wavelength behind beam A. For directions between 1 and 2, beam B lags behind beam A by a fraction of one wavelength, and *destructive interference* occurs. For a certain direction, 3, B is exactly half a wavelength behind A and complete destructive interference or *cancellation* occurs. For other directions between 1 and 2, the destructive interference is only partial. Thus, directions 1 and 2 have maximum light intensity and this falls off gradually to zero as the angle changes to direction 3. In the optical grating, however, there are not just two parallel diffracted beams A and B but several hundred or thousand, one for each line on the grating. This causes the resultant diffracted beams to sharpen enormously after interference so that intense beams occur in directions 1 and 2 with virtually no intensity over the whole range of directions between 1 and 2.

The directions, such as 2, in which constructive interference occurs are governed by the wavelength of the light, λ, and the separation, a, of the lines on the grating. Consider the diffracted beams 1 and 2 (Fig. 3.4) which are at an angle, ϕ, to the direction of the incident beam. If 1 and 2 are in phase with each other, the distance AB must equal a whole number of wavelengths; i.e.

$$AB = \lambda, 2\lambda, \ldots, n\lambda$$

But

$$AB = a \sin \phi$$

Therefore,

$$a \sin \phi = n\lambda \tag{3.2}$$

This equation gives the conditions under which constructive interference occurs and relates the spacing of the grating to the light wavelength and the diffraction order, n. Hence, depending on the value of $a \sin \phi$, one or more diffraction orders, corresponding to $n = 1, 2$, etc., may be observed.

We can now understand why the separation of the lines on the grating must be of the same order of magnitude as, but somewhat larger than, the wavelength of light. The condition for the first-order diffracted beam to occur is $a \sin \phi = \lambda$. The maximum value of $\sin \phi$ is 1, corresponding to $\phi = 90°$ but realistically, in order to observe first-order diffraction, $\sin \phi < 1$ and, therefore, $a > \lambda$. If $a < \lambda$, only the zero order direct beam is observable.

If, on the other hand, $a \gg \lambda$, individual diffraction orders ($n = 1, 2, 3, \ldots$, etc.) are

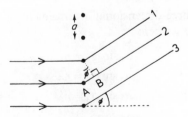

Fig. 3.4 Diffraction of light by an optical grating

so close together as to be unresolved and, effectively, a diffraction continuum results. This is because, for large values of a, $\sin \phi$ and, hence, ϕ must be very small. therefore $\phi_{n=1} \gtrsim 0$ and the first-order beam is not distinguishable from the primary beam. Visible light has wavelengths in the range 4,000 to 7,000 Å and so, in order to observe well-separated spectra, grating spacings are usually 10,000 to 20,000 Å.

The other condition to be observed in the construction of an optical grating is that the lines should be accurately parallel. If this were not so, ϕ would vary over the grating and the diffraction spectra would be blurred or irregular and of poor quality generally.

Crystals and diffraction of X-rays

By analogy with the diffraction of light by an optical grating, crystals, with their regularly repeating structures, should be capable of diffracting radiation that has a wavelength similar to the interatomic separation, ~ 1 Å. Three types of radiation are used for crystal diffraction studies: X-rays, electrons and neutrons. Of these, X-rays are the most useful but electron and neutron diffraction both have important specific applications and are discussed later.

The X-ray wavelength commonly employed is the characteristic $K\alpha$ radiation, $\lambda = 1.5418$ Å, emitted by copper. When crystals diffract X-rays, it is the atoms or ions which act as secondary point sources and scatter the X-rays; in the optical grating, it is the lines scratched or ruled on the glass surface which cause scattering.

Historically, two approaches have been used to treat diffraction by crystals. These are as follows.

The Laue equations

Diffraction from a hypothetical 1-dimensional crystal, constituting a row of atoms, may be treated in the same way as diffraction of light by an optical grating because, in projection, the grating is a row of points. An equation is obtained which relates the separation, a, of the atoms in the row, the X-ray wavelength, λ, and the diffraction angle, ϕ; i.e.

$$a \sin \phi = n\lambda$$

A real crystal is a three-dimensional arrangement of atoms for which three *Laue equations* may be written:

$$a_1 \sin \phi_1 = n\lambda$$
$$a_2 \sin \phi_2 = n\lambda$$
$$a_3 \sin \phi_3 = n\lambda$$

The equations correspond to each of the three crystallographic axes needed to represent the atomic arrangement in the crystal. For a diffracted beam to occur, these three equations must all be satisfied simultaneously.

The Laue equations provide a rigorous and mathematically correct way to describe diffraction by crystals. The drawback is that they are cumbersome to use. The alternative theory of diffraction, based on Bragg's law, is much simpler and is used almost universally in solid state chemistry. No further discussion of the Laue equations is given in this book.

Bragg's law

The Bragg approach to diffraction is to regard crystals as built up in layers or planes such that each acts as a semi-transparent mirror. Some of the X-rays are reflected off a plane with the angle of reflection equal to the angle of incidence, but the rest are transmitted to be subsequently reflected by succeeding planes.

The derivation of Bragg's law is shown in Fig. 3.5. Two X-ray beams, 1 and 2, are reflected from adjacent planes, A and B, within the crystal and we wish to know under what conditions the reflected beams 1′ and 2′ are in phase. Beam 22′ has to travel the extra distance xyz as compared to beam 11′, and for 1′ and 2′ to be in phase, distance xyz must equal a whole number of wavelengths. The perpendicular distance between pairs of adjacent planes, the *d-spacing*, d, and the angle of incidence, or *Bragg angle*, θ, are related to the distance xy by

$$xy = yz = d \sin \theta$$

Thus

$$xyz = 2d \sin \theta$$

But

$$xyz = n\lambda$$

Therefore

$$2d \sin \theta = n\lambda \qquad Bragg's\ law \qquad (3.3)$$

When Bragg's law is satisfied, the reflected beams are in phase and interfere constructively. At angles of incidence other than the Bragg angle, reflected beams are out of phase and destructive interference or cancellation occurs. In real crystals, which contain thousands of planes and not just the two shown in Fig. 3.5, Bragg's law imposes a stringent condition on the angles at which reflection may occur. If the incident angle is incorrect by more than a few tenths of a degree, cancellation of the reflected beams is usually complete.

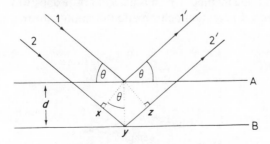

Fig. 3.5 Derivation of Bragg's law for X-ray diffraction

For a given set of planes, several solutions of Bragg's law are usually possible, for $n = 1, 2, 3$, etc. It is customary, however, to set n equal to 1 and for situations where, say, $n = 2$, the d-spacing is instead halved by doubling up the number of planes in the set; hence n is kept equal to 1. (Note that $2\lambda = 2d \sin \theta$ is equivalent to $\lambda = 2(d/2) \sin \theta$).

It is difficult to give an explanation of the nature of the semi-transparent layers or planes that is immediately convincing. This is because they are a concept rather than a physical reality. Crystal structures, with their regularly repeating patterns, may be referred to a three-dimensional grid and the repeating unit of the grid, the *unit cell*, can be found. The grid may be divided up into sets of planes in various orientations and it is these planes which are considered in the derivation of Bragg's law. In some cases, with simple crystal structures, the planes do correspond to layers of atoms, but this is not generally the case.

Some of the assumptions upon which Bragg's law is based may seem to be rather dubious. For instance, it is known that diffraction occurs as a result of interaction between X-rays and atoms. Further, the atoms do not *reflect* X-rays but scatter or diffract them in all directions. Nevertheless, the highly simplified treatment that is used in deriving Bragg's law gives exactly the same answers as are obtained by a rigorous mathematical treatment. We therefore happily use terms such as reflexion (often deliberately with this alternative spelling) and bear in mind that we are fortunate to have such a simple and picturesque, albeit inaccurate, way to describe what in reality is a very complicated process.

The X-ray diffraction experiment

When reduced to basic essentials, the X-ray diffraction experiment, Fig. 3.6, requires an X-ray source, the sample under investigation and a detector to pick up the diffracted X-rays. Within this broad framework, there are three variables which govern the different X-ray techniques:

(a) radiation—monochromatic or of variable λ;
(b) sample—single crystal, powder or a solid piece;
(c) detector—radiation counter or photographic film.

These are summarized for the most important techniques in Fig. 3.7. With the exception of the Laue method, which is used almost exclusively by metallurgists and is not discussed here, monochromatic radiation is nearly always used.

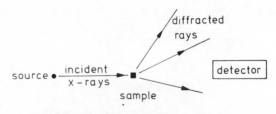

Fig. 3.6 The X-ray diffraction experiment

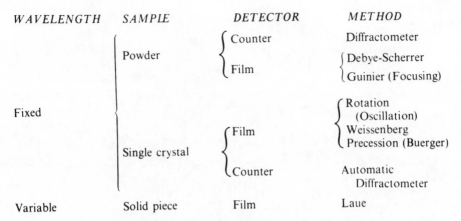

WAVELENGTH	SAMPLE	DETECTOR	METHOD
Fixed	Powder	Counter	Diffractometer
		Film	Debye-Scherrer / Guinier (Focusing)
	Single crystal	Film	Rotation (Oscillation) / Weissenberg / Precession (Buerger)
		Counter	Automatic Diffractometer
Variable	Solid piece	Film	Laue

Fig. 3.7 The different X-ray diffraction techniques

The powder method—principles and uses

The principles of the powder method are shown in Fig. 3.8. A monochromatic beam of X-rays strikes a finely powdered sample that, ideally, has crystals randomly arranged in every possible orientation. In such a powder sample, the various lattice planes are also present in every possible orientation. For each set of planes, therefore, at least some crystals must be oriented at the Bragg angle, θ, to the incident beam and thus, diffraction occurs for these crystals and planes. The diffracted beams may be detected either by surrounding the sample with a strip of photographic film (Debye–Scherrer and Guinier focusing methods) or by using a movable detector, such as a Geiger counter, connected to a chart recorder (diffractometer).

The original powder method, the *Debye–Scherrer method*, is little used nowadays, but since it is simple it is instructive to consider its mode of operation. For any set of lattice planes, the diffracted radiation forms the surface of a cone, as shown in Fig. 3.9. The only requirement for diffraction is that the planes be at angle θ to the incident beam; no restriction is placed on the angular orientation of the planes about the axis of the incident beam. In a finely powdered sample, crystals are present at every possible angular position about the incident beam

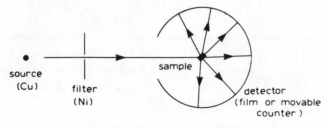

Fig. 3.8 The powder method

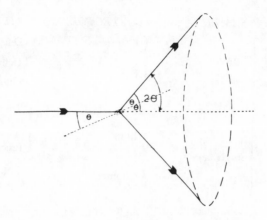

Fig. 3.9 The formation of a cone of diffracted
radiation in the powder method

and the diffracted beams that result appear to be emitted from the sample as
cones of radiation. (Each cone is in fact a large number of closely spaced diffracted
beams). If the Bragg angle is θ, then the angle between diffracted and undiffracted
beams is 2θ and the angle of the cone is 4θ. Each set of planes gives its own cone of
radiation. The cones are detected by a thin strip of film wrapped around the
sample (Fig. 3.8); each cone intersects the film as two short arcs (Fig. 3.10) which
are symmetrical about the two holes in the film (these allow entry and exit of
incident and undiffracted beams). In a well-powdered sample, each arc appears as
a continuous line, but in coarser samples the arcs may be spotty due to the
relatively small number of crystals present.

To obtain d-spacings from the Debye–Scherrer film, the separation, S, between
pairs of corresponding arcs is measured. If the camera (film) radius, R, is known,
then

$$\frac{S}{2\pi R} = \frac{4\theta}{360} \tag{3.4}$$

from which 2θ and therefore d may be obtained for each pair of arcs. The
disadvantages of this method are that exposure times are long (6 to 24 hours) and
that closely spaced arcs are not well resolved. This is because, although the

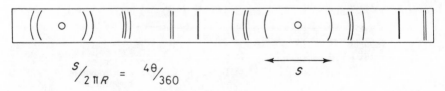

$$S\big/2\pi R = 4\theta\big/360$$

$$S$$

Fig. 3.10 Schematic Debye–Scherrer photograph

incident beam enters the camera through a pinhole slit and collimator tube, the beam is somewhat divergent and the spread increases in the diffracted beams. If, in an effort to increase the resolution, a finer collimator is used, the resulting diffracted beams have much less intensity and longer exposure times are needed. Apart from considerations of the extra time involved, the amount of background radiation detected by the film (as fogging) increases with exposure time and, consequently, weak lines may be lost altogether in the background.

In modern film methods (*Guinier focusing methods*) a *convergent*, intense incident beam is used with the result that excellent resolution of lines is obtained and exposure times are much reduced (10 min to 1 hr). Methods for obtaining a convergent beam of X-rays are discussed in the next section.

The other modern powder technique is *diffractometry*, which gives a series of peaks on a strip of chart paper. A convergent incident beam is again used to give fairly good resolution of peaks. Both peak positions and intensities (peak heights) are readily obtained from the chart to make this a very useful and rapid method of phase analysis.

The most important use of the powder method is in the qualitative identification of crystalline phases or compounds. While most chemical methods of analysis give information about the *elements* present in a sample, powder diffraction is very different and perhaps unique in that it tells which *crystalline compounds* or *phases* are present but gives no direct information about their chemical constitution.

Each crystalline phase has a characteristic powder pattern which can be used as a fingerprint for identification purposes. The two variables in a powder pattern are peak position, i.e. *d*-spacing, which can be measured very accurately if necessary, and intensity, which can be measured either qualitatively or quantitatively. It is rare but not unknown that two materials have identical powder patterns. More often, two materials have one or two lines with common *d*-spacings, but on comparing the whole patterns, which may contain between ~ 5 and 100 observed lines, the two are found to be quite different. In more extreme cases, two substances may happen to have the same unit cell parameters and, therefore, the same *d*-spacings, but since different elements are probably present in the two, their intensities are quite different. The normal practice in using powder patterns for identification purposes is to pay most attention to the *d*-spacings but, at the same time, check that the intensities are roughly correct.

For the identification of unknown crystalline materials, an invaluable reference source is the *Powder Diffraction File* (Joint Committee on Powder Diffraction Standards, Swarthmore, USA), previously known as the ASTM file, which contains the powder patterns of about 35,000 materials; new entries are added at the current rate of $\sim 2,000$ p.a. In the search indices, materials are classified either according to their most intense peaks or according to the first eight lines in the powder pattern in order of decreasing *d*-spacing. Identification of an unknown is usually possible within 30 min of obtaining its measured powder pattern. Problems arise if the material is not included in the file (obviously!) or if the material is not pure but contains lines from more than one phase.

Powder diffractometers and the focusing of X-rays

The powder diffractometer has a proportional, scintillation or Geiger counter as the detector which is connected to a chart recorder or sometimes to a means of digital output. In normal use, the counter is set to scan over a range of 2θ values at a constant angular velocity (it is common practice to refer to the angle 2θ between diffracted and undiffracted beams (Fig. 3.9), rather than to the Bragg angle, θ). Usually, the range 10 to 80° 2θ is sufficient to cover the most useful part of the powder pattern. A typical diffractometer trace is shown in Fig. 3.11a for SiO_2. The scale is linear in 2θ and d-spacings of the peaks may be calculated from Bragg's law or obtained from standard tables of d versus 2θ. The scanning speed of the counter is usually $2°\ 2\theta\ min^{-1}$ and, therefore, about 30 min are needed to obtain a trace. Intensities are taken as peak heights, unless very accurate work is being done, in which case areas may be measured; the most intense peak is given intensity of 100 and the rest are scaled accordingly.

If very accurate d-spacings or intensities are desired, slower scanning speeds (e.g. $\frac{1}{8}°\ 2\theta\ min^{-1}$) are used. To obtain accurate d-spacings, an internal standard (a pure material, such as KCl, whose d-spacings are known accurately) is mixed in with the sample. A correction factor, which may vary with 2θ, is obtained from the discrepancy between observed and true d-spacings of the standard and is then applied to the pattern that is being measured. Accurate intensities are obtained from peak areas by cutting out the peaks and weighing them, by measuring their area with a device such as a planimeter or by using an automatic counter fitted to the diffractometer.

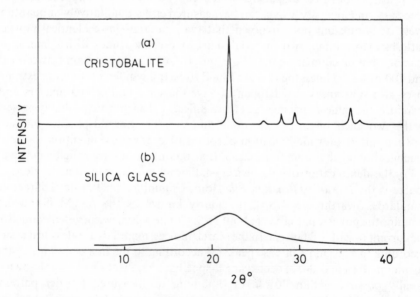

Fig. 3.11   X-ray powder diffraction pattern of (a) cristobalite and (b) glassy SiO_2, Cu $K\alpha$ radiation

Samples for diffractometry take various forms: they include thin layers of the fine powder sprinkled onto a glass slide smeared with vaseline and thin flakes pressed onto a glass slide. Different people prefer different methods of sample preparation and the objective is always to obtain a sample which contains a random arrangement of crystal orientations. If the crystal arrangement is not random, then *preferred orientation* exists and can introduce errors, sometimes very large, into the measured intensities. Preferred orientation is a serious problem for materials that crystallize in a characteristic, very non-spherical shape, e.g. clay minerals which usually occur as thin plates or some cubic materials which crystallize as cubes and, on crushing, break up into smaller cubes. In a powder aggregate of such materials, the crystals tend to sit on their faces, resulting in a far from random average orientation.

The big disadvantage of early Debye–Scherrer cameras is that incident and diffracted beams are, inevitably, somewhat divergent and of low intensity. In diffractometers and modern focusing cameras, a convergent X-ray beam is used; this gives a dramatic improvement in resolution and, because much more intense beams may be used, exposure times are greatly reduced. It is not possible to focus or converge X-rays using the X-ray equivalent of an optical lens; instead, use is made of certain geometric properties of the circle in order to obtain a convergent X-ray beam. These properties are illustrated in Fig. 3.12(a). The arc XY forms part of a circle and all angles subtended on the circumference of this circle by the arc XY are equal, i.e. $XCY = XC'Y = XC''Y = \alpha$. Suppose that X is a source of X-rays and XC, XC′ represent the extremities of a divergent X-ray beam emitted from X. If the beam is diffracted by a sample which covers the arc between C and C′ such that the diffracting planes are tangential to the circle, then the diffracted beam, represented by CY and C′Y, will focus to a point at Y. The principle of the focusing method is therefore to arrange that the source of X-rays, the sample and the detector all lie on the circumference of a circle (Fig. 32b).

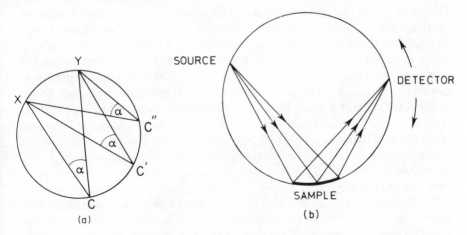

Fig. 3.12 (a) Theorem of a circle used to focus X-rays. (b) Arrangement of sample, source and detector on the circumference of a circle

Focusing (Guinier) cameras and crystal monochromators

The same focusing principle that is basic to the construction of diffractometers is also used in focusing cameras, although several different arrangements are found in commercial instruments. An additional feature of focusing cameras is the inclusion of a *crystal monochromator* which serves two functions: to give highly monochromatic radiation and to produce an intense, convergent X-ray beam. There are several sources of background scattering in diffraction experiments, one of which is the presence of radiation of wavelength different from that of the $K\alpha$ radiation. $K\alpha$ radiation may be separated from the rest by the use of filters or, better, by a crystal monochromator.

A crystal monochromator consists of a large single crystal of, for example, quartz, oriented such that one set of planes which diffracts strongly ($10\bar{1}1$ for quartz) is at the Bragg angle to the incident beam. This Bragg angle is calculated for $\lambda_{K\alpha_1}$ and so only the $K\alpha_1$ rays are diffracted, giving monochromatic radiation. If a flat crystal monochromator were used, much of the $K\alpha$ radiation would be lost since the X-ray beam emitted from a source is naturally divergent; only a small amount of the $K\alpha$ component would therefore be at the correct Bragg angle to the monochromator. To improve the efficiency, the crystal monochromator is bent, in which case a divergent X-ray beam may be used which is diffracted by the crystal monochromator to give a beam that is intense, monochromatic and convergent.

The arrangement of a focusing or Guinier camera which uses a crystal monochromator M and also makes use of the theorem of the circle described above is shown in Fig. 3.13(a). The convergent beam of monochromatic radiation passes through the sample at X. Radiation that is not diffracted comes to a focus at A, where a beam stop is placed in front of the film to prevent its blackening. Various beams diffracted by the sample focus at B, C, etc. We know from the theorem of the circle that A, B, C and X must lie on the circumference of a circle. The film is placed in a cassette which is in the form of a short cylinder and lies on the circle ABC. The scale of the film is linear in 2θ, as is the chart output from a diffractometer. A schematic film is as shown in Fig. 3.13(b) except that instead of peaks of different height, lines of different intensity or different degrees of blackness are seen. Film dimensions are $\sim 1 \times 15\,\text{cm}$ which makes them very convenient to handle. The line at $0°\ 2\theta$ or ∞ d-spacing corresponds to the undiffracted beam at A in (a). This is the reference position on the film. The mark is made by removing the beam stop for a fraction of a second while the X-rays are switched on. If required, a scale may be printed onto the film and the positions of the lines, relative to A, may be measured with a travelling microscope or, better, by microdensitometry; 2θ values and d-spacings may then be computed or obtained from tables.

The Guinier method is capable of giving accurate d-spacings, if desired, and the results are comparable to those obtained by diffractometry using very slow scanning speeds. Intensities of the lines on the films are either estimated visually or may be measured quantitatively using microdensitometry. Sample sizes are

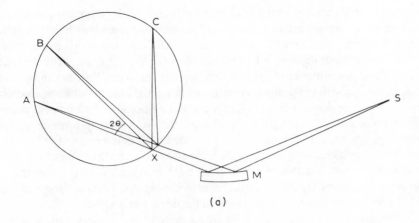

(a)

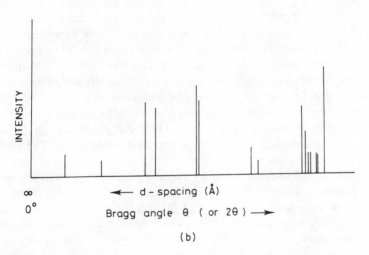

∞
0°

← d – spacing (Å)

Bragg angle θ (or 2θ) →

(b)

Fig. 3.13 (a) Crystal monochromator M, source S and sample X, in a focusing camera. (b) Schematic Guinier X-ray powder diffraction pattern

very small, 1 mg or less, and necessary exposure times vary between 5 min and 1 hr, depending on factors such as the crystallinity of the sample and the presence or absence of heavy elements which absorb X-rays.

A powder pattern is a crystal's 'fingerprint'

The powder X-ray diffraction method is very important and useful in qualitative phase analysis because every crystalline material has its own characteristic powder pattern; indeed, the method is often called the powder fingerprint method. There are two main factors which determine powder patterns: (a) the size and shape of the unit cell and (b) the atomic number and position of the various atoms in the cell. Thus, two materials may have the same crystal structure but almost certainly they will have quite distinct powder patterns. For example, KF, KCl and KI all have the rock salt structure and should show the same set of lines in their powder patterns, but, as can be seen from Table 3.2, both the positions and intensities of the lines are different in each. The positions or d-spacings of the lines are shifted because the unit cells are of different size and, therefore, the a parameter in the d-spacing formula varies. Intensities are different because different anions with different atomic numbers and therefore different scattering powers are present in the three materials, even though the atomic coordinates are the same for each (i.e. cations at corner and face centre positions, etc.). KCl is a rather extreme example because the intensities of 111 and 311 reflections are too small to measure, but it serves to illustrate the importance of scattering power of the atoms present. Intensities are discussed in more detail in the next section.

The powder pattern has two characteristic features, therefore: the d-spacings of the lines and their intensity. Of the two, the d-spacing is far more useful and capable of precise measurement. The d-spacings should be reproducible from sample to sample unless impurities are present to form a solid solution or the material is in some stressed, disordered or metastable condition. On the other hand, intensities are more difficult to measure quantitatively and often vary from sample to sample. Intensities can usually be measured only semi-quantitatively and may show variation of, say, 20 per cent from sample to sample (much more if preferred crystal orientation is present). Thus, the differences in tabulated intensities for, say, the (220) reflection of the three materials in Table 3.2 are probably not very significant.

The likelihood of two materials having the small cell parameters and d-spacings decreases considerably with decreasing crystal symmetry. Thus, cubic materials have only one variable, a, and there is a fair chance of finding two materials with the same a value. On the other hand, triclinic powder patterns have six variables, a, b, c, α, β and γ, and so accidental coincidences are far less

Table 3.2 *X-ray powder diffraction patterns for potassium halides.* (Data from Joint Committee on Powder Diffraction Standards, Swarthmore)

(*hkl*)	KF, $a = 5.347$ Å		KCl, $a = 6.2931$ Å		KI, $a = 7.0655$ Å	
	d (Å)	I	d (Å)	I	d (Å)	I
111	3.087	29		—	4.08	42
200	2.671	100	3.146	100	3.53	100
220	1.890	63	2.224	59	2.498	70
311	1.612	10	—	—	2.131	29
222	1.542	17	1.816	23	2.039	27
400	1.337	8	1.573	8	1.767	15

likely. Problems of identification, if they occur at all, are most likely to be experienced with high symmetry, especially cubic, materials.

Intensities

Intensities of X-ray reflections are important for two main reasons. First, quantitative measurements of intensity are necessary in order to determine unknown crystal structures. Second, qualitative or semi-quantitative intensity data are needed in using the powder fingerprint method to characterize materials and especially in using the *Powder Diffraction File* to identify unknowns. Although this book is not concerned with the methods of crystal structure determination, it is important that the factors which control the intensity of X-ray reflections be understood. The topic falls into two parts: the intensity scattered by individual atoms and the resultant intensity scattered from the large number of atoms that are arranged periodically in a crystal.

Scattering of X-rays by an atom

Atoms diffract or scatter X-rays because an incident X-ray beam, which can be described as an electromagnetic wave with an oscillating electric field, sets each electron of an atom into vibration. A vibrating charge such as an electron emits radiation and this radiation is in phase or *coherent with* the incident X-ray beam. The electrons of an atom therefore act as secondary point sources of X-rays. Coherent scattering may be likened to an elastic collision between the wave and the electron: the wave is deflected by the electron without loss of energy and, therefore, without change of wavelength. The intensity of the radiation scattered coherently by 'point source' electrons has been treated theoretically and is given by the Thomson equation:

$$I_p \propto \tfrac{1}{2}(1 + \cos^2 2\theta) \tag{3.5}$$

where I_p is the scattered intensity at any point, P, and 2θ is the angle between the directions of the incident beam and the diffracted beam that passes through P. From this equation it can be seen that the scattered beams are most intense when parallel or antiparallel to the incident beam and are weakest when at 90° to the incident beam. The Thomson equation is also known as the *polarization factor* and is one of the standard angular correction factors that must be applied during the processing of intensity data (for use in structure determination).

At this point, it is worth mentioning that X-rays can interact with electrons in a different way to give *Compton scattering*. Compton scattering is rather like an inelastic collision in that the X-rays lose some of their energy on impact and so the scattered X-rays are of longer wavelength than the incident X-rays. They are also no longer in phase with the incident X-rays; nor are they in phase with each other. A close similarity exists between Compton scattering and the generation of white radiation in an X-ray tube; both are examples of incoherent scattering that are sources of background radiation in X-ray diffraction experiments. As Compton scattering is caused by interaction between X-rays and the more loosely held outer valence electrons, it is an important effect with the lighter elements and can have a particularly deleterious effect on the powder patterns of organic materials such as polymers.

The X-rays that are scattered by an atom are the resultant of the waves scattered by each electron in the atom. The electrons may be regarded as particles that occupy different positions in an atom and interference occurs between their scattered waves. For scattering in the direction of the incident beam (Fig. 3.14a) beams 1' and 2', all electrons scatter in phase irrespective of their position. The scattered intensity is, then, the sum of the individual intensities. The *scattering factor*, or *form factor*, f, of an atom is proportional to its atomic number, Z, or, more strictly, to the number of electrons possessed by that atom.

For scattering at some angle 2θ to the direction of the incident beam, a phase difference, corresponding to the distance XY, exists between beams 1″ and 2″. This phase difference is usually rather less than one wavelength (i.e. XY < 1.5418 Å for Cu $K\alpha$ X-rays) because distances between electrons within an atom are short. As a result, only partial destructive interference occurs between 1″ and 2″. The net effect of interference between the beams scattered by all the electrons in the atom is to cause a gradual decrease in scattered intensity with increasing angle, 2θ. For example, the scattering power of copper is proportional to 29 (i.e. Z) at $2\theta = 0°$, to 14 at 90° and to 11.5 at 120°. It should also be apparent that for a given angle, 2θ, the net intensity decreases with decreasing X-ray wavelength. The form factors of atoms are given in *International Tables for X-ray Crystallography*, Vol. 3 (1952). They are tabulated against $(\sin \theta/\lambda)$ to include the effect of both angle and X-ray wavelength; examples are shown in Fig. 3.14(b).

Two consequences of the dependence of form factors on $\sin \theta/\lambda$ and atomic number are as follows. First, the powder patterns of most materials contain only weak lines at high angles (above ~ 60 to $70° 2\theta$). Second, in crystal structure determinations using X-rays, it is difficult to locate light atoms because their diffracted radiation is so weak. Thus hydrogen atoms cannot usually be located unless all the other elements present are also extremely light (e.g. in boron hydride crystals). Atoms that have as many electrons as oxygen can usually be located easily unless a very heavy atom such as uranium is present. Structures that are particularly difficult to solve are those in which a considerable number or all of the atoms present have similar atomic number, e.g. large organic molecules with carbon, nitrogen and oxygen atoms. In such cases, a common ploy is to make a derivative of the compound of interest which contains a heavy metal atom. The heavy atoms can be detected very readily and because they determine the phase of diffracted beams, a lead is given towards placing the remaining atoms. Because of their similar atomic numbers, aluminium and silicon are very difficult to distinguish, which causes problems in determinations of aluminosilicate structures. One of the advantages of using neutrons instead of (or as well as) X-rays for crystallographic work is that the neutron form factors are not a simple function of atomic number. Light atoms, e.g. hydrogen and lithium, are often strong neutron scatterers.

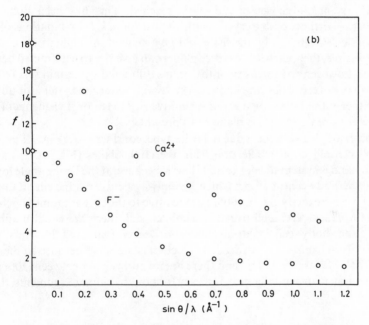

Fig. 3.14   (a) Scattering of X-rays by electrons in an atom. (b) Form factors of Ca^{2+} and F^-

Scattering of X-rays by a crystal—systematic absences

Earlier, we treated diffraction by crystals in terms of Bragg's law, with the crystals divided up into sets of lattice planes, defined with reference to the shape and dimensions of the unit cell. For each set of planes, with its characteristic d-spacing, the Bragg angle can be evaluated for a given wavelength. The number of possible sets of planes is limited since, using equation (1.1) for orthogonal crystals,

Table 3.3 *Calculated d-spacings for an ortho-*
rhombic cell, for a = 3.0, b = 4.0, c = 5.0 Å

hkl	d(Å)
001	5.00
010	4.00
011	3.12
100	3.00
101	2.57
110	2.40
111	2.16

h, k and l must be integers. It is possible to calculate all the possible d-spacing values from either equation (1.1) or the appropriate equations for other unit cell shapes, although the calculation is usually terminated when either a minimum d-spacing or maximum set of indices is reached. This has been done for a hypothetical orthorhombic crystal, with all possible h, k, l combinations of 0 and 1, in Table 3.3. Obviously, the list could be extended for higher indices.

In principle, then, each set of lattice planes can give rise to a diffracted beam. In practice, however, the intensity of the beams diffracted by certain sets of lattice planes may be zero. These are known as *systematic absences*. Systematic absences arise either if the lattice type is non-primitive (I, F, etc.) or if elements of space symmetry (screw axes, glide planes) are present.

As an example of absences due to lattice type, consider α-Fe (Fig. 3.15a), which is body centred cubic. Reflection from the (100) planes (Fig. 3.15b) has zero intensity and is systematically absent. This is because, at the Bragg angle for these planes, the body centre atoms which lie midway between adjacent (100) planes diffract X-rays exactly 180° out of phase relative to the corner atoms which lie on the (100) planes. Averaged over the whole crystal, there are equal numbers of corner and body centre atoms and the beams diffracted by each cancel completely. In contrast, a strong 200 reflection is observed because all the atoms lie on (200) planes (Fig. 3.15c) and there are no atoms lying between (200) planes to cause destructive interference. It is easy to show, by similar arguments, that the

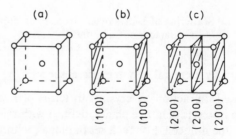

Fig. 3.15 (a) Body centred cubic α-Fe, (b) (100) planes, (c) (200) planes

Table 3.4   *Systematic absences due to lattice type*

Lattice type	Rule for reflection to be observed*
Primitive, P	None
Body centred, I	hkl; $h + k + l = 2n$
Face centred, F	hkl; h, k, l either all odd or all even
Side centred, e.g. C	hkl; $h + k = 2n$
Rhombohedral, R	hkl; $-h + k + l = 3n$
	or $(h - k + l = 3n)$

*If space symmetry elements are present, additional rules limiting the observable reflections may apply.

110 reflection is observed whereas 111 is systematically absent in α-Fe. For each non-primitive lattice type there is a simple characteristic formula for systematic absences (Table 3.4). For a body centred cell reflections for which $(h + k + l)$ is odd are absent, e.g. reflections such as 100, 111, 320, etc., are systematically absent.

Systematic absences caused by the presence of space symmetry elements are rather complicated and difficult to describe and will not be dealt with here.

The occurrence of systematic absences is an extreme case of destructive interference between the X-ray beams diffracted by individual atoms. In particular, systematic absences arise when one set of atoms diffracts X-rays that are exactly out of phase with those diffracted by a second set of atoms of the same type. Two conditions must be met for systematic absences, therefore: the diffracted beams must be out of phase (by $\lambda/2$ or π) and of the same amplitude (determined by scattering powers, f). In the more general case, where the destructive interference is not complete and intensities are actually observed, then one or both of these conditions may not be met.

Let us consider now the rock salt structure. It is face centred cubic and

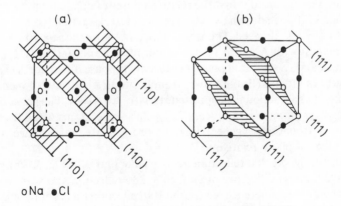

Fig. 3.16   (a) (110) and (b) (111) planes in NaCl

therefore, only those reflections for which h, k, l are either all odd or all even may be observed (Table 3.4). From this rule, for instance, 110 is systematically absent. This may be seen from Fig. 3.16a: (110) planes have Na^+ and Cl^- ions lying on the planes but equal numbers of the same ions lying midway between the planes. Both conditions specified above are met and complete cancellation occurs. For the (111) planes, however, only the first condition is met and this gives rise to an allowable reflection. The (111) planes are shown in Fig. 3.16b. Na^+ ions lie on the planes and Cl^- ions lie exactly midway between them. Hence, the Na^+ and Cl^- ions scatter exactly 180° out of phase with each other for these planes, but since they have different scattering powers the destructive interference that occurs is only partial. The intensity of the 111 reflection in materials that have the rock salt structure is, therefore, related to the difference in atomic number of anion and cation. For the potassium halides, the 111 intensity is zero for KCl, since K^+ and Cl^- are isoelectronic, and its intensity should increase in the order.

$$KCl < KF < KBr < KI$$

Some data which confirm this are given in Table 3.2.

Similar effects may be found in other simple crystal structures. In primitive cubic CsCl, if the difference between caesium and chlorine is ignored the atomic positions are the same as in body centred α-Fe (Fig. 3.15). The 100 reflection is systematically absent in α-Fe but is an observed reflection with CsCl because the scattering powers of Cs^+ and Cl^- are different, i.e. $F_{Cs^+} \neq f_{Cl^-}$.

Intensities—general formulae and a model calculation for CaF$_2$

Each atom in a crystal scatters X-rays by an amount related to the scattering power, f, of that atom. In summing the individual waves to give the resultant diffracted beam, both the *amplitude* and *phase* of each wave are important. If we know the atomic positions in the structure, the amplitude and phase appropriate to each atom in the unit cell may be calculated and the summation carried out by various mathematical methods, therefore simulating what happens during diffraction. Let us consider first the relative phases of different atoms in the unit cell. In Fig. 3.17(a) are drawn two (100) planes of a crystal that has an orthogonal (i.e. $\alpha = \beta = \gamma = 90°$) unit cell. The atoms A, B, C, A' lie on the a axis (perpendicular to (100) planes) with A and A' at the origin of adjacent unit cells. For the 100 reflection, A and A' scatter in phase because their phase difference is exactly one wavelength, 2π radians (Bragg's law). Atom B, situated halfway between adjacent (100) planes, has a fractional x coordinate (relative to A) of $\frac{1}{2}$. The phase difference between (waves diffracted from) A and B is $\frac{1}{2} \cdot 2\pi = \pi$, i.e. atoms A and B are exactly out of phase. Atom C has a general fractional coordinate x (at distance xa from A) and, therefore, a phase relative to A of $2\pi x$.

Consider, now, the 200 reflection for the same unit cell (Fig. 3.17b). Atoms A and B have a phase difference of 2π for the 200 reflection and scatter in phase, whereas their phase difference is π for the 100 reflection (in order to obey Bragg's law, if d is halved, $\sin \theta$ must double; thus $\theta_{200} \gg \theta_{100}$). Comparing the Bragg

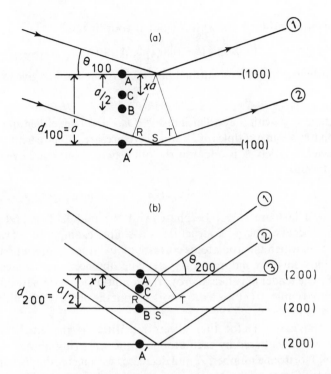

Fig. 3.17 (a) (100) planes for an orthogonal unit cell ($\alpha = \beta = \gamma = 90°$). Atoms A, B, C, A' lie on the a cell edge. (b) (200) planes for the same unit cell as in (a)

diffraction conditions for the (100) and (200) planes, the effect of halving d is to double the relative phase difference between pairs of atoms such as A and B; therefore, A and C have a phase difference of $(2x \cdot 2\pi)$ for the (200) reflection.

For the general case of an $h00$ reflection, the d-spacing between adjacent ($h00$) planes is $(1/h)a$ and the phase difference, δ, between A and C is given by

$$\delta = 2\pi hx \qquad (3.6)$$

The phase difference between atoms depends, therefore, on two factors: the Miller indices of the reflection that is being considered and the fractional coordinates of the atoms in the unit cell. The above reasoning may be extended readily to a general three-dimensional situation. For reflection from the set of planes with indices (hkl), the phase difference, δ, between atoms at the origin and a position with fractional coordinates (x, y, z) is given by

$$\delta = 2\pi(hx + ky + lz) \qquad (3.7)$$

This is an important formula and is applicable to all unit cell shapes. Let us use it on a simple structure, γ-Fe, which is face centred cubic with atoms at the corner and face centre positions, i.e. with fractional coordinates:

$$(0,0,0); \quad (\tfrac{1}{2},\tfrac{1}{2},0); \quad (\tfrac{1}{2},0,\tfrac{1}{2}); \quad (0,\tfrac{1}{2},\tfrac{1}{2})$$

These coordinates may be substituted into the formula for δ to give four phases:

$$0, \quad \pi(h+k), \quad \pi(h+l), \quad \pi(k+l)$$

How do these vary with the Miller indices? If h, k and l are either all even or all odd, the phases are in multiples of 2π and, therefore, are in phase with each other.

If, however, one, say h, is odd and the other two, k and l, are even, the four phases reduce to

$$0, \quad (2n+1)\pi, \quad (2n+1)\pi, \quad 2n\pi$$

The first and last are π out of phase with the middle two and complete cancellation occurs. The γ-Fe structure is a simple example of a face centred cubic lattice in which the iron atoms correspond to lattice points and, in fact, we have just proved the condition for systematic absences due to face centring (Table 3.4). The reader may like to prove the condition for systematic absences in a body centred cubic structure, e.g. by working out the phases of the atoms for the structure of α-Fe.

The second major factor that affects intensities is the amplitude of the individual waves scattered by each atom, i.e. the scattering power, f, which is proportional to atomic number, Z, and decreases with increasing Bragg angle, θ.

We now wish to generalize the treatment to consider any atom in the unit cell. For atom j, the diffracted wave of amplitude f_j and phase δ_j may be represented by a sine wave of the form

$$F_j = f_j \sin(\omega t - \delta_j) \tag{3.8}$$

The waves diffracted from each atom in the cell have the same angular frequency, ω, but may differ in f and δ. The resultant intensity is obtained from the summation of the individual sine waves. Mathematically, addition of waves may be carried out by various methods, including vector addition and by the use of complex numbers. In complex notation, wave j may be written as

$$F_j = f_j(\cos\delta_j + i\sin\delta_j) \tag{3.9}$$

or as

$$F_j = f_j e^{i\delta_j} \tag{3.10}$$

The intensity of a wave is proportional to the square of its amplitude; i.e.

$$I \propto f_j^2 \tag{3.11}$$

and is obtained by multiplying the equation for the wave by its complex conjugate; i.e.

$$I \propto (f_j e^{i\delta_j})(f_j e^{-i\delta_j})$$

and therefore

$$I \propto f_j^2$$

Alternatively,

$$[f_j(\cos \delta_j + i \sin \delta_j)][f_j(\cos \delta_j - i \sin \delta_j)] = f_j^2(\cos^2 \delta_j + \sin^2 \delta_j) = f_j^2$$

Substituting the expression for δ, the equation of a diffracted wave becomes

$$F_j = f_j \exp 2\pi i (hx_j + ky_j + lz_j)$$
$$= f_j[\cos 2\pi(hx_j + ky_j + lz_j) + i \sin 2\pi(hx_j + ky_j + lz_j)] \qquad (3.12)$$

When written in these forms, the summation over the j atoms in the unit cell may be carried out readily, to give the *structure factor* or *structure amplitude*, F_{hkl}, for the hkl reflection; i.e.

$$F_{hkl} = \sum_{j=1 \to n} (f_j e^{i\delta_j})$$

or

$$F_{hkl} = \sum_j f_j(\cos \delta_j + i \sin \delta_j) \qquad (3.13)$$

The intensity of the diffracted beam I_{hkl} is proportional to $|F_{hkl}|^2$ and is obtained from

$$I_{hkl} \propto |F_{hkl}|^2 = \left[\sum_j f_j(\cos \delta_j + i \sin \delta_j)\right]\left[\sum_j f_j(\cos \delta_j - i \sin \delta_j)\right]$$
$$= \sum_j (f_j \cos \delta_j)^2 + \sum_j (f_j \sin \delta_j)^2 \qquad (3.14)$$

This latter is a very important formula in crystallography because by using it the intensity of any hkl reflection may be calculated from a knowledge of the atomic coordinates in the unit cell. Let us see one example of its use. Calcium fluoride, CaF_2, has the fluorite structure with atomic coordinates in the face centred cubic unit cell:

Ca	$(0,0,0)$	$(\tfrac{1}{2},\tfrac{1}{2},0)$	$(\tfrac{1}{2},0,\tfrac{1}{2})$	$(0,\tfrac{1}{2},\tfrac{1}{2})$
F	$(\tfrac{1}{4},\tfrac{1}{4},\tfrac{1}{4})$	$(\tfrac{1}{4},\tfrac{1}{4},\tfrac{3}{4})$	$(\tfrac{1}{4},\tfrac{3}{4},\tfrac{1}{4})$	$(\tfrac{3}{4},\tfrac{1}{4},\tfrac{1}{4})$
	$(\tfrac{3}{4},\tfrac{3}{4},\tfrac{1}{4})$	$(\tfrac{3}{4},\tfrac{1}{4},\tfrac{3}{4})$	$(\tfrac{1}{4},\tfrac{3}{4},\tfrac{3}{4})$	$(\tfrac{3}{4},\tfrac{3}{4},\tfrac{3}{4})$

Substitution of these coordinates into the structure factor equation (3.14), yields

$$F_{khl} = f_{Ca}[\cos 2\pi(0) + \cos \pi(h+k) + \cos \pi(h+l) + \cos \pi(k+l)]$$
$$+ if_{Ca}[\sin 2\pi(0) + \sin \pi(h+k) + \sin \pi(h+l)$$
$$+ \sin \pi(k+l)] + f_F[\cos \pi/2(h+k+l)$$
$$+ \cos \pi/2(h+k+3l) + \cos \pi/2(h+3k+l)$$
$$+ \cos \pi/2(3h+k+l) + \cos \pi/2(3h+3k+l)$$
$$+ \cos \pi/2(3h+k+3l) + \cos \pi/2(h+3k+3l)$$
$$+ \cos \pi/2(3h+3k+3l)] + if_F[\sin \pi/2(h+k+l)$$
$$+ \sin \pi/2(h+k+3l) + \sin \pi/2(h+3k+l) + \sin \pi/2(3h+k+l)$$
$$+ \sin \pi/2(3h+3k+l) + \sin \pi/2(3h+k+3l)$$
$$+ \sin \pi/2(h+3k+3l) + \sin \pi/2(3h+3k+3l)]$$

Since the fluorite structure is face centred cubic, h, k and l must be either all odd or all even for an observed reflection; for any other combination, $F = 0$ (try it!). Consider the reflection 202:

$$F_{202} = f_{Ca}(\cos 0 + \cos 2\pi + \cos 4\pi + \cos 2\pi)$$
$$+ if_{Ca}(\sin 0 + \sin 2\pi + \sin 4\pi + \sin 2\pi)$$
$$+ f_F(\cos 2\pi + \cos 4\pi + \cos 2\pi + \cos 4\pi + \cos 4\pi + \cos 6\pi$$
$$+ \cos 4\pi + \cos 6\pi) + if_F(\sin 2\pi + \sin 4\pi + \sin 2\pi + \sin 4\pi + \sin 4\pi$$
$$+ \sin 6\pi + \sin 4\pi + \sin 6\pi)$$

That is,

$$F_{202} = f_{Ca}(1 + 1 + 1 + 1) + if_{Ca}(0 + 0 + 0 + 0)$$
$$+ f_F(1 + 1 + 1 + 1 + 1 + 1 + 1 + 1)$$
$$+ if_F(0 + 0 + 0 + 0 + 0 + 0 + 0 + 0)$$

or

$$F_{202} = 4f_{Ca} + 8f_F$$

The 202 reflection in CaF_2 has a d-spacing of $1.929\,\text{Å}$ $(a = 5.464\,\text{Å})$. Therefore

$$\theta_{202} = 23.6° \quad \text{and} \quad \sin\theta/\lambda = 0.259 \quad \text{for } \lambda = 1.5418\,\text{Å}(Cu\,K\alpha)$$

Form factors for calcium and fluorine are given in Fig. 3.14(b); for $\sin\theta/\lambda = 0.259$, by interpolation,

$$f_{Ca} = 12.65 \quad \text{and} \quad f_F = 5.8$$

Therefore,

$$F_{202} = 97$$

This calculation may be made for a series of hkl reflections and the results, after scaling, may be compared with the observed values (Table 3.5). In solving unknown crystal structures, the objective is always to obtain a model structure for which the calculated structure factors, F_{hkl}^{calc}, are in good agreement with those obtained from the experimental intensities, i.e. F_{hkl}^{obs}.

An important feature which simplifies the above calculations is that all the sine terms are zero. This is because the origin of the unit cell is also a centre of symmetry. For each atom at position (x, y, z) there is a centrosymmetrically related atom at $(-x, -y, -z)$ (e.g. F at $(\frac{1}{4}, \frac{1}{4}, \frac{1}{4})$ and $(-\frac{1}{4} -\frac{1}{4} -\frac{1}{4})$, i.e. $(1 - \frac{1}{4}, 1 - \frac{1}{4}, 1 - \frac{1}{4})$ or $(\frac{3}{4}, \frac{3}{4}, \frac{3}{4})$) and since $\sin(-\delta) = -\sin\delta$, the summation of the sine terms over the unit cell contents is zero. If, on the other hand, one of the F atoms was taken as the origin of the cell, the sine terms would be non-zero because, F, with its immediate coordination environment of 4Ca arranged tetrahedrally, does not lie on a centre of symmetry. Many structures, of course, belong to non-centric space groups, in which case the complete calculation of F using both cosine and sine terms cannot be avoided.

Table 3.5 *Structure factor calculations for CaF$_2$: X-ray powder diffraction*

d(Å)	hkl	I	Multiplicity*	I/(multiplicity × L$_p$)†	F^{obs}	F^{calc}	F^{obs}scaled	‖F^{obs}| − |F^{calc}‖
3.143	111	100	8	0.409	0.640	67	90	23
1.929	202	57	12	0.476	0.690	97	97	0
1.647	311	16	24	0.098	0.313	47	44	3
1.366	400	5	6	0.193	0.439	75	62	13
1.254	331	4	24	0.047	0.217	39	31	8

$\sum F^{obs}$ scaled = 324

$\sum \| F^{obs}| - |F^{calc} \| = 47$

$$R = \frac{\sum |\Delta F|}{\sum F^{obs}} = \frac{47}{324} = 0.15$$

*The multiplicity of an X-ray powder line is given by the number of equivalent sets of planes that diffract at the same Bragg angle and therefore overlap. Thus, overlapping with 111 are $\bar{1}11$, $1\bar{1}1$, $11\bar{1}$, $\bar{1}\bar{1}1$, $\bar{1}1\bar{1}$, $1\bar{1}\bar{1}$ and $\bar{1}\bar{1}\bar{1}$, where a negative index indicates that a negative crystallographic direction was used in defining it.

†The Lorentz polarization factor, L$_p$, is an angular correction factor that includes the effect of equation (3.5) and certain instrumental factors. Available from standard tables.

R-factors and structure determination

It was shown above how the structure factor, F_{hkl}^{calc}, may be calculated for any *hkl* reflection from a knowledge of the coordinates of the atoms in the unit cell. The values of F_{hkl}^{calc} for the first five lines in the powder pattern of CaF_2 are given in Table 3.5, column 7. The experimental intensities are given in column 3 and the intensities after correction for the L_p factor and *multiplicities* in column 5. The observed structure factor, F_{hkl}^{obs}, is related to the corrected intensities by the relation: $F_{hkl}^{obs} = \sqrt{I_{corr}}$, and these values are given in column 6. In order to be able to compare the values of F_{hkl}^{obs} and F_{hkl}^{calc}, they must be scaled such that $\sum F_{hkl}^{obs} = \sum F_{hkl}^{calc}$. Multiplication of each F_{hkl}^{obs} value by 141 gives the scaled values in column 8. The measure of agreement between the individual, scaled F_{hkl}^{obs} and F_{hkl}^{calc} values is given by the *residual factor* or *R-factor*, defined as follows:

$$R = \frac{\sum \|F^{obs}| - |F^{calc}\|}{\sum |F^{obs}|}$$ (3.15)

Values of the numerator are listed in column 9 and an *R*-factor of 0.15 (or 15 per cent after multiplying by 100) is obtained.

In solving unknown crystal structures, one is guided, among other things, by the value of *R*; the lower it is, the more likely is the structure to be correct. The calculation given for CaF_2 is rather artificial since only five reflections were used (one normally uses hundreds or thousands of reflections), but it serves as an illustration. It is not possible to give hard and fast rules about the relation between the magnitude of *R* and the likely correctness of the structure, but, usually, when *R* is less than 0.1 to 0.2, the proposed structure is essentially correct. A structure which has been solved fully using good quality intensity data has *R* typically in the range 0.02 to 0.06.

Electron density maps

An electron density map is a plot of the variation of electron density throughout the unit cell. During the processes of solving an unknown structure it is often useful to construct electron density maps (Fourier maps) in order to try and locate atoms. As the structure refinement proceeds, the quality of the electron density map usually improves: the background electron density decreases and, at the same time, more peaks due to individual atoms become resolved. We are concerned here, not with the methods of structure refinement but only with the results, and the final electron density map obtained at the end of a structure determination is an important piece of information. Electron density maps usually take the form of sections through the structure at regular intervals; by superposing these, a three-dimensional picture of the electron density distribution may be obtained. In Fig. 3.18 is shown the electron density distribution for a section through a very simple structure, NaCl. The section is parallel to one face of the unit cell and passes through the centres of the Na^+, Cl^- ions. It has the following features.

Fig. 3.18 Electron density map for NaCl

An electron density map resembles a geographical contour map. The contours represent lines of constant electron density throughout the structure. Peaks of electron density maxima may be distinguished clearly and these correspond to atoms; the coordinates of the atoms in the unit cell are given by the coordinates of the peak maxima. The assignment of peaks to particular atoms is made from the relative heights of the electron density peaks: the peak height is proportional to the number of electrons possessed by that atom, which apart from very light atoms is approximately equal to the atomic number of that atom. In Fig. 3.18, two types of peak of relative heights 100 and 50, are seen; these are assigned to chlorine and sodium, respectively. (The atomic numbers of chlorine and sodium are 17 and 11; for ions, the number of electrons are 18 (Cl^-) and 10(Na^+). The experimental maxima are therefore in fair agreement with the expected values).

Electron density maps also show that our mental picture of atoms as spheres is essentially correct, at least on a time average. The electron density drops to almost zero at some point along the line connecting pairs of adjacent atoms in Fig. 3.18 and this supports the model of ionic bonding in NaCl. In other structures which have covalent bonding, there is residual electron density between atoms on the electron density map. However, in other than very simple structures, such as the alkali halides, there is one serious difficulty in using an electron density map to determine quantitatively the distribution of valence electrons. In most structure refinements, both the position and thermal vibration factors of atoms are allowed to vary in order to achieve the best agreement between measured and calculated structure factors and intensities. The final parameters may represent a compromise, therefore, and the electron density map, which is greatly influenced by the thermal vibration factors, is not necessarily a true representation of the distribution of valence electrons. In refining simple structures, the atomic coordinates are usually known accurately. This then gives more accuracy to the thermal vibration factors (or temperature factors) and hence to the electron density map.

X-ray crystallography and structure determination—what's involved?

For determining the structures of crystalline materials, X-ray crystallography reigns supreme. For molecular materials, it complements the use of spectroscopic techniques—NMR, mass spectroscopy, etc.—and often, one may use either crystallography or spectroscopy to determine molecular structures. For non-molecular materials, however, or for molecular materials whose arrangement in the crystalline state is important, or whose bond lengths and angles must be determined, then X-ray crystallography is by far the most important technique for structure determination.

Nowadays, solving structures is mathematically complex but usually highly automated with computer-controlled collection and processing of diffraction data. It requires large, expensive instruments to do the diffraction experiments and is time-consuming: given a reasonable-sized single crystal, several days are required to collect the X-ray diffraction data and at least a few days of computation to solve the structure. It is not the kind of experiment that is carried out in an afternoon in a laboratory, therefore, but nevertheless, it is important to have at least a passing familiarity with the processes and problems involved. In fact, we have already covered much of the groundwork earlier in this chapter.

Solving an unknown crystal structure is a bit like solving a set of simultaneous equations. The unknowns in the equations are the atomic coordinates and the equations are the experimental intensity data. Obviously, we must have at least as many equations (i.e. intensities) as variables but in practice, we need many more intensities than variables in order to obtain good quality structure determinations. This is partly because the intensity data may have considerable errors, partly because the computational methods involved in structure determination often involve statistical analyses of data which can function properly only with large data sets, and partly because the Fourier methods used in structure determination are effective only with a large number of coefficients (intensities). Let us see some examples with structures of varying complexity, because the number of variables to be determined dictates the type of diffraction experiment that is required.

A simple crystal structure is that of rutile, TiO_2. Suppose that we have prepared a new material, MX_2, suspected to have the rutile structure: how many variables must be determined in order to confirm the structure? From the rutile structure, Fig. 1.32, atoms M are in fixed positions, at the corner and body centre and hence have no positional variables. The atoms, X, have a single positional variable, x, since the coordinates of the four atoms in the unit cell are given by:

$$x, x, 0; \qquad 1 - x, 1 - x, 0; \qquad \tfrac{1}{2} - x, \tfrac{1}{2} + x, \tfrac{1}{2}; \qquad \tfrac{1}{2} + x, \tfrac{1}{2} - x, \tfrac{1}{2}$$

Hence, once x has been determined, all four atoms have been located. (Note, for TiO_2, Fig. 1.32, $x = 0.3$).

In addition to the positional variables, atoms are vibrating, either isotropically or anisotropically and the so-called *temperature factors* are additional variables which must be determined in any good quality structure refinement. The first stage is to determine *isotropic temperature factors*, B_{iso}, which assumes that the atoms are vibrating isotropically; there are two of these for rutile, one for M and one for X. In many cases, determination of positional coordinates and B_{iso} values is all that is required; for our MX_2 compound this involves a total of three variables. In order to determine these accurately, intensity data for at least 10–20 reflections would normally be required; since the powder X-ray pattern may contain at least this number of lines, the structure could probably be determined satisfactorily from powder data.

A moderately complex structure is that of $YBa_2 Cu_3 O_x$, one of the new family of recently discovered ceramic superconductors. It is orthorhombic with a structure related to but rather more complex than that of perovskite. The unit cell contains one formula unit with $x = 7.0$. There are 5 positional variables to be determined to locate all the atoms and 8 B_{iso} values (4 for oxygen, since there are four crystallographically distinct oxygens in the structure, 2 for Cu and 1 each for Ba, Y). One of the oxygens is present in variable amounts (x is variable in the formula) and the fractional occupancy of its site is a variable. This gives a total of 14 variables and for a good structure determination, 200–300 intensities would be required. Routine powder methods do not give this number of well-resolved reflections; either special powder techniques, such as high-resolution neutron or X-ray diffraction, or single crystal methods must be used. In these, the data set is increased both by collecting data to lower d-spacings using shorter-wavelength radiation and, with single crystals, by recording data that are too weak to appear in powder patterns.

For yet more complex structures, such as many silicates and complex organic molecules, especially if they are of low symmetry, the number of variables may be 50–100 and single crystal data, giving perhaps 2,000 to 3,000 intensities, are essential.

Once the intensity data have been collected, corrected for factors such as Lorentz polarization and converted to observed structure factors using, $F^{obs} = \sqrt{I^{obs}}$, the process of solving the structure can begin. The problem is essentially to determine the values of the atomic coordinates x, y, z that, when substituted into equation (3.12), yield calculated F values that match the observed ones.

In cases where both the signs ($+$ or $-$) and magnitudes of the F^{obs} values are known, there are standard mathematical procedures, based on Fourier series, for attacking this problem. This is because the observed diffracted X-ray beams may be combined mathematically, as a Fourier transform, to give the crystal structure in the form of an electron density map, as in Fig. 3.18, for example. The relevant equation is:

$$\rho(u, v, w) = \frac{1}{V} \sum_h \sum_k \sum_l F_{hkl} \cos 2\pi(hu + kv + lw) \qquad (3.16)$$

from which, for each point u, v, w in the unit cell, the electron density, ρ, is calculated by substituting into the summation all the observed F and h, k, l values. In order to obtain a good quality map, there must be a large number of terms in the summation and this is the main reason why the number of intensity values required for a structure determination is much greater than the number of variables.

There is a clear analogy between this mathematical transform and the production of images in optical and electron microscopes. In microscopes, the diffraction pattern obtained by shining a beam of either light or electrons on the sample, is combined to give an image using either an optical lens or an electromagnetic lens. There is, unfortunately, no material that can act as an 'X-ray lens' and we must resort to mathematical methods to perform this combination. This brings us to the key problem in crystallography: while the amplitudes of the F^{obs} values are determined directly and unequivocally from the intensities, their phases are not. Thus, while the I values must be positive, the F values, given by $\sqrt{I}$, may be positive or negative. Many means have been devised to attack this 'phase problem'. The most successful ones are outlined next.

The Patterson method

This method uses a Fourier summation rather similar to equation (3.16) but in which intensity (or F_{hkl}^2) data form the coefficients, i.e.

$$P(u, v, w) = \frac{1}{V} \sum_h \sum_k \sum_l |F_{hkl}|^2 \cos 2\pi(hu + kv + lw) \qquad (3.17)$$

The resulting Patterson map looks similar to a Fourier map, but the regions of high electron density correspond to vectors between pairs of atoms. The peak heights are proportional to the products of the atomic numbers; hence, vectors formed from pairs of the heaviest atoms in the unit cell give rise to the largest peaks. The peak positions give the position apart, vectorially, of these two atoms in the structure. Thus, two atoms at x_1, y_1, z_1 and x_2, y_2, z_2 will give rise to a Patterson peak, relative to the origin of the Patterson map, at $x_1 - x_2, y_1 - y_2, z_1 - z_2$. This does not give the positions of heavier atoms directly, but it does give their relative positions in the unit cell and this is often of great assistance in getting started: thus, if one atom can be located with confidence, then the Patterson map may suggest the location of others. Once such a start has been made, other Fourier methods may be used to complete the structure determination.

Fourier methods

These may be used even though the complete list of F^{obs} values, with their signs, is not available. Use is made of composite F values, whose magnitudes are taken from the F^{obs} listing and whose signs are taken from a partial structure factor calculation based on the heavy atom positions only. Since the heavy atoms scatter X-rays most strongly, they are likely to dominate the intensities and, in particular, control the phases (i.e. whether $+$ or $-$). Hence, most of the signs can probably be determined correctly at this stage, especially for the larger F values. A Fourier map, made from equation (3.16) and these composite F values, should then reveal considerably more electron density peaks and enable lighter atoms to be located. This process may be repeated: a more accurate structure factor calculation is performed using the more extensive list of atomic coordinates obtained from the Fourier map; consequently, the signs of more of the F values are determined correctly and an improved Fourier map can be calculated.

Once most of the atoms have been located approximately, *least squares refinement* procedures may be used. The objective is to improve the agreement between F^{obs} and F^{calc} and thereby, minimize the R-factor, equation (3.15). In these procedures, atom positions are permitted to vary somewhat, the effect on R is noted and the optimum positions (for minimum R) are found by trial and error. It is also useful to construct a *difference Fourier map* using $(F^{obs} - F^{calc})$ values as the coefficients:

$$\Delta(u, v, w) = \frac{1}{V}\sum_{h}\sum_{k}\sum_{l}(F^{obs} - F^{calc})\cos 2\pi(hu + kv + lw) \qquad (3.18)$$

A difference map may show regions of low, positive electron density associated with previously undetected light atoms such as H. Or, it may show regions of negative electron density, indicating that an atom has been wrongly placed at that position. Or it may be completely featureless, indicating that the structure refinement is correct and complete.

Direct methods

These are very useful for getting a structure determination started and work best when all the component atoms have similar atomic number, thus complementing the Patterson method which works best when a small number of heavy atoms are present. Direct methods are used to determine phases. They are based on the statistical probabilities of phases being either $+$ or $-$. For example, for centrosymmetric structures, the *Sayre probability relationship* indicates that for three reflections h, k, l, h', k', l' and h'', k'', l'' that are related by $h'' = h - h', k'' = k - k'$ and $l'' = l - l'$, then the sign of one phase is likely to be the same as the product of the signs of the other two. Thus, if 312 and 111 are both $-$, then 201 is probably $+$. By choosing three reflections, whose phases are not known, there are eight possible combinations of $+$ and $-$. For each combination, the phases of other reflections may be predicted. A range of methods are available for predicting phases and optimising the value of the predictions. The end result is an *E-map*, similar to an electron density map and from which atomic positions may be determined.

Electron diffraction

Electrons have wave characteristics which allow them also to be used for diffraction experiments. Their wavelength is related to velocity, which is governed by the voltage through which they are accelerated in an 'electron gun' and is usually about 0.04 Å in conventional electron microscopes. The two techniques of electron diffraction and electron microscopy are closely related. The latter is used more widely and is discussed in Chapter 4; here we shall see briefly the general characteristics of electron diffraction and how it compares with X-ray diffraction.

Electrons interact strongly with matter and intense diffraction patterns are obtained from very small samples; indeed, for transmission studies, the samples should not be thicker than about 100 nm (1 nm = 10 Å), otherwise the electrons may not pass through the sample. This contrasts markedly with X-ray diffraction: the efficiency with which X-rays are diffracted by matter is low and, for single crystal studies, relatively large crystals, 0.05 mm or greater in dimensions are required.

One disadvantage of electron diffraction is that *secondary diffraction* commonly occurs. Because the scattering efficiency of electrons is high, the diffracted beams are strong. Secondary diffraction occurs when these diffracted beams effectively become the incident beam and are diffracted by another set of lattice planes. There are two undesirable consequences of secondary diffraction. First, under certain circumstances, extra reflections may appear in the diffraction pattern; care is therefore needed in the interpretation of diffraction patterns. Second, the intensities of diffracted beams are unreliable and cannot be used quantitatively for crystal structure determination.

In spite of these disadvantages, electron diffraction is very useful and complements the various X-ray techniques. Thus with X-rays, the scattering efficiency is small, secondary diffraction is rarely a problem and intensities are reliable, but (relatively) large samples are needed. On the other hand, with electrons, the scattering efficiency is high and although intensities are unreliable very small samples can be studied. The technique is very useful for obtaining unit cell and space group information for crystals smaller than 0.01 to 0.02 mm in diameter; it is, in fact, the only reliable method for obtaining such information. The other methods that are sometimes used, such as graphical or computer-assisted indexing of X-ray powder patterns, give results that are not always 100 per cent reliable. In spite of its utility, electron diffraction is very much underused in solid state chemistry for unit cell and space group determination.

Electron diffraction is unsuitable as a routine method of phase identification in relatively large (e.g. 10 mg or more) samples. It is useful, however, (a) when only very small quantities are available, (b) for thin film samples, and (c) for detecting small amounts of impurity phases. In all these cases, there would be insufficient material to show up in X-ray diffraction.

Neutron diffraction

Neutron diffraction is a very expensive technique. In order to get a sufficiently intense source of neutrons, a nuclear reactor is needed. Few laboratories have their own neutron facility and, instead, experiments are carried out at central laboratories which provide a user service (e.g. at Grenoble, France). In spite of its high cost, neutron diffraction is a valuable technique and can provide information, especially on magnetic materials, that is not attainable with other techniques. Clearly, it is never used when alternative techniques, such as X-ray diffraction, can solve a particular problem.

Neutron beams are usually of low intensity and therefore the size of sample required for neutron diffraction work is relatively large, at least $1 \, mm^3$. Since crystals of this size are often not available, crystallographic studies are usually carried out on polycrystalline samples. A powder neutron diffraction pattern looks very much like an X-ray one.

There are several characteristic differences between neutron and X-ray diffraction. First, the neutrons obtained from a nuclear reactor give a continuous spectrum of radiation, without the intense characteristic peaks that are present in X-ray spectra (as in Fig. 3.1b). The neutrons that are used for diffraction are so-called thermal neutrons that have had their energy moderated by heavy water; their wavelength is of the order 0.5 to 3 Å. In order to have monochromatic neutrons, it is necessary to select a particular wavelength and filter out the remainder using a crystal monochromator. Most of the available neutron energy is wasted, therefore, and the beam that is used is weak and not particularly monochromatic.

A recent advance which offers exciting prospects is the use of pulsed neutron sources coupled with 'time of flight' analysis. The neutrons are obtained using particle accelerators to bombard a heavy metal target with high-energy particles such as protons. The efficiency of the *spallation* process, yielding about 30 neutrons per proton, gives a high neutron flux suitable for diffraction experiments. In the time of flight method, the entire neutron spectrum (variable wavelength) is used with a fixed diffraction angle, θ. Neutron wavelength depends on velocity, given by the de Broglie relation, $\lambda = h/mv$, where m is the mass of the neutron, 1.675×10^{-27} kg. Hence, the diffracted radiation arriving at the detector is separated according to its time of flight and wavelength. The fundamental law of diffraction is Bragg's law, $n\lambda = 2d \sin \theta$. In the time of flight method, λ and d (the d-spacing) are the variables at fixed θ. This compares with conventional diffraction techniques in which d and θ are the variables at fixed λ. The pulsed method gives much more rapid data collection. It has the additional advantage that it may be used for studies of short time relaxation phenomena, especially in experiments where samples are subjected to pulsed magnetic fields.

A second difference between neutron and X-ray diffraction is that the scattering powers of atoms towards neutrons are quite different from those towards X-rays. In the latter, scattering power is a simple function of atomic number and light atoms such as hydrogen diffract X-rays only weakly. With neutrons, the atomic nuclei, rather than the extranuclear electrons, are responsible for the scattering and in fact, hydrogen is a strong scatterer of neutrons. There is no simple dependence of neutron scattering power on atomic number and additionally, some atoms cause a change of phase of π(or $\lambda/2$) in the diffracted neutron beam.

Crystal structure determination

Neutron diffraction may be used for crystallographic work in cases where X-ray diffraction is inadequate. It has been much used to locate light atoms, especially hydrogen in hydrides, hydrates and organic structures. Usually, the main part of the structure is solved by X-ray methods and neutron diffraction is subsequently used to provide the finishing touches and locate the light atoms. Neutron diffraction is also used to distinguish between atoms that have similar X-ray scattering powers, such as manganese and iron or cobalt and nickel. The neutron scattering powers of these atoms are different and, for instance, superlattice phenomena, associated with Mn/Fe ordering in alloys, are readily observed by neutron diffraction.

Magnetic structure analysis

Magnetic properties depend on the presence of unpaired electrons, especially in d or f orbitals. Since neutrons possess a magnetic dipole moment, they interact with unpaired electrons giving rise to an additional scattering effect. This forms the basis of a powerful technique for studying the magnetic structure of materials. A simple example of magnetic structure and order is shown by NiO. By X-ray

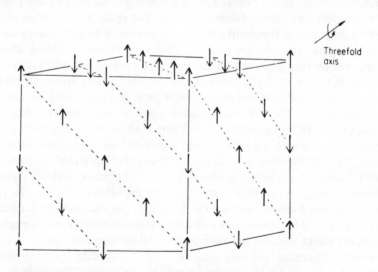

Fig. 3.19 Antiferromagnetic superstructure in MnO, FeO and NiO, showing pseudo-cubic unit cell for which a (supercell) $= 2a$ (subcell). Oxygen positions are not shown

diffraction, NiO has the face centred cubic rock salt structure. When examined by neutron diffraction, however, extra peaks are observed which indicate the presence of a superstructure. This arises because the unpaired d electrons (in the e_g orbitals) are arranged so as to be antiparallel in alternate layers of nickel atoms (Fig. 3.19). Neutrons detect this ordering of spins whereas X-rays do not. The unit cell of 'antiferromagnetic' NiO, which is stable below 250 °C, has eight times the volume of paramagnetic NiO, stable above 250 °C.

Neutron diffraction is widely used to study ferro-, ferri- and antiferromagnetic phenomena in materials. Such magnetic effects depend on the interactions between electrons on different atoms and neutron diffraction provides an invaluable insight into the nature and strength of these exchange interactions. Further details are given in Chapter 8.

Inelastic scattering, soft modes and phase transitions

'Slow' neutrons possess kinetic energy that is comparable to the thermal energy levels in a solid. Such neutrons are inelastically scattered by phonons (i.e. vibrational modes) in the solid. From an analysis of the energy of the scattered neutrons, information on phonons and interatomic forces is obtained. For magnetic materials, further information on their electron exchange energy is obtained.

Displacive phase transitions are believed to be associated with the instability of a lattice vibration. A certain type of vibrational mode in the low temperature structure effectively collapses at the critical temperature. Such modes are called soft modes. They may be studied by IR and Raman spectroscopy, provided the vibrations involved are spectroscopically active, and also by neutron scattering. The latter technique is useful since it is not limited by the spectroscopic selection rules; by measuring the inelastic scattering about a number of 'Bragg reflections', the atomic displacements that are responsible for the soft mode and the phase transition may be determined. For instance, the mechanism of the displacive transition in quartz, SiO_2, at 573 °C has been analysed by recording the neutron spectra at several temperatures below and above 573 °C.

Chapter 4

Other Techniques: Microscopy, Spectroscopy, Thermal Analysis

A wide variety of spectroscopic, microscopic and other techniques may be used to obtain information on the identity, structure and properties of solids. The main objective of this chapter is to give a glossary of these techniques and especially to indicate the kind of useful information that they can give about solids. Lack of space precludes a full description of each technique; instead a concise summary of the basic principles is given. There is a considerable amount of specialized terminology; such words and phrases are given in italics, explanations are not usually given, but the meanings should be clear from the text.

Microscopic techniques

As a first step in examining a solid, it is usually well worthwhile to have a look at it under magnification. This may involve no more than a brief look with a polarizing microscope or it may involve a more in-depth study using one or more instruments. Materials that visually appear to be somewhat similar, such as fine powders of white sand or table salt, may look quite different under the microscope. Thus, the crystals in these two particular examples have a different morphology (i.e. shape) and their optical properties, as studied in plane polarized light, are quite distinct.

Various kinds of microscope are available and they can be divided into two groups: optical and electron. With optical microscopes, particles down to a few micrometres ($1\,\mu m = 10^4\,\text{Å} = 10^{-3}\,mm$) in diameter may be seen under high magnification. The lower limit is reached when the particle size approaches the wavelength of visible light, 0.4 to 0.7 μm. For submicrometre-sized particles it is essential to use electron microscopy. With this technique it is possible to image features that have diameters as small as a few angstroms. Both optical and electron microscopy use two types of instrument in that the sample can be viewed in transmission (i.e. the beam of light or electrons passes through the sample) or in reflection (the beam of light or electrons is reflected off the sample surface).

Optical microscopy

Optical microscopes are of two basic types, as indicated above. The *polarizing* or *petrographic microscope* is a transmission instrument. It is widely used by geologists and mineralogists and can be used profitably in solid state chemistry. Samples are usually in the form of either fine powders or thin slices cut off a solid piece. The *metallurgical* or *reflected light microscope* is suitable for looking at the surfaces of materials, especially opaque ones. It is much used in metallurgy, mineralogy and ceramics.

Samples for the polarizing microscope are often fine powders with particle sizes in the range 10 to $100 \mu m$. On this scale, substances may often be transparent whereas they would be opaque in bulk form. The sample is immersed in a liquid whose refractive index is close to that of the sample. If this is not done and the solid is examined in air, much of the light is scattered off the surface of the sample rather than transmitted through it. It is then difficult, if not impossible, to measure the various optical properties of the solid.

Polarizing microscope

The basic components of a polarizing microscope are indicated in Fig. 4.1. The source may give either white or monochromatic light. This shines on the *polarizer* and only that component whose vibration direction is parallel to that of the polarizer is permitted to pass through. The resulting *plane polarized light* passes through lenses, apertures and accessories and on to the sample which is mounted on the *microscope stage*. Light transmitted by the sample and immersion liquid is picked up by the *objective lens*. The instrument usually has several of these, of different magnification, which are readily interchangeable. The *analyser* may be placed in or out of the path of light. It is similar to the polarizer but is oriented so that its vibration direction is at 90° to that of the polarizer. When the analyser is 'in', only light vibrating in the correct direction is permitted to pass through and on to the *eyepiece*. When the analyser is 'out', the microscope behaves as a simple magnifying microscope. Various other attachments may also be present between the sample and the eyepiece.

EYEPIECE LENS
↑
ANALYSER (in or out)
↑
OBJECTIVE LENS (interchangeable, to vary magnification)
↑
SAMPLE
↑
POLARIZER
↑
LIGHT SOURCE (white or monochromatic light)

Fig. 4.1 Basic components of a polarizing microscope

In practice, samples are usually examined with the analyser alternately 'out' and 'in'. With the analyser 'out', the instrument can be focused and a first examination made, noting for instance the size and shape of the particles. With the analyser 'in', the sample is viewed between *crossed polars*, and one can immediately tell whether the sample is *isotropic* (dark) or *anisotropic* (bright or coloured). By rotating the sample and stage (in some instruments, the polars rotate and the stage is fixed), the *extinction directions* can be seen if the crystals are anisotropic; from the nature of the extinction, information on the quality of the crystals may be obtained. If it is desired to measure the *refractive index* (or indices) of the sample, the analyser is again taken out and various immersion liquids, of different refractive index, are tried until one is found in which the sample is effectively invisible. The refractive index of the sample then matches that of the liquid. It is possible to do this systematically; using the *Becke line method*, it is possible to tell whether the refractive index of the sample is larger or smaller than that of the immersion liquid.

The above measurements are simple and fairly rapid. Often the information obtained is all that is required from microscopy. If necessary, further measurements can be made. The variation of refractive index with crystal orientation may be studied using the concept of the *optical indicatrix*. Anisotropic crystals may be classed into *uniaxial* or *biaxial*, depending on whether they have one or two *optic axes*. When viewed down an optic axis, anisotropic crystals appear to be isotropic, i.e. they are dark between crossed polars. If a convergent beam of light is shone onto the sample, a *conoscopic examination* may be made and *interference figures* are seen; from these, further information on uniaxial and biaxial crystals is obtained.

Reflected light microscope

The reflected light microscope is similar to the transmission one except that the source and objective lens are on the same side of the sample. It is used to examine solid lumps of opaque material such as metals, minerals and ceramics. The amount of information that can be obtained depends very much on the care and skill with which the sample is prepared. It is best to have a flat polished surface which has been treated chemically to cause preferential etching of part of the sample. Some phases etch away more quickly than others and this gives relief to the initially flat surface; other phases take on a coloured hue after etching. The information that can be obtained by this technique primarily concerns the texture of the solid, i.e. the phases present, their identification, the number, size and distribution of particles.

Applications

Crystal morphology and symmetry. Before the advent of X-ray diffraction ($\sim$ 1910), morphological data provided an important means of classifying crystals. The method used (and still uses) a goniometric stage on which a crystal is mounted. The crystal can be rotated on the stage and effectively viewed from all angles under the microscope. From the shape of the crystal, information on the internal symmetry of the crystal structure may be obtained. With well-formed crystals, the number and disposition of the crystal faces can be determined, which may indicate the class (i.e. point group) of the crystal. Nowadays, goniometry is little used for crystallographic work. It is possible, however, to obtain quite a lot of useful information from a cursory examination of powdered samples using the polarizing microscope. For instance:

(a) The crystal fragments often have some kind of characteristic shape, especially if the crystals cleave easily and in a particular orientation.

(b) By examining between crossed polars, it is possible to classify substances into *isotropic* and *anisotropic*. Isotropic crystalline substances are limited to those of cubic symmetry although amorphous solids such as glasses and gels are also isotropic. Isotropic substances appear 'dark' between crossed polars since plane-polarized light passes through them without modification. Anisotropic substances cause a partial rotation of the plane of polarization of light as it passes through them. The emergent beam has a component that is vibrating parallel to the vibration direction of the analyser and hence anisotropic substances appear 'light'. Anisotropic substances include all non-cubic crystalline solids. The vast majority of crystalline solids come into this category. It is often possible to distinguish between uniaxial crystals (of hexagonal, trigonal or tetragonal symmetry) and biaxial crystals (ortho-rhombic, monoclinic and triclinic symmetry). In uniaxial crystals, the optic axis is parallel to the unique symmetry axis (six-, three- or fourfold, respectively) and when viewed down this axis such crystals appear to be isotropic. In a powdered sample it is quite common to find crystals in this orientation, especially if they exhibit preferred orientation. For instance, crystals of hexagonal symmetry often exist as thin hexagonal plates and these naturally tend to lie on their flat faces. Although biaxial crystals have two optic axes, these are not usually parallel to any pronounced edge or feature of the crystal. The chance of finding biaxial crystals oriented with an optic axis parallel to the light beam is quite small.

(c) When anisotropic substances are viewed between crossed polars they usually appear 'light'. On rotating the sample on the microscope stage, the sample becomes 'dark' in a certain position. This is known as *extinction*. On rotating a sample, extinction occurs every $90°$; at the $45°$ position, the sample exhibits *maximum brightness*. If the sample has *parallel extinction*, i.e. the extinction occurs when a pronounced feature of the crystal, such as an edge, is parallel to the direction of vibration of the polarized light, then it is likely that the crystal possesses some symmetry, i.e. it is not triclinic but is of monoclinic symmetry, or higher.

By using steps (a) to (c), outlined above, it is often possible to make a good guess at the symmetry and unit cell of a crystalline substance. This can be a useful prelude to making more time-consuming X-ray diffraction studies. If crystals have a recognizable morphology, use can be made of this to save time in orienting crystals for X-ray work or in measuring some physical property of the crystal as a function of crystal orientation.

Phase identification, purity and homogeneity. Crystalline substances may be identified according to their optical properties, refractive indices, optic axes, etc. Much use is made of this in mineralogy. Standard tables are available with which optical data for an unidentified mineral may be compared. The method is little used outside mineralogy for the identification of complete unknowns. However, where only a limited number of substances are possible, microscopy can be a very powerful method of identification. For instance, in the synthesis of new compounds or phase diagram studies on a specific system, only a limited number of phases are likely to appear. Provided the general appearance and optical properties of most or all are known, then optical microscopy may provide an extremely rapid method of phase analysis. In the space of a few minutes, a sample can be prepared for examination and analysed in the microscope. Often, this gives sufficient information to answer a particular question regarding phase content or purity and may avoid the necessity of making more time-consuming analyses by other techniques.

The purity of a sample may be checked rapidly provided the impurities form a separate crystalline or amorphous phase. Low impurity levels are easily detected, especially if the optical properties of the impurity are markedly different from those of the major phase.

The quality and homogeneity of glasses may be checked readily with a polarizing microscope. For instance, in the preparation of silicate glasses using sand (SiO_2) as one of the raw materials, it may take a long time before all the SiO_2 crystals dissolve during the melting operation. The presence of undissolved silica may not be noticed from a visual inspection of a glass sample, whereas it would be obvious immediately under the microscope. The compositional homogeneity of glass, once all the crystalline raw materials have melted, may be checked by measuring the refractive index of a random selection of fragments of crushed glass. The method works because the refractive index of glass is usually composition dependent. Although glass is isotropic, sometimes glasses that have not been annealed properly show *stress birefringence*. They appear light and apparently anisotropic between crossed polars.

The quality of single crystals may be assessed with a polarizing microscope. Good quality crystals should show a *sharp extinction*; i.e. on rotating the crystal relative to the polars, extinction should occur simultaneously throughout the entire crystal. Crystal aggregates may show an extinction that is *wavy* or irregular. If the crystal divides into strips which extinguish alternately as the crystal rotates, then the crystal is likely to be *twinned*. Other optical effects are also associated with twinning such as striation and crosshatching. Twinning may occur either (a) as a result of the crystal growth mechanism or (b) as a consequence of a phase transition from a high symmetry to a lower symmetry phase, as in paraelectric–ferroelectric or paramagnetic–ferromagnetic transitions. Optical microscopy is a powerful method for studying the latter, especially if a hot stage is fitted so that the sample temperature can be varied under the microscope. A twinned crystal is not a single crystal but contains crystal domains in two, or more, symmetry-related orientations.

Crystal defects—grain boundaries and dislocations. Reflected light microscopy on polished and etched materials can provide much information on internal crystal surfaces (e.g. grain boundaries) and line defects (dislocations). Such defects are always present, even in good quality single crystals. They can be detected because, for example, in the region of an emergent dislocation at the crystal surface, the crystal structure is in a stressed condition; if the surface is treated with a suitable chemical reagent etching may occur preferentially at such stressed sites, resulting in etch pits. Dislocation densities may be determined by counting the number of etch pits per unit area; this has been applied to, for example, metals and alkali halide crystals. Grain boundaries also etch away preferentially and low angle grain boundaries are seen as rows of dislocation etch pits.

Electron microscopy

Electron microscopy is an extremely versatile technique capable of providing structural information over a wide range of magnification. At one extreme, *scanning electron microscopy* (SEM) complements optical microscopy for studying the texture, topography and surface features of powders or solid pieces; features up to tens of micrometres in size can be seen and, because of the depth of focus of SEM instruments, the resulting pictures have a definite three-dimensional quality. At the other extreme, *high-resolution electron microscopy* (HREM) is capable, under favourable circumstances, of giving information on an atomic scale, by direct *lattice imaging*. Resolution of $\sim 2\,\text{Å}$ has been achieved, which means that it is now becoming increasingly possible to 'see' individual atoms. However, lest any one should think that HREM is on the verge of solving all remaining problems concerning the structure of materials, it must be emphasized that there are still formidable obstacles to be overcome before this goal is achieved; there is no immediate prospect of redundancy for more conventional crystallographers!

Electron microscopes are of either transmission or reflection design. For examination in transmission, samples should usually be thinner than $\sim 2000\,\text{Å}$. This is because electrons interact strongly with matter and are completely absorbed by thick particles. Sample preparation may be difficult, especially if it is not possible to prepare thin foils. Thinning techniques, such as ion bombardment, are used, but not always satisfactorily, especially with, for example, polycrystalline ceramics. There is also a danger that ion bombardment may lead to structural modification of the solid in question or that different parts of the material may be etched away preferentially in the ion beam. One possible solution is to use higher voltage instruments, e.g. 1 MV. Thicker samples may then be used since the beam is more penetrating; in addition, the amount of background scatter is reduced and higher resolution may be obtained. Alternatively, if the solid to be examined can be crushed into a fine powder then at least some of the resulting particles should be thin enough to be viewed in transmission.

With reflection instruments, sample thickness is no longer a problem and special methods of sample preparation are not required. It is usually necessary to coat the sample with a thin layer of metal, especially if the sample is a poor electrical conductor, in order to prevent the build-up of charge on the surface of the sample. The main reflection instrument is the SEM. It covers the magnification range between the lower resolution limit of optical microscopy ($\sim 1\,\mu m$) and the upper practical working limit of *transmission electron microscopy* (TEM) ($\sim 0.1\,\mu m$) although, in fact, SEM can be used to study structure over a much wider range, from $\sim 10^{-2}$ to $\sim 10^{2}\,\mu m$. The approximate working ranges of different kinds of microscope are summarized in Fig. 4.2.

Some SEM and TEM instruments have the very valuable additional feature of providing an elemental analysis of sample composition. There are various names for the technique including *electron probe microanalysis* (EPMA), *electron microscopy with microanalysis* (EMMA) and *analytical electron microscopy* (AEM). The main mode of operation makes use of the fact that when a sample is placed in the microscope and bombarded with high-energy electrons, many things can happen (Fig. 4.3), including the generation of X-rays (Fig. 3.1a). These X-rays are characteristic emission spectra of the elements present in the sample (Fig. 3.1b). By scanning either the wavelength (*wavelength dispersive*, WD) or the energy (*energy dispersive*, ED) of the emitted X-rays, it is possible to identify the elements present. If a suitable calibration procedure has been adopted, a quantitative elemental analysis may be made. At present, only elements heavier than and including sodium can be determined; lighter elements do not give suitable X-ray spectra. For lighter elements, however, there are alternative techniques, *Auger spectroscopy* and *electron energy loss spectroscopy* (EELS).

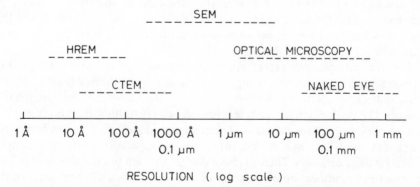

Fig. 4.2 Working ranges of various techniques used for viewing solids. CTEM = conventional transmission electron microscopy; HREM = high-resolution electron microscopy; SEM = scanning electron microscopy

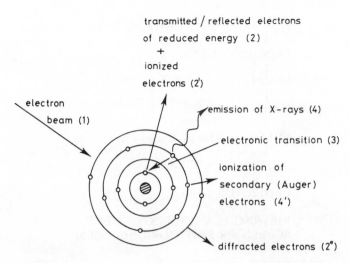

transmitted / reflected electrons
of reduced energy (2)
+
ionized
electrons (2')

electron
beam (1)

emission of X-rays (4)

electronic transition (3)

ionization of
secondary (Auger)
electrons (4')

diffracted electrons (2″)

Fig. 4.3 Some of the processes that occur on bombarding a sample with electrons (in, for example, an electron microscope)

Generation of low-energy Auger electrons is another process (4') that occurs on electron bombardment of the sample in the microscope (Fig. 4.3). The energy of Auger electrons is characteristic of the atom from which they are emitted. Auger emission occurs when electrons, and not X-rays, are ejected from the sample as an energy release mechanism.

EELS detects the transmitted/reflected electrons (2) which were responsible for the initial ionization of electrons (2'). Since energy is required to ionize the atoms (2') the EELS electrons are of reduced energy compared to the energy of the incident electron beam (1). EELS and Auger techniques are discussed further in the next section.

The basic components of a TEM instrument are listed in Fig. 4.4. Electrons emitted from a tungsten filament (*electron gun*) are accelerated through a high voltage (50 to 100 kV). Their wavelength is related to the accelerating voltage, V, by

$$\lambda = h(2meV)^{-1/2} \tag{4.1}$$

where m and e are the mass and charge of the electron. At high voltage, as the velocity of the electrons approaches the velocity of light, m is increased by relativistic effects. The electron wavelengths are much smaller than the X-ray wavelengths used in diffraction experiments, e.g. $\lambda \sim 0.04$ Å at 90 kV accelerating voltage. Consequently, the Bragg angles for diffraction are small and the diffracted beams are concentrated into a narrow cone centred on the undiffracted beam.

In order to use electrons, instead of light, in a microscope it is necessary to be able to focus them. As yet, no one has found a substance that can act as a lens to focus electrons but, fortunately, electrons may be focused by an electric or

ELECTRON SOURCE (GUN)
↓
CONDENSER LENSES
↓
SAMPLE
↓
OBJECTIVE LENS
↓
(DIFFRACTION PLANE)
↓
(INTERMEDIATE IMAGE)
↓
PROJECTOR LENS
↓
VIEWING PLANE (FLUORESCENT
SCREEN OR PHOTOGRAPHIC FILM)

Fig. 4.4 Basic components of a transmission electron microscope

magnetic field. Electron microscopes contain several *electromagnetic lenses*. The *condenser lenses* are used to control the size and angular spread of the electron beam that is incident on the sample. Transmitted electrons then pass through a sequence of lenses—*objective, intermediate* and *projector*—and form a magnified image of the sample on a fluorescent viewing screen. Photographs may be taken if so desired. By changing the relative position of the viewing screen, the diffraction pattern, rather than the image, of the specimen is seen and can also be photographed. The region of the sample that is chosen for imaging may be controlled by an aperture placed in the *intermediate image plane*. This is particularly important in examining polycrystalline materials that contain more than one phase. The technique is then known as *selected area electron diffraction* (SAD). The quality of photographs, especially where the particles of interest are difficult to pick out from the background, may be improved by *dark field imaging*. In this, only the diffracted beams from the particle of interest are allowed to recombine to form the image.

In the scanning electron microscope, electrons from the electron gun are focused to a small spot, 50 to 100 Å in diameter, on the surface of the sample. The electron beam is scanned systematically over the sample, rather like the spot on a television screen. Both X-rays and secondary electrons are emitted by the sample; the former are used for chemical analysis and the latter are used to build up an image of the sample surface which is displayed on a screen. A limitation with SEM instruments is that the lower limit of resolution is ∼ 100 Å. A recent advance is the development of the scanning transmission electron microscope (STEM). This combines the scanning feature of the SEM with the intrinsically higher resolution obtainable with TEM. Some applications of electron microscopes that primarily involve electron diffraction are given in Chapter 3. Other uses are as follows.

Applications

Particle size and shape, texture, surface detail. Electron microscopy, especially SEM, is invaluable for surveying materials under high magnification and providing information on particle sizes and shapes. The results complement those obtained from optical microscopy by providing information on submicrometre-sized particles.

Crystal defects. Using the dark field technique on, for example, thin foils with TEM, crystal defects such as dislocations, stacking faults, antiphase boundaries and twin boundaries may be seen directly. The domain structure of ferromagnetic and ferroelectric materials may be observed. With HREM, it is now possible to see detail on an atomic scale although, as yet, clear images of atoms are not usually obtained. Variations in local structure such as site occupancies and vacancies can be observed directly; this, therefore, complements the results obtained by conventional X-ray crystallography which yield an average structure.

Precipitation and phase transitions. Many glasses exhibit a liquid immiscibility texture on a submicrometre scale and this can be seen by electron microscopy. Precipitation phenomena in crystalline systems are widespread and often technologically important. Hardening processes in alloys often involve precipitation; e.g. the formation of cementite, Fe_3C, precipitates in the hardening of steels. Similar processes are increasingly being studied in inorganic (i.e. non-metallic) and ceramic systems. Many minerals have undergone precipitation (or exsolution) reactions over geological timescales. In all such cases, EM can be used to study the nature of the precipitates, i.e. their crystal structures (by electron diffraction), their crystallographic orientation relative to the parent structure, the overall texture of the solid and compositional variations throughout the solid.

Chemical analysis. The use of AEM and related techniques enables the composition of small regions of solid, ~ 1000 Å in diameter, to be determined. This may be applied to small individual particles or to larger masses, in which case compositional variations across the solid may be determined. This is becoming extremely valuable (a) as a straightforward means of elemental analysis and (b) to study inhomogeneous materials such as complex mineral aggregates, cements and concretes, electronic ceramics, etc.

Structure determination. In favourable circumstances, HREM may give direct information on crystal structures. A certain number of the diffracted beams are allowed to recombine and form a direct image of the crystal structure.

Difficulties may arise, however, because first, the images are only a projection of the structure and second, the image obtained varies greatly with the

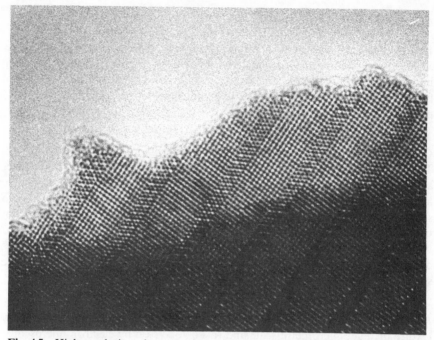

Fig. 4.5 High resolution electron micrograph of an intergrowth tungsten bronze, $Rb_{0.1}WO_3$. Black dots represent WO_6 octahedra. The structure may be regarded as an intergrowth of primitive cubic WO_3 (strips of black dots based on a square grid) and hexagonal WO_3 containing Rb (narrow strips of black dots on a hexagonal grid). Photograph provided by Dr M. Sundberg, University of Stockholm, Department of Chemistry. Further details including related Cs phases are given in L. Kihlborg *et al.*, *Chemica Scripta*, March, 1988

instrumental focusing conditions. It is possible to calculate images for different models of the crystal structure and, by comparison with the observed images, one can find structural models that are at least consistent with the experimental data. This is clearly not as useful as solving crystal structures without making prior assumptions about structural models. Nevertheless, it is a valuable technique especially for studying complex inorganic materials which have pronounced layer, shear or block structures. Images may be obtained from which, for instance, layer spacings or block widths may be determined. An example of an HREM image is shown in Fig. 4.5.

Spectroscopic techniques

There are many different spectroscopic techniques but all work on the same basic principle. This is that, under certain conditions, materials are capable of absorbing or emitting energy. The energy can take various forms. Usually it is electromagnetic radiation but it also can be sound waves, particles of matter, etc.

The experimental results, or spectra, take the form of a plot of intensity of absorption or emission (y axis) as a function of energy (x axis). The energy axis is often expressed in terms of either frequency, f, or wavelength, λ, of the appropriate radiation. The various terms are interrelated by the classic equation

$$E = hf = hc\lambda^{-1} \tag{4.2}$$

where h is Planck's constant (6.6×10^{-34} J sec), c is the velocity of light (2.998×10^{10} cm sec^{-1}), f is frequency (in hertz, cycles per second) and λ is the wavelength (in centimetres). Units of λ^{-1} are cm^{-1} or wavenumbers and E is expressed in joules. Chemists usually prefer joules per mole for units of E, in which case the above equation is multiplied by Avogadro's number, N. Some useful interconversions between the different energy units, obtained by substituting for the constants into equation (4.2) are

$$E(\text{in J mol}^{-1}) = 3.991 \times 10^{-10} f \simeq 4 \times 10^{-10} f \quad (f \text{ in sec}^{-1})$$
$$E(\text{in J mol}^{-1}) = 11.97\lambda^{-1} \simeq 12\lambda^{-1}(\lambda \text{ in cm}) \tag{4.3}$$
$$E(\text{in sec}^{-1}) = 3 \times 10^{10} \lambda^{-1} \quad (\lambda \text{ in cm})$$
$$E(\text{eV}) \simeq 96E(kJ \text{ mol}^{-1})$$

The electromagnetic spectrum covers an enormous span of frequency, wavelength and, therefore, energy. The different spectroscopic techniques operate over different, limited frequency ranges within this broad spectrum, depending on the processes and magnitudes of the energy changes that are involved (Fig. 4.6). At the low frequency, long wavelength end, the associated energy changes are small, < 1 J mol^{-1}, but may be sufficient to cause the reversal of spins of either nuclei or electrons in an applied magnetic field. Thus, the nuclear magnetic resonance (NMR) technique operates in the radiofrequency region at, for example, 100 MHz (10^8 Hz) and detects changes in nuclear spin state.

At higher frequencies and shorter wavelengths, the amount of associated energy increases and, for instance, the vibrational motions of atoms in molecules or solids may be altered by the absorption or emission of infrared (IR) radiation. At still higher frequencies, electronic transitions within atoms may occur from one energy level to another. For electronic transitions involving outer (valence) shells, the associated energy usually lies in the visible and ultraviolet regions; however, for inner shell transitions, much larger energies are involved and fall in the X-ray region.

Although many of the spectroscopic techniques were initially developed for and applied to molecular materials, often liquids and gases, they are finding many applications in solid state studies. Spectroscopic measurements on solids complement well the results obtained from (X-ray) diffraction since spectroscopy gives information on local order whereas diffraction is concerned primarily with long range order. Spectroscopic techniques may be used to determine coordination numbers and site symmetries; variations in local order can be detected, as can impurities and imperfections; amorphous materials, such as glasses and gels, can be studied just as easily as crystalline materials. By contrast, the long-range

170

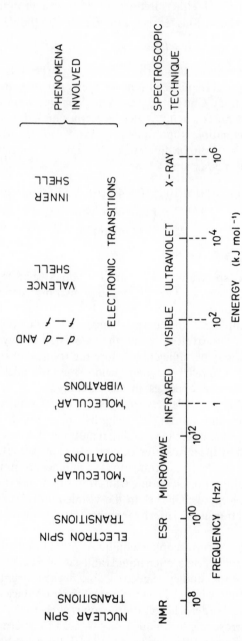

Fig. 4.6 Principal regions of the electromagnetic spectrum and the associated spectroscopic techniques

periodic structures of crystals can be determined only by diffraction techniques (and occasionally by HREM); however, this results in an *average* picture of the local structure in which information on defects, impurities and subtle variations in local order may effectively be lost. Also, diffraction techniques are of limited use for studying amorphous materials.

Some of the principal spectroscopic techniques and their uses in solid state chemistry are as follows.

Vibrational spectroscopy: IR and Raman

Atoms in solids vibrate at frequencies of approximately 10^{12} to 10^{13} Hz. Vibrational modes, involving pairs or groups of bonded atoms, can be excited to higher energy states by absorption of radiation of appropriate frequency. IR spectra and closely related Raman spectra are plots of intensity of absorption (IR) or scattering (Raman) as a function of frequency or wavenumber. In the IR technique, the frequency of the incident radiation is varied and the quantity of radiation absorbed or transmitted by the sample is obtained. In the Raman technique, the sample is illuminated with monochromomatic light, usually generated by a laser. Two types of scattered light are produced by the sample. *Rayleigh scatter* emerges with exactly the same energy and wavelength as the incident light. *Raman scatter*, which is usually much less intense than Rayleigh scatter, emerges at either longer or shorter wavelength than the incident light. Photons of light from the laser, of frequency v_0, induce transitions in the sample and the photons gain or lose energy as a consequence. For a vibrational transition of frequency v_1, associated Raman lines of frequency $v_0 \pm v_1$ appear in the scattered beam. This scattered light is detected in a direction perpendicular to the incident beam.

IR and Raman spectra of solids are usually complex with a large number of peaks, each corresponding to a particular vibrational transition. A complete assignment of all the peaks to specific vibrational modes is possible with molecular materials and, in favourable cases, is possible with non-molecular solids. The IR and Raman spectra of a particular solid are usually quite different since the two techniques are governed by different selection rules. The number of peaks that are observed with either technique tends to be considerably less than the total number of vibrational modes and different modes may be active in the two techniques. For instance, in order for a particular mode to be IR active, the associated dipole moment must vary during the vibrational cycle. Consequently, centrosymmetric vibrational modes are IR inactive. The principal selection rule for a vibrational mode to be Raman active is that the nuclear motions involved must produce a change in polarizability.

IR and Raman spectra are much used for the straightforward identification of specific functional groups, especially in organic molecules. In inorganic solids, covalently bonded linkages such as hydroxyl groups, trapped water and oxyanions—carbonate, nitrate, sulphate, etc.—give rise to intense IR and Raman peaks. Some examples are shown in Fig. 4.7. The three spectra are quite

172

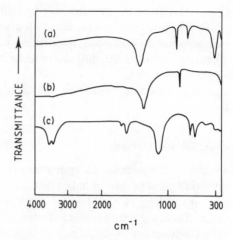

Fig. 4.7 Infrared absorption spectra of
(a) calcite, $CaCO_3$, (b) $NaNO_3$, (c) gypsum,
$CaSO_4 \cdot 2H_2O$. (Data taken from P. A.
Estep-Barnes, p. 532 in Zussman, *Physical
Methods in Determinative Mineralogy,*
Academic Press, 1977)

different in the region 300 to 1500 cm^{-1}, largely due to the different oxyanions present in each sample. Spectrum (c) contains an additional absorption doublet at ~ 3500 cm^{-1} caused by the H_2O molecules in the gypsum. Peaks that occur in the region ~ 3000 to 3500 cm^{-1} are usually characteristic of OH groups in some form: the frequencies of the peak maxima depend on the O—H bond strength and information may be obtained on, for instance, the location of the OH group, whether it belongs to a water molecule and whether or not hydrogen bonding is present.

Peaks associated with the vibrational modes of covalently bonded groups, such as oxyanions, usually occur at relatively high frequencies, above ~ 300 cm^{-1}. At lower frequencies, in the far infrared region, lattice vibrations give rise to peaks; e.g. alkali halides give broad lattice absorption bands in the region 100 to 300 cm^{-1}. The peak positions depend inversely on the mass of the anions and cations, as shown in the following sequences (positions in cm^{-1}): LiF(307), NaF(246), KF(190), RbF(156), CsF(127); and LiCl(191), NaCl(164), KCl(141), RbCl(118), CsCl(99).

Since inorganic solids tend to give characteristic vibrational spectra, these may be used for identification purposes; by setting up a file of reference spectra it should be possible to identify unknowns. This has already been done, with considerable success, for organic molecules, but has not yet been adequately done for inorganic solids. Perhaps, with the comprehensive coverage and success achieved by the JCPDS X-ray powder diffraction file, there is less demand for a second method of fingerprinting inorganic solid materials.

An interesting example of the use of laser Raman spectroscopy for fingerprint-

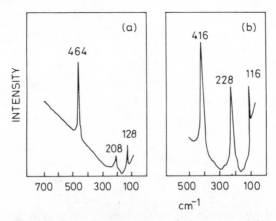

Fig. 4.8 Laser Raman spectra of (a) quartz and (b) cristobalite. (From Farwell and Gage, *Anal. Chem.*, **53**, 1529, 1981)

ing crystalline solids is in the distinction between two of the polymorphs of silica, quartz and cristobalite (Fig. 4.8). In the region 100 to 500 cm^{-1}, each polymorph gives three sharp peaks, but at different positions in the two cases. Use has been made of these spectral differences in the analysis of ash from the recent Mount St Helens volcanic eruption in the United States.

Apart from possible uses in routine sample identification, vibrational spectra may be used to characterize solids and provide some structural information. For these purposes, a deeper understanding of the spectra is necessary and, in particular, the assignment of peaks to specific vibrational modes is desirable. Methods are available which can be used to carry out this assignment; they will not be discussed here since they are rather complicated and as yet have been used only for simple crystal structures.

Visible and ultraviolet spectroscopy

Transitions of electrons between outermost energy levels are associated with energy changes in the range $\sim 10^4$ to 10^5 cm^{-1} or $\sim 10^2$ to 10^3 kJ mol^{-1}. These energies span the range from the near infrared through the visible to the ultraviolet (Fig. 4.6) and are therefore often associated with colour. Various types of electronic transition occur and may be detected spectroscopically; some are shown schematically in Fig. 4.9. The two atoms A and B are neighbouring atoms in some kind of solid structure; they may be, for instance, an anion and a cation in an ionic crystal. The inner electron shells are localized on the individual atoms. The outermost shells may overlap to form delocalized bands of energy levels. Four basic types of transition are indicated in Fig. 4.9:

(i) Promotion of an electron from a localized orbital on one atom to a higher energy but still localized orbital on the same atom. The spectroscopic absorption band associated with this transition is sometimes known as an

174

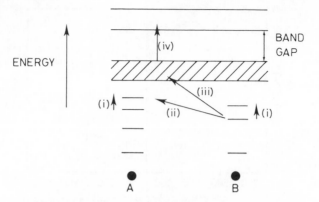

Fig. 4.9 Possible electronic transitions in a solid. These involve electrons in localized orbitals and/or delocalized bands

exciton band. Transitions in this category include (a) $d–d$ and $f–f$ transitions in transition metal compounds, (b) outer shell transitions in heavy metal compounds, e.g. $6s–6p$ in lead (II) compounds, (c) transitions associated with defects such as trapped electrons or holes, e.g. colour centres (F, H, etc.) in alkali halides, and (d) transitions involving, for example, silver atoms in photochromic glasses: colloidal silver is precipitated initially on photo-irradiation and subsequent electronic transitions occur within the reduced silver atoms.

(ii) Promotion of an electron from a localized orbital on one atom to a higher energy but still localized orbital on an adjacent atom. The associated absorption bands are known as *charge transfer spectra*. The transitions are usually 'allowed transitions' according to the spectroscopic selection rules and hence the absorption bands are intense. Charge transfer processes are, for example, responsible for the intense yellow colour of chromates; an electron is transferred from an oxygen atom in a $(CrO_4)^{2-}$ tetrahedral complex anion to the central chromium atom. In mixed valence transition metal compounds, e.g. in magnetite, Fe_3O_4, charge transfer processes also occur.

(iii) Promotion of an electron from a localized orbital on one atom to a delocalized energy band, the conduction band, which is characteristic of the entire solid. In many solids the energy required to cause a transition such as this is very high but in others, especially those containing heavy elements, the transition occurs in the visible/ultraviolet region and the materials are photoconductive, e.g. some chalcogenide glasses are photoconductive.

(iv) Promotion of an electron from one energy band (the valence band) to another band of higher energy (the conduction band). The magnitude of the band gap in semiconductors (Si, Ge, etc.) may be determined spectroscopically; a typical semiconductor has a band gap of 1 eV, 96 kJ mol^{-1}, which lies between the visible and UV regions.

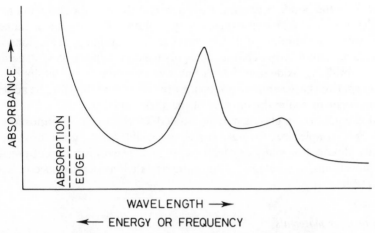

Fig. 4.10 Schematic typical UV/visible absorption spectrum

The appearance of a typical UV and visible absorption spectrum is shown schematically in Fig. 4.10. It contains two principal features. Above a certain energy or frequency known as the *absorption edge*, intense absorption occurs. Since the transmittance of the sample drops to essentially zero at the absorption edge, this places a high frequency limit on the spectral range that can be investigated. If it is desired to go to frequencies above the absorption edge, then reflectance techniques must be used. Transitions of types (ii) and (iii) are responsible for the appearance of the absorption edge. Its position varies considerably between different materials. In electronically insulating ionic solids it may occur in the ultraviolet, but in photoconducting and semiconducting materials it may occur in the visible or even in the near infrared spectral regions.

The second feature is the appearance of broad absorption peaks or bands at frequencies below that of the absorption cut-off. These are generally associated with type (i) transitions.

Visible and UV spectroscopy has a variety of applications associated with the local structure of materials. This is because the positions of the absorption bands are sensitive to coordination environment and bond character.

Structural studies on glass

Information on the local structure of amorphous materials may be obtained by UV and visible spectroscopy. Often it is necessary to add a small quantity of a spectroscopically active species to the raw materials. This may be a transition metal compound, in which case the *d–d* spectra of the transition metal ion are recorded. Alternatively, heavy metal cations such as Tl^+, Pb^{2+} may be added. From the nature of the spectra, coordination numbers can usually be deduced and, hence, information obtained on the availability of such sites in the glass structure. The spectra of ions such as Pb^{2+} are sensitive to the degree of covalent character in the bonds between a Pb^{2+} ion and the neighbouring anions. This provides a means for studying the basicity of glasses. Basicity is associated with electron-donating ability and, for instance, non-bridging oxide ions which contain a net unit negative charge are highly basic in a silicate glass structure. By contrast, bridging oxide ions, which link two silicon atoms, contain very little excess negative charge and are not basic. Pb^{2+} ions may therefore be added to a glass in order to probe the basicity of the sites available to it.

Redox equilibria in glasses may be studied. One of the most important is the $Fe^{2+}-Fe^{3+}$ couple since the Fe^{3+} ion is responsible for the green/brown colour of many glasses. For many applications, such as glasses for use in optical fibre communications, the presence of these ions and their associated spectra is highly undesirable.

Study of laser materials

Laser materials often contain a transition metal ion as the active species, e.g. the ruby laser is essentially Al_2O_3 doped with a small amount of Cr^{3+}; the neodymium glass laser consists of a glass doped with Nd^{3+} ions. Information on the Cr^{3+}, Nd^{3+} ions and their coordination environment in the host structure may be obtained from the UV and visible spectra. Laser action results from a population inversion in which a large number of electrons are promoted into a higher energy level. These subsequently drop back into a lower level and, at the same time, emit the laser beam. It is therefore very important to have a detailed knowledge of the available energy levels and possible electronic transitions.

Nuclear magnetic resonance (NMR) spectroscopy

NMR spectroscopy has made an enormous impact on the determination of molecular structure over the last two or three decades, but until recently has yielded only limited structural information about solids. Molecular substances, especially organic molecules in the liquid state, give NMR spectra which at high resolution are composed of a number of sharp peaks. From the positions and relative intensities of the peaks it is often possible to tell which atoms are bonded together, coordination numbers, next nearest neighbours, etc. In the solid state, by contrast, broad featureless peaks are observed from which little direct structural information can be obtained. Strenuous efforts have been made to sharpen these broad peaks that are observed in solids and recently, considerable success has been achieved. In the '*magic angle spinning*' (MAS) technique, the sample is rotated at a high velocity at a critical angle of 54.74° to the applied magnetic field; the resulting spectra contain groups of sharp peaks, from which, much structural information may be obtained.

NMR spectroscopy is a technique that involves the magnetic spin energy of atomic nuclei. For elements that have a non-zero nuclear spin, such as ^1H, ^2H, ^6Li, ^7Li, ^{13}C and ^{29}Si, but not, for example, ^{12}C, ^{16}O, or ^{28}Si, an applied magnetic field will influence the energy of the nuclei. The magnetic energy levels split into two groups, depending on whether the nuclear spins are aligned parallel or antiparallel with the applied magnetic field. The magnitude of the energy difference between parallel and antiparallel spin states is small, $\sim 0.01 \, J \, mol^{-1}$ for an applied magnetic field of $10^4 \, G$ (1 T). This amount of energy is associated with the radiofrequency region of the electromagnetic spectrum. NMR spectrometers operating at, for example, 50 MHz can therefore induce nuclear spin transitions. The magnitude of the energy change and the associated frequency of absorption depends not only on the particular element involved but also on its chemical environment. Thus, in organic molecules, hydrogen atoms bonded to different types of carbon atom or different functional groups may be distinguished since they absorb at slightly different frequencies. Note, however, that NMR instruments are usually operated at fixed frequency, e.g. 220 MHz, and the magnitude of the energy difference between the parallel and antiparallel spin states is varied, by varying the applied magnetic field strength.

Conventional NMR measurements on solids give broad, featureless bands which are of little use for structural work. By using refinements such as the MAS technique, the broad bands collapse to reveal a fine structure. Lippmaa and others have applied this technique to crystalline silicates and found that the ^{29}Si NMR spectra give peaks whose positions depend on the nature of the silicate anion. In particular, the spectra can distinguish between isolated SiO_4 tetrahedra and SiO_4 tetrahedra linked by sharing common corners (oxygen atoms) to one, two, three or four other tetrahedra. It is customary to assign to each silicon (or its SiO_4 tetrahedron) a 'Q value' which represents the number of adjacent SiO_4 tetrahedra to which it is directly bonded. Q values range from zero (as in orthosilicates such as Mg_2SiO_4 with isolated tetrahedra) to four (as in three-dimensional framework structures such as SiO_2, in which all four corners are shared). The positions of the ^{29}Si NMR peaks, i.e. the *chemical shifts* relative to the internal standard (tetramethylsilane, $(CH_3)_4Si$, or TMS), depend approximately on the Q value, as indicated in Fig. 4.11. For each Q value, a range of chemical shifts is observed, depending on other features of the crystal structure.

The ^{29}Si NMR spectrum of a calcium silicate, xonotlite, is shown in Fig. 4.12. The silicate anion of xonotlite is an infinite double chain or ladder with a cross link or rung at every third tetrahedron in each chain, as shown schematically in the diagram. Two types of silicon are present, therefore—Q^2 and Q^3—and in the

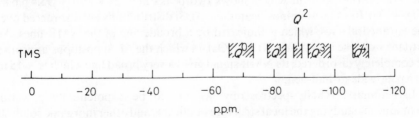

Fig. 4.11 Positions of ^{29}Si NMR peaks in silicates as a function of the degree of condensation, Q, of the silicate anion. (From E. Lippmaa *et al.*, *J. Amer. Chem. Soc.*, **102**, 4889, 1980)

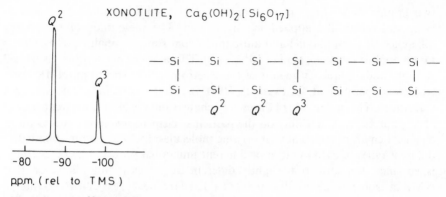

Fig. 4.12 The ^{29}Si NMR spectrum of xonotlite (from Lippmaa *et al.*, *J. Amer. Chem. Soc.*, **102**, 4889, 1980). Also shown in a highly schematic manner is the structure of the silicate double chain anion and the Q values of the silicon atoms involved. Oxygen atoms are omitted

relative amounts of 2:1. The NMR peaks appear in the appropriate positions for Q^2 and Q^3 silicon atoms and their intensities are in the ratio of 2:1, as expected.

NMR spectroscopy can also be used to probe the fine structure of aluminosilicate crystal structures. In these, the aluminium atoms can play two possible rôles: either they occupy octahedral sites and do not really form part of the aluminosilicate framework of linked tetrahedra; or, more often, they occupy tetrahedral sites, similar to silicon, and do form part of the aluminosilicate framework. In the latter case, the chemical shift of a particular silicon atom depends on how many aluminium atoms are in its second coordination sphere (Fig. 4.13). For instance, a Q^4 silicon atom in an aluminosilicate framework is surrounded by four other tetrahedral atoms and, of these, any number between zero and four may be aluminium atoms (i.e. AlO_4 tetrahedra). It is found that the chemical shift increases with decreasing number of aluminium neighbours, from -84 p.p.m. with four aluminium neighbours, as in nepheline, $KNa_3(AlSiO_4)_4$, to ~ -108 p.p.m. with no aluminium neighbours (as in SiO_2). The crystal structure of natrolite, $Na_2(Al_2Si_3O_{10})\cdot 2H_2O$, contains two types of Q^4 silicon atoms with three and two neighbouring aluminium atoms, respectively. This is reflected in the ^{29}Si NMR spectrum which shows two peaks at -87.7 and -95.4 p.p.m. (Fig. 4.13c). In some aluminosilicates, the Al, Si distributions are disordered over the tetrahedral sites, which is indicated by a broadening of the NMR lines. An extreme example is sanidine, $K(AlSi_3O_8)$, in which the Al, Si positions appear to be completely disordered; its NMR signal gives a very broad line which is ~ 15 to 20 p.p.m. wide at half-height.

In summary, NMR spectroscopy appears to be a potentially powerful technique for studying the local structure of silicates and other inorganic solids. It has been used for such work only since about 1980 and, therefore, considerable future developments are hoped for. The information that is obtained complements very well the 'average' structural information that is obtained by X-ray

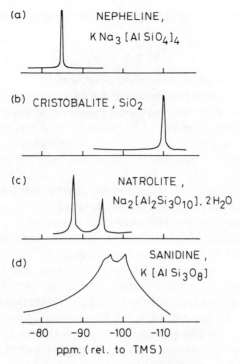

Fig. 4.13 Schematic ^{29}Si NMR spectra of silicates containing different Q^4 silicon atoms. The number of aluminium atoms in the second coordination sphere are (a) 4, (b) 0, (c) 3 and 2, (d) 2 and 1. (From Lippmaa *et al.*, *J. Amer. Chem. Soc.*, **102**, 4889, 1980)

diffraction. Thus the nature of the Al, Si distribution in silicate structures has long been a thorny problem for mineralogists and crystallographers and only in favourable cases have X-ray measurements fully resolved the structure.

Electron spin resonance (ESR) spectroscopy

The ESR technique is closely related to NMR; it detects changes in electron spin configuration. ESR depends on the presence of permanent magnetic dipoles, i.e. unpaired electrons in the sample, such as those which occur in many transition metal ions. The reversal of spin of these unpaired electrons in an applied magnetic field is recorded. The magnitude of the energy change is again small, $\sim 1.0 \, \mathrm{J \, mol^{-1}}$, although somewhat larger than the corresponding NMR energy changes (Fig. 4.6). ESR spectrometers operate at microwave frequencies, e.g. at 2.8×10^{10} Hz (28 GHz), with an applied magnetic field of, for example, 3000 G. In practice, the spectra are obtained by varying the magnetic field at constant frequency. Absorption of energy associated with the spin transition occurs at the resonant condition:

$$\Delta E = hf = g\beta_e H \qquad (4.4)$$

where β_e is a constant, the *Bohr magneton* ($\beta_e = eh/4\pi mc = 9.723 \times 10^{-12}$ J G^{-1}), and H is the strength of the applied magnetic field. The factor g, the *gyromagnetic ratio*, has a value of 2.0023 for a free electron but varies significantly for paramagnetic ions in the solid state. The value of g depends on the particular paramagnetic ion, its oxidation state and coordination number. It is therefore responsible for the position of an absorption peak and is analogous to the 'chemical shift' of NMR spectra.

ESR spectra of solids often show broad absorption peaks (as do NMR spectra) and, therefore, certain conditions must be met in order to get sharp peaks from which useful information may be obtained. One source of line broadening is spin–spin interactions between neighbouring unpaired electrons. This is overcome by having only a low concentration of unpaired electrons, e.g. 0.1 to 1 per cent of a paramagnetic transition metal ion dissolved in a diamagnetic host structure. A second source of line broadening is the occurrence of low-lying excited states near to the ground state energy of the paramagnetic ion. This leads to frequent electron transitions, short relaxation times and broad peaks. To overcome this, and reduce the frequency of the transitions, the spectra are recorded at low temperatures, often at the liquid helium temperature, 4.2 K. The interpretation of ESR spectra is greatly simplified if the paramagnetic species has only one unpaired electron, as in, for example, d^1 transition metal ions, V^{4+}, Cr^{5+}, etc.

ESR spectra are usually presented as the first derivative of the absorption (Fig. 4.14b), rather than as absorption itself, a. ESR spectra often comprise a set of closely spaced peaks. The 'hyperfine splitting' which causes multiple peaks arises because, in addition to the applied external magnetic field, there are internal nuclear magnetic fields present associated with the transition metal ion itself or with surrounding ligands. For instance, ^{53}Cr has a nuclear spin moment, I, of $\frac{3}{2}$ and the spectrum of $^{53}Cr^{5+}$ (d^1 ion) is split into $2I + 1 = 4$ hyperfine lines. An example is shown in Fig. 4.14(c) for the ESR spectrum of Cr^{5+}, as CrO_4^{3-}, dissolved in apatite, Ca_2PO_4Cl, at 77 K. Naturally occurring chromium is a mixture of isotopes, the main one being ^{52}Cr which has a nuclear spin moment of zero. Unpaired electrons on the $^{52}Cr^{5+}$ ions are responsible for the very intense central line, 3, in the spectrum. The four small, equally spaced lines 1, 2, 4 and 5 are associated with unpaired electrons on the small percentage of $^{53}Cr^{5+}$ ions which have $I = \frac{3}{2}$. In some cases, additional line structure, 'superhyperfine splitting' may occur due to interactions of an unpaired electron with the nuclear moment of neighbouring ions.

From ESR spectra such as Fig. 4.14, one can (with appropriate experience!) obtain information on the paramagnetic ion and its immediate environment in the host structure. Specifically, one may determine:

(a) the oxidation state, electronic configuration and coordination number of the paramagnetic ion,

(b) the ground state d orbital configuration of the paramagnetic ion and any

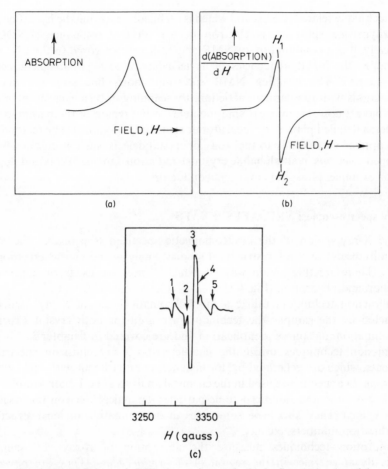

Fig. 4.14 (a) Schematic ESR absorption peak, (b) the first derivative of the absorption and (c) the ESR spectrum of CrO_4^{3-} in a Ca_2PO_4Cl at 77 K. (From Greenblatt, *J. Chem. Educ.*, **57**, 546, 1980)

structural distortions arising from, for example, a Jahn–Teller effect,

(c) the extent, if any, of covalency in the bonds between the paramagnetic ion and its surrounding anions or ligands.

Since the paramagnetic ion is present only in small amounts, it is assumed that its site symmetry is identical to that of the host ion that it substitutionally replaces. For instance, it has been shown that in chromium-doped apatites, $M_5(PO_4)_3X$: M = Ca, Sr, Ba; X = Cl, F, the CrO_4^{3-} tetrahedron is distorted in such a way as to be squashed or compressed along one of the fourfold inversion axes of the tetrahedron. It is very likely, therefore, that the PO_4 tetrahedra have a similar distortion. On the other hand, in $BaSO_4$, the SO_4 tetrahedra are distorted by stretching along a fourfold inversion axis. This was shown by doping $BaSO_4$ with MnO_4^{2-} containing Mn^{6+} (also a d^1 ion) and recording the ESR spectra.

A technique related to ESR and which is useful for observing the hyperfine and superhyperfine splittings is electron nuclear double resonance (ENDOR). Basically, it is a combination of NMR and ESR spectroscopy. In the ENDOR technique, the NMR frequencies of nuclei adjacent to a paramagnetic centre are scanned. The resulting NMR spectrum shows fine structure due to interactions with the paramagnetic ion. By scanning in turn each of the nuclei (that have a non-zero nuclear spin moment) in the region of the paramagnetic species, a detailed plan of the local atomic structure is obtained. The method has been applied successfully to the study of crystal defects, such as colour centres (trapped electrons in alkali halide crystals), radiation damage effects and doping in, for example, phosphors and semiconductors.

X-ray spectroscopies: XRF, AEFS, EXAFS

The X-ray region of the electromagnetic spectrum is probably the most generally useful region for structural studies, analysis and characterization of solids. There are three main ways in which X-rays are used: for diffraction, emission and absorption (Fig. 4.15a).

Diffraction techniques utilize a monochromatic beam of X-rays which is diffracted by the sample; the techniques are used for both crystal structure determination and phase identification and are covered in Chapter 3.

Emission techniques utilize the characteristic X-ray emission spectra of elements, which are generated by, for instance, bombardment with high-energy electrons. The spectra are used in the chemical analysis of both bulk samples (*X-ray fluorescence*) and submicroscopic particles (analytical electron microscopy, EPMA, etc.). They also have some uses in determination of local structure, coordination numbers, etc.

Absorption techniques measure the absorption of X-rays by samples, especially at energies in the region of *absorption edges*. They are powerful techniques for studying local structure but are not accessible to most people since a *synchrotron radiation source* is needed.

Emission techniques

On bombardment of matter with, for instance, high-energy electrons, inner shell electrons may be ejected from atoms. Outer shell electrons then drop into the vacancies in the inner levels and the excess energy is released in the form of electromagnetic radiation, often X-radiation (Figs. 3.1a and 4.3). Each element gives a characteristic X-ray emission spectrum (see Fig. 3.1b) composed of a set of sharp peaks. The spectra are different for each element since the peak positions depend on the difference in energy between electron levels, e.g. $2p$ and $1s$, which, in turn, depend on atomic number (Moseley's law, equation (3.1)). X-ray emission spectra may, therefore, be used for elemental analysis, both qualitatively by looking for peaks at certain positions or wavelengths, and quantitatively by measuring peak intensities and comparing them against a calibration chart. *X-*

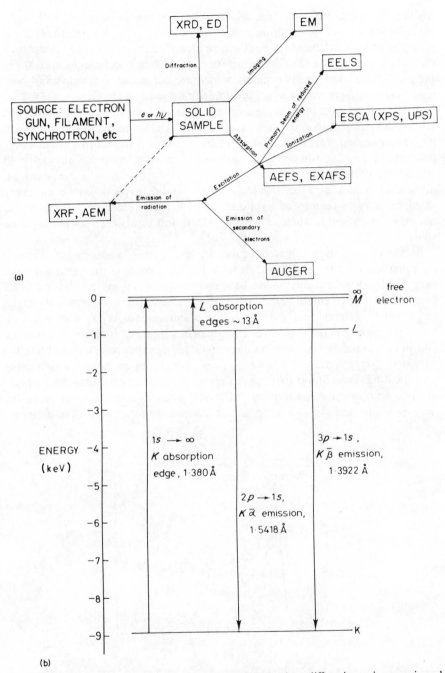

Fig. 4.15 (a) Flowsheet showing relations between various diffraction, microscopic and spectroscopic techniques associated with electrons and X-rays. (b) Electronic transitions responsible for emission and absorption X-ray spectra. Wavelength values are for copper

ray fluorescence (XRF) does just this: a solid sample is bombarded with high-energy electrons and the resulting emission spectrum recorded. From the spectral peak positions the elements present can be identified and from their intensities a quantitative analysis made. XRF is an important analytical technique, especially in industry. With modern instruments, which are computer controlled and fully automated, large numbers of solid samples can be analysed.

Analytical electron microscopy and related techniques such as electron probe microanalysis operate in the same way as XRF. The sample is bombarded with high-energy electrons; the diffracted electrons are used to record both electron micrographs and electron diffraction patterns whereas the emitted X-rays are used for elemental analysis. Analytical electron microscopy is rapidly becoming an invaluable technique for the chemical analysis of solids, in particular because very small particles or regions of a solid may be characterized. The heterogeneous nature of many solids, such as cements, steels and catalysts, may therefore be studied.

X-ray emission spectra also have uses in determining local structure such as coordination numbers and bond distances. This is because the peak positions vary slightly, depending on the local environment of the atoms in question. As an example, the Al $K\beta$ emission bands (corresponding to the $3p \rightarrow 1s$ transition) for three widely different materials, metallic aluminium, α-Al_2O_3 and sanidine, $KAlSi_3O_8$, are shown in Fig. 4.16. The peaks clearly occur in different positions and, in the case of Al_2O_3 and sanidine, appear to be poorly resolved doublets. By comparing the spectra for a wide range of aluminium-containing oxide compounds, it has been found that approximate correlations exist between (a) peak position and coordination number of the aluminium (it is usually 4 or 6) and (b) peak position and Al—O bond distance for a given coordination number.

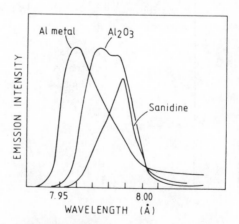

Fig. 4.16 Al $K\beta$ emission spectra of three aluminium-containing materials. (From White and Gibbs, *Amer. Mineral.*, **54**, 931, 1969)

Similar studies on silicates (in which silicon is almost always tetrahedrally coordinated) have shown a correlation between the Si $K\beta$ peak position and Si—O bond length; since the Si—O bond length varies according to whether the oxygen bridges two silicate tetrahedra or is non-bridging, this may be used to study polymerization of silicate anions in, for example glasses, gels and crystals.

For elements that can exist in more than one valence state or oxidation state, a correlation usually exists between peak position and oxidation state. Thus, XRF may be used to distinguish the different oxidation states of, for example, sulphur ($-$ II to $+$ VI) in a variety of sulphur-containing compounds.

Absorption techniques

Atoms give characteristic X-ray absorption spectra as well as characteristic emission spectra. These arise from the various ionization and intershell transitions that are possible, as shown in Fig. 4.15(b). In order to ionize a $1s$ electron in copper metal an energy of almost 9 keV, corresponding to a wavelength of 1.380 Å, is needed. Much less energy ($\sim$ 1 keV) is needed to ionize L shell ($2s$, $2p$) electrons. The X-ray absorption spectrum of copper is shown in Fig. 4.17. It takes the form of a smooth curve which increases rapidly at low X-ray energies and

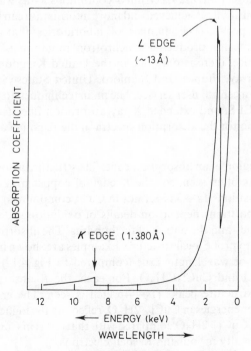

Fig. 4.17 Variation of X-ray absorption coefficient with wavelength for copper metal. (From Stern, *Sci. Amer.*, **234**, 96, 1976)

superposed on which are K, L, etc., absorption edges. The K absorption edge represents the minimum energy that is required to ionize a $1s$ electron in copper. The absorption coefficient therefore undergoes an abrupt increase as the energy is increased to this value. The L absorption edge (in fact, three closely spaced L edges are usually seen) represents ionization of $2s$, $2p$ electrons. At intermediate energies of the incident X-ray beam, e.g. at 4 keV, electrons may be ionized from L or outer shells and they leave the atom with a net kinetic energy, E, given by

$$E = hv - E_0 \qquad (4.5)$$

where hv is the energy of the incident X-ray photon (4 keV in this case) and E_0 is the critical ionization energy required to free the electron from the atom.

The wavelengths at which absorption edges occur depend on the relative separation of the atomic energy levels in atoms which, in turn, depend on atomic number (Moseley's law, equation (3.1)). They are, therefore, characteristic for each element and may be used for identification purposes in a similar manner to emission spectra.

Although X-ray absorption techniques have been in use since the 1930s, a recent development involving the use of synchrotrons and storage rings as a source of X-rays has given them a new impetus. This is because synchrotron radiation, produced when charged particles, electrons, are accelerated in a magnetic field, is a very intense spectrum of continuous X-ray wavelengths. Using it, much higher sensitivity is achieved and more information can be obtained from the absorption spectra. The number of laboratories that have a particle accelerator suitable for producing synchrotron radiation is, of course, very limited; for instance, there is only one in the United Kingdom, at Daresbury. Others are at Orsay, France, and Stanford, United States. Such laboratories therefore provide a central user service. The main techniques are absorption edge fine structure (AEFS) and extended X-ray absorption fine structure (EXAFS), both of which monitor the absorption spectra in the region of absorption edges.

AEFS. In the region of an absorption edge, fine structure associated with inner shell transitions is often seen, e.g. the K edge of copper may show additional peaks due to transitions $1s \rightarrow 3d$ (but not in Cu^{+1} compounds), $1s \rightarrow 4s$, $1s \rightarrow 4p$. The exact peak positions depend on details of oxidation state, site symmetry, surrounding ligands and the nature of the bonding. The absorption spectra may therefore be used to probe local structure. Examples are shown in Fig. 4.18, on an expanded energy or wavelength scale (compared to Fig. 4.17), for two copper compounds, CuCl and $CuCl_2 \cdot 2H_2O$. For each, the K absorption edge, with superposed fine structure peaks ($1s \rightarrow 4p$ etc.) is seen. The entire spectrum is displaced to higher energies in $CuCl_2 \cdot 2H_2O$, reflecting the higher oxidation state of copper ($+2$) in $CuCl_2 \cdot 2H_2O$ compared with that ($+1$) in CuCl and, therefore, the increased difficulty in ionizing K shell electrons.

EXAFS. Whereas the AEFS technique examines at high resolution the details

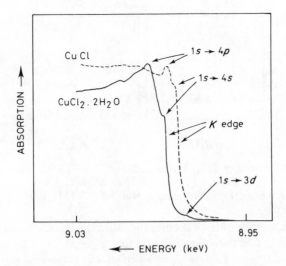

Fig. 4.18 AEFS spectra of CuCl and $CuCl_2 2H_2O$.
(From Chan, Hu and Gamble, *J. Mol. Str.*, **45**, 239,
1978)

of the fine structure in the region of an absorption edge, the EXAFS technique examines the variation of absorption with energy (or wavelength) over a much wider range, extending out from the absorption edge to higher energies by up to ∼ 1 keV. The absorption usually shows a ripple, known also as the *Kronig fine structure* (Fig. 4.19) from which, with suitable data processing, information on local structure and, especially, bond distances may be obtained. Explanations of the origin of the ripple will not be attempted here. Suffice it to say that it is related

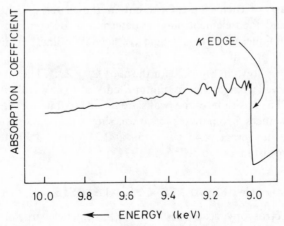

Fig. 4.19 EXAFS spectrum of copper metal. (From
E. A. Stern, *Sci. Amer.*, **234**, 96, 1976)

188

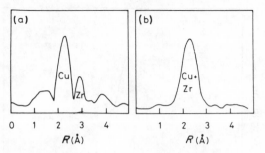

Fig. 4.20 EXAFS-derived partial RDFs for an amorphous $Cu_{46}Zr_{54}$ alloy: Zr K edge, (b) Cu K edge. (From Gurman, *J. Mat. Sci.*, **17**, 1541, 1982)

to the wave properties of the electron; the ionized photoelectrons interact with neighbouring atoms in the solid which then act as secondary sources of scattering for the photoelectrons. Interference between adjacent scattered waves may occur and this influences the probability of absorption of an incident X-ray photon. The degree of interference depends on the wavelength of the photoelectron (and hence on the wavelength of the incident X-ray photons) and the local structure, including interatomic distances, in the region of the emitting atom. EXAFS is therefore a kind of *in situ* electron diffraction in which the source of the electron is the actual atom which participates in the X-ray absorption event. Using Fourier transform techniques, it is possible to analyse the ripple pattern and obtain information on coordination numbers and bond distances.

EXAFS is a technique for determining local structure and is equally suitable for non-crystalline as well as crystalline materials. It is particularly valuable for studying disordered and amorphous materials such as glasses, gels and amorphous metals since structural information on them is generally hard to obtain. This is because EXAFS has one great advantage: by tuning in to the absorption edge of each element present in the material in turn, the local structure around each element may be determined. By contrast, conventional diffraction techniques give only a single averaged coordination environment for all the elements present.

An example is shown in Fig. 4.20 for the alloy $Cu_{46}Zr_{54}$. The plots are Fourier transforms derived from (a) the zirconium K edge at 18 keV and (b) the copper K edge at 9 keV. The positions of the peaks are related to, but not directly equal to, interatomic distances. From these plots it was shown that each zirconium atom is surrounded by an average of 4.6 Cu atoms at 2.74 Å and 5.1 Zr atoms at 3.14 Å: copper–copper distances are 2.47 Å.

Electron spectroscopies: ESCA, XPS, UPS, AES, EELS

Electron spectroscopy techniques measure the kinetic energy of electrons that are emitted from matter as a consequence of bombarding it with ionizing radiation or high-energy particles. Various processes take place when atoms are exposed to ionizing radiation (Fig. 4.15a). The simplest is the direct ionization of an electron from either a valence or an inner shell. The kinetic energy, E, of the ionized electron is equal to the difference between the energy, hv, of the incident radiation and the binding energy or ionization potential, E_b, of the electron, i.e. $E = hv - E_b$. For a given atom, a range of E_b values is possible, corresponding to the ionization of electrons from different inner and outer valence shells, and these E_b values are characteristic for each element. Measurement of E, and therefore E_b, provides a means of identification of atoms and forms the basis of the ESCA (*electron spectroscopy for chemical analysis*) technique developed by Siegbahn and coworkers in Uppsala (1967). The ionizing radiation that is used in ESCA is usually either X-rays (Mg $K\alpha$, 1254 eV or Al $K\alpha$, 1487 eV monochromatic radiation) or ultraviolet light (He discharge, 21.4 and 40.8 eV for the $2p \rightarrow 1s$ transitions in He and He$^+$, respectively) and the techniques are then also known as XPS (*X-ray photoelectron spectroscopy*) and UPS (*ultraviolet photoelectron spectroscopy*), respectively. The main difference between XPS and UPS is in the electron shells that are accessible for ionization. Thus, inner shell electrons may be ionized in XPS but only outer electrons—in valence shells, molecular orbitals or energy bands—may be ejected in UPS.

A related technique is *Auger electron spectroscopy* (AES). In this, the electrons that are ejected and detected are not the primary ionized electrons, as in ESCA, but are electrons produced by secondary processes involving the decay of ionized atoms from excited states to lower energy states (Fig. 4.21).

The ESCA and Auger processes may be visualized as involving:

$$\text{Atom A} \xrightarrow{\text{radiation}} A^{+*} + e^-$$

where A^{+*} refers to an ionized atom which is in an excited state and e^- is the ionized electron which is detected in ESCA. The excited state condition A^{+*} arises either if the electron is ejected from an inner shell, leaving a vacancy, or if other electrons in the atom have been promoted to higher, normally empty levels, during irradiation. Either way, the excited atom decays when electrons drop into vacancies in lower energy levels. Energy is consequently released by one of two

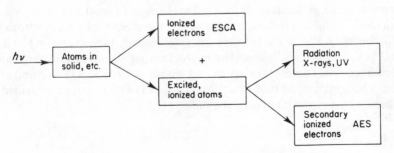

Fig. 4.21 Origins of ESCA and Auger spectra

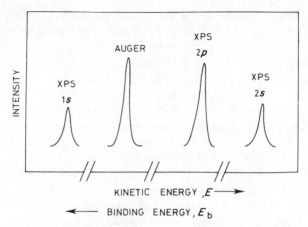

Fig. 4.22 Schematic XPS, AES spectrum of Na$^+$ in a sodium-containing solid. The Auger peak arises from an initial $1s$ vacancy which is filled by a $2p$ electron, causing ionization of another $2p$ (Auger) electron

methods:

$$A^{+*} \rightarrow A^+ + h\nu \quad \text{(X-rays, UV)}$$

or

$$A^{+*} \rightarrow A^{++} + e^- \quad \text{(Auger electrons)}$$

The energy may be emitted as electromagnetic radiation: this is, in fact, the normal method by which X-rays are produced, although for lighter atoms, UV photons are generated instead. Alternatively, the energy may be transferred to another (outer shell) electron in the same atom which is then ejected. Such secondary ionized electrons are known as Auger electrons.

AES spectra are usually observed at the same time as ESCA spectra. A schematic example is shown in Fig. 4.22 and is a plot of intensity of ionized electrons against their energy. Often, AES spectra are complex and difficult to interpret. Although AES is not, as yet, a widely used technique, this may change in the near future.

In the last decade or so, ESCA has proved to be a powerful technique for determining energy levels in atoms and molecules. In solid state sciences, it is particularly useful as a technique for studying surfaces because the electrons that are produced in ESCA are not very energetic (usually their energy is much less than 1 keV) and are rapidly absorbed by solid matter. Consequently, they cannot escape from solids unless they are ejected within ~ 20 to $50\,\text{Å}$ (2 to 5 nm) of the surface. Information on bulk structure of solids is also obtained, provided the surface layer is representative of the bulk.

Applications

Chemical shifts and local structure. Some limited success has been achieved in

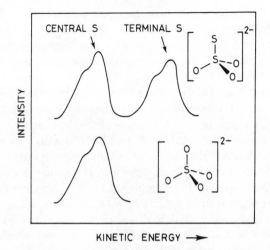

Fig. 4.23 Schematic XPS $2p$ spectra of sodium thiosulphate and sodium sulphate. Note that each peak is a doublet depending on the spin–orbit states, 1/2 and 3/2, of the $2p$ electron

using XPS to probe the local structure of solids. This has been possible because the binding energies of electrons in a particular atom may show a small variation, depending on the immediate environment of the atom and its charge or oxidation state. The idea is to measure the 'chemical shift' of an atom, relative to a standard, and thereby obtain information on local structure (as, in for example, NMR). Examples of chemical shift effects are shown in Figs 4.23 and 4.24. In sodium thiosulphate, $Na_2S_2O_3$, the two types of sulphur atom may be distinguished (Fig. 4.23). Peaks are of equal height, indicating equal numbers of each. Assignment of the peaks of higher kinetic energy to the terminal S atom is made

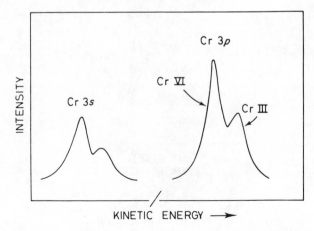

Fig. 4.24 XPS spectrum of Cr $3s$, $3p$ electrons in KCr_3O_8

on the basis that this atom carries more negative charge than the central atom and is, therefore, easier to ionize. Hence E_b is less and $(h\nu - E_b)$ greater for the terminal S atom than for the central S atom. By comparison, sodium sulphate, Na_2SO_4, shows a single sulphur $2p$ peak at the same energy as that for the central sulphur atom in $Na_2S_2O_3$.

The compound KCr_3O_8 is a mixed valence compound better written as $KCr^{III}(Cr^{VI}O_4)_2$. Its XPS spectrum (Fig. 4.24) shows doubled peaks for chromium both for $3s$ and $3p$ electrons. The intensities are in the ratio of 2:1 and the peaks are assigned to the oxidation states Cr^{VI} and Cr^{III}. This fits with the formula and also with the expectation that E_b is greater for Cr^{VI} than for Cr^{III}.

Although the two examples above show clearly the influence of local structure effects in the ESCA spectra, these examples are by no means typical. In many other cases, the chemical shifts associated with different oxidation states or local environments may be quite small and the spectra insensitive to local structure. The promise of ESCA as a local structure probe, applicable to virtually all elements of the periodic table, is therefore as yet only partly fulfilled.

Bonding and band structure. ESCA may be used to investigate the band structures of solids, especially in metals and semiconductors. An interesting example is shown in Fig. 4.25 for a series of sodium tungsten bronzes. These have

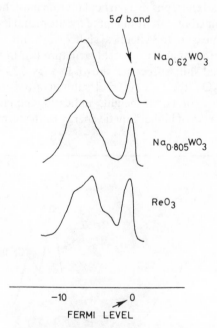

Fig. 4.25 XPS spectra for tungsten bronzes. (From Campagna *et al.*, *Phys. Rev. Letts.*, **34**, 738, 1975)

the ReO_3 structure but contain variable amounts of intercalated sodium. W^{VI} is d^0 and hence, ideally, WO_3 should have no $5d$ electrons. On addition of sodium to WO_3, to form Na_xWO_3, electrons from sodium enter the tungsten $5d$ band (leaving Na^+). This shows up in the XPS spectrum as a peak representing ionization from the $5d$ band and whose intensity increases with increasing sodium content. The spectrum of ReO_3 is shown for comparison. Re^{VI} is d^1 and hence its spectrum is similar to that which might be expected for $Na_{1.0}WO_3$.

Surface studies. The principal use and potential of ESCA in solid state sciences is probably for studying surfaces, mainly because it is one of the few techniques that is surface selective. It may be used purely as an analytical method for detecting which elements are present at a surface and in particular for detecting layers of absorbed molecules on, for example, the surfaces of metals. This is clearly of great relevance to catalysis. Coatings may be analysed; for instance, a GeO_2 layer on the surface of germanium metal may be detected, since a chemical shift effect is seen. Sometimes, surface states in the electron band structure of a metal can be detected. These appear as electron orbitals localized at or near the surface, rather than delocalized throughout the bulk of the metal. In order to detect them, very clean surfaces are needed.

Electron energy loss spectroscopy

EELS or ELS is a technique associated with analytical electron microscopy. It can be used for elemental analysis, for light elements such as carbon and nitrogen and for studying energy levels in surface layers of solids. In the EELS technique, a monochromatic incident beam of electrons is used which causes ionization of inner shell electrons in atoms of the sample (see Fig. 3.1a). These ionizations are a necessary precursor to the generation of X-rays. The electrons that are responsible for these ionizations suffer an energy loss as a consequence. An EELS spectrum is a plot of intensity of these electrons against energy loss. An intense peak occurs at zero energy loss. This corresponds to electrons that are either scattered elastically or do not interact with the sample. The other peaks in the EELS spectrum correspond to the electrons responsible for inner shell ionizations. The peaks are usually weak and broad and the spectra increase in complexity with increasing atomic number. EELS is a particularly useful technique for analysing light atoms and therefore complements X-ray fluorescence which is more useful with heavier elements. A derivative of EELS is *extended energy loss fine structure spectroscopy* (EXELFS), which is concerned with fine structure in the EELS spectrum. EXELFS is the electron analogue of the X-ray technique EXAFS.

Mössbauer spectroscopy

Mössbauer or γ-ray spectroscopy is akin to NMR spectroscopy in that it is concerned with transitions that take place inside atomic nuclei. The incident radiation that is used is a highly monochromatic beam of γ-rays whose energy may be varied by making use of the Doppler effect. The absorption of γ-rays by the sample is monitored as a function of energy and a spectrum is obtained which usually consists of a number of poorly resolved peaks. From an appropriate analysis of the spectrum, information on local structure—oxidation states, coordination numbers and bond character—may be obtained.

The γ-rays that are used in Mössbauer spectroscopy are produced by decay of radioactive elements such as $^{57}Fe^*_{29}$ or $^{119}Sn^*_{50}$. The γ-rays appear in the electromagnetic spectrum (Fig. 4.6) to the high-energy (right-hand) side of X-rays. The γ-emission is associated with a change in population of energy levels in the nuclei responsible, rather than with a change in atomic mass or number. Under certain conditions of 'recoilless emission', all of the energy change in the nuclei is transmitted to the emitted γ-rays and this gives rise to a highly monochromatic beam of radiation. This radiation may then be absorbed by a sample that contains similar atoms to those responsible for the emission. However, the nuclear energy levels of the absorbing atoms do vary somewhat, depending on the oxidation state of the element, its coordination number, etc., and hence some means of modulating either the energy of the incident radiation or the energy levels within the nuclei of the sample is required. In practice, the energy of the γ-rays is modified by making use of the Doppler effect. With the sample placed in a fixed position, the γ-ray source is moved at a constant velocity either towards or away from the sample. This has the effect of either increasing or decreasing the energy of the γ-rays incident upon the sample. In this way, the

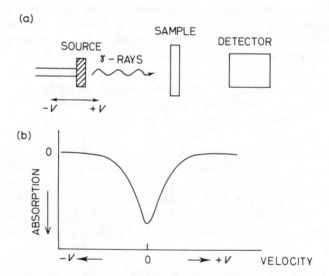

Fig. 4.26 (a) The Mössbauer effect. The energy of the γ-rays is modified when emitted from a moving source (Doppler effect). (b) Typical single line spectrum obtained when source and sample are identical

energy of the γ-rays may be varied and the γ-ray absorption spectrum of the sample determined. Only a limited number of isotopes can be induced to emit γ-rays suitable for Mössbauer work. The most widely used isotopes are ^{57}Fe and ^{119}Sn and, hence, Mössbauer spectroscopy has been most used on iron- and tin-containing substances. Other elements and isotopes which have been profitably studied include ^{129}I, ^{99}Ru and ^{121}Sb. A schematic illustration of the technique is given in Fig. 4.26.

Several types of information may be obtained from Mössbauer spectra. In the simplest case, in which both the emitter and sample are identical, the resonant absorption peak occurs when the source is stationary (Fig. 4.26b). When the emitter and sample are not identical, however, the absorption peak is shifted. This *chemical shift*, δ, arises because the nuclear energy levels in the atoms concerned have been modified by changes in the extra-nuclear density distribution in the atoms. In particular, chemical shifts may be correlated with the density at the nucleus of the outer shell s electrons. Chemical shifts are controlled principally by oxidation state, coordination number and the type of bonding. A summary given in Fig. 4.27 of the chemical shifts observed in various iron-containing ionic solids shows the influence of both charge and coordination number on the chemical shift. Use may be made of this for diagnostic purpose, e.g. to determine the nature of iron in inorganic compounds, minerals, etc.

For those nuclei that have a nuclear spin quantum number $I > \frac{1}{2}$, the distribution of positive charge inside the nucleus is non-spherical and a quadrupole moment, Q, results. The net effect of this is to cause splitting of the nuclear energy levels and hence splitting of the peaks in the Mössbauer spectrum. For both ^{57}Fe and ^{119}Sn, the peaks are split into doublets. The separation of the doublets, known as the *quadrupole splitting*, Δ, is, like the chemical shift, δ, sensitive to local structure and oxidation state.

A second type of line splitting which is important in the study of magnetic interactions is the *magnetic hyperfine Zeeman splitting*. This arises when a nucleus, of spin I, is placed in a magnetic field; each nuclear energy level splits into $(2I + 1)$

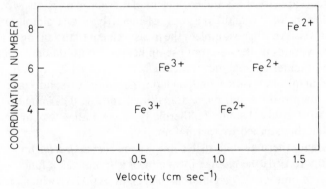

Fig. 4.27 Chemical shifts in iron-containing compounds. (From Bancroft, *Mössbauer Spectroscopy*, McGraw-Hill, 1973)

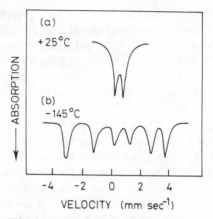

Fig. 4.28 Mössbauer spectrum of $KFeS_2$ (a) above and (b) below the Néel temperature. (From Greenwood, *Chem. Brit.*, **3** 56, 1967)

sublevels. The magnetic field arises either from magnetic exchange effects in ferro-, antiferro- and paramagnetic samples or from an externally applied field.

The study of hyperfine splitting and, especially, its temperature dependence provides information on magnetic ordering. For instance, $KFeS_2$ is antiferromagnetic below 245 K and hyperfine splitting results in a spectrum containing six peaks. Above 245 K only quadrupole coupling occurs and the spectrum reduces to a doublet (Fig. 4.28).

Thermal analysis

Thermal analysis may be defined as the measurement of physical and chemical properties of materials as a function of temperature. In practice, however, the term thermal analysis is used to cover certain specific properties only. These are enthalpy, heat capacity, mass and coefficient of thermal expansion. Measurement of the coefficient of thermal expansion of metal bars is a simple example of thermal analysis. Another example is the measurement of the change in weight of oxysalts or hydrates as they decompose on heating. With modern equipment, a wide range of materials may be studied. Uses of thermal analysis in solid state science are many and varied and include the study of solid state reactions, thermal decompositions and phase transitions and the determination of phase diagrams. Most kinds of solid are 'thermally active' in one way or another and may be profitably studied by thermal analysis.

The two main thermal analysis techniques are *thermogravimetry* (TG), which automatically records the change in weight of a sample as a function of either temperature or time, and *differential thermal analysis* (DTA), which measures the difference in temperature, ΔT, between a sample and an inert reference material as a function of temperature; DTA therefore detects changes in heat content. A technique that is closely related to DTA is *differential scanning calorimetry* (DSC). In DSC, the equipment is designed to allow a quantitative measure of the enthalpy changes that occur in a sample as a function of either temperature or time. A fourth thermal analysis technique is *dilatometry*, in which the change in linear dimension of a sample as a function of temperature is recorded. Dilatometry has long been used to measure coefficients of thermal expansion of metals; recently, it has acquired a new name, *thermomechanical analysis* (TMA), and has been applied to more diverse materials and problems, e.g. to the quality control of polymers.

With modern, automatic thermal analysis equipment it is possible to do TG, DTA and DSC using the same instrument; with some models, TG and DTA may be carried out simultaneously. Thermal analysis equipment is necessarily rather complicated and expensive, in order that a wide variety of thermal events and properties may be studied, both rapidly and with high sensitivity and accuracy. The basic principles of operation of each technique are simple, however.

In this section, the basic principles of TG, DTA and DSC are described, together with applications of each; instrumental descriptions are omitted.

Thermogravimetry (TG)

Thermogravimetry is a technique for measuring the change in weight of a substance as a function of temperature or time. The results usually appear as a continuous chart record; a schematic, typical, single step decomposition reaction is shown in Fig. 4.29. The sample, usually a few milligrams in weight, is heated at a constant rate, typically in the range 1 to $20\,°C\,min^{-1}$, and has a constant weight W_i, until it begins to decompose at temperature T_i. Under conditions of dynamic heating, decomposition usually takes place over a range of temperatures, T_i to T_f, and a second constant-weight plateau is then observed above T_f, which corresponds to the weight of the residue W_f. The weights W_i, W_f and the difference in weight ΔW are fundamental properties of the sample and can be used for quantitative calculations of compositional changes, etc. By contrast, the

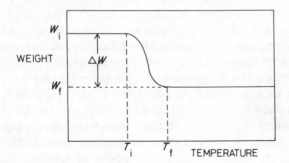

Fig. 4.29 Schematic TG curve for a single step decomposition reaction

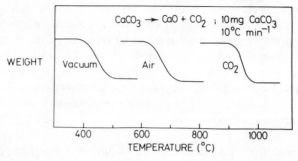

Fig. 4.30 Decomposition of $CaCO_3$ in different atmospheres

temperatures T_i and T_f depend on variables such as heating rate, the nature of the solid (e.g. its particle size) and the atmosphere above the sample. The effect of atmosphere can be dramatic, as shown in Fig. 4.30 for the decomposition of $CaCO_3$: in vacuum, the decomposition is complete by $\sim 500\,°C$, but in CO_2 at one atmosphere pressure, decomposition does not even commence until above $900\,°C$. T_i and T_f pertain to the particular experimental conditions, therefore, and do not necessarily represent equilibrium decomposition temperatures.

Differential thermal analysis (DTA) and differential scanning calorimetry (DSC)

Differential thermal analysis is a technique in which the temperature of a sample is compared with that of an inert reference material during a programmed change of temperature. The temperature of sample and reference should be the same until some thermal event, such as melting, decomposition or change in crystal structure, occurs in the sample, in which case the sample temperature either lags behind (if the change is endothermic) or leads (if the change is exothermic) the reference temperature.

The reason for having both a sample and a reference is shown in Fig. 4.31. In (a), a sample is shown heating at a constant rate and its temperature, T_s, is monitored continuously with a thermocouple. The temperature of the sample as a function of time is shown in (b); the plot is linear until an endothermic event occurs in the sample, e.g. melting at temperature T_c. The sample temperature remains constant at T_c until the event is completed; it then increases rapidly to catch up with the temperature required by the programmer. The thermal event in the sample at T_c therefore appears as a rather broad deviation from the sloping baseline in (b). Such a plot is insensitive to small heat effects since the time taken to complete such processes may be short and hence the deviation from the baseline is small. Further, any spurious variations in the baseline, caused by, for example, fluctuations in the heating rate, would appear as apparent thermal events. Because of its insensitivity, this technique has only limited applications; its main use historically has been in the 'method of cooling curves' which was used to

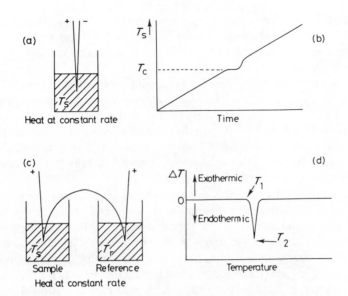

Fig. 4.31 The DTA method. Graph (b) results from the set-up shown in (a) and graph (d), a typical DTA trace, results from the arrangement shown in (c)

determine phase diagrams: the sample temperature was recorded on cooling rather than on heating and since the heat effects associated with solidification and crystallization are usually large, they could be detected by this method.

In (c), the arrangement normally used in DTA is shown. Sample and reference are placed side by side in a heating block which is either heated or cooled at a constant rate; identical thermocouples are placed in each and are connected 'back to back'. When the sample and reference are at the same temperature, the net output of this pair of thermocouples is zero. When a thermal event occurs in the sample, a temperature difference, ΔT, exists between the sample and reference which is detected by the net voltage of the thermocouples. A third thermocouple (not shown) is used to monitor the temperature of the heating block and the results are presented as ΔT against temperature (d). A horizontal baseline, corresponding to $\Delta T = 0$, occurs and superposed on this is a sharp peak due to the thermal event in the sample. The temperature of the peak is taken either as the temperature at which deviation from the baseline begins, T_1, or as the peak temperature, T_2. While it is probably more correct to use T_1, it is often not clear where the peak begins and, therefore, it is more common to use T_2. The size of the ΔT peak may be amplified so that events with very small enthalpy changes may be detected. Figure (d) is clearly a much more sensitive and accurate way of presenting data than (b) and is the normal method for presenting DTA results.

Commercial DTA instruments are available which enable the temperature range -190 to $1600\,^{\circ}\mathrm{C}$ to be covered. Sample sizes are usually small, a few milligrams, because then there is less trouble with thermal gradients within the sample which could lead to reduced sensitivity and accuracy. Heating and cooling rates are usually in the range 1 to $50\,°C\,min^{-1}$. With the slower rates, the sensitivity is reduced because ΔT for a particular event decreases with decreasing heating rate.

DTA cells are usually designed for maximum sensitivity to thermal changes, but this is often at the expense of losing a calorimetric response; thus peak areas or peak heights are only qualitatively related to the magnitude of the enthalpy changes occurring. It is possible to calibrate DTA equipment so that quantitative enthalpy values can be obtained from the peak areas, but the calibration is usually tedious. If calorimetric data are required it is usually better and easier to use differential scanning calorimetry (DSC).

DSC is very similar to DTA. A sample and an inert reference are also used in DSC but the cell is designed differently. In some DSC cells, the sample and reference are maintained at the same temperature during the heating programme and the extra heat input to the sample (or to the reference if the sample undergoes an exothermic change) required in order to maintain this balance is measured. Enthalpy changes are therefore measured directly. In other DSC cells, the difference in temperature between the sample and reference is measured, as in DTA, but by careful attention to cell design the response of the cell is calorimetric.

Applications

Uses of thermal analysis in solid state science are many and varied. Generally, DTA is more versatile than TG: TG detects effects which involve weight changes only. DTA also detects such effects, but, in addition, detects other effects, such as polymorphic transitions, which do not involve changes in weight. For many problems, it is advantageous to use both DTA and TG, because the DTA events can then be classified into those which do and those which do not involve weight change. An example is the decomposition of kaolin, $Al_4(Si_4O_{10})(OH)_8$ (Fig. 4.32). By TG a change in weight occurs at ~ 500 to $600\,°C$, which corresponds to dehydration of the sample; this dehydration also shows up on DTA as an endothermic event. A second DTA effect occurs at 950 to $980\,°C$, which has no counterpart in the TG trace: it corresponds to a recrystallization reaction in the dehydrated kaolin. This latter DTA effect is exothermic, which is unusual; it means that the structure that is obtained between ~ 600 and $950\,°C$ is metastable and the DTA exotherm signals a decrease in enthalpy of the sample and therefore a change to a more stable structure. The details of the structural changes that occur at this transition have still not been fully resolved.

Another useful ploy is to follow the thermal changes on cooling as well as on heating. This enables a separation of reversible changes, such as melting/solidification, from irreversible changes, such as most decomposition reactions. A schematic DTA sequence illustrating reversible and irreversible changes is shown in Fig. 4.33. Starting with a hydrated material, dehydration is the first event that occurs on heating and appears as an endotherm. The dehydrated material undergoes a polymorphic transition, which is also endo-

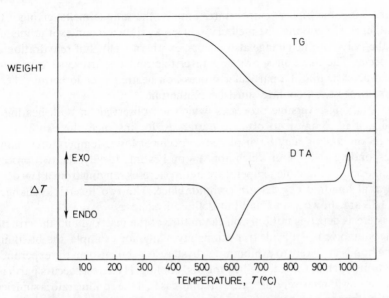

Fig. 4.32 Schematic TG and DTA curves for kaolin minerals. Curves vary depending on the sample structure and composition, e.g. the TG weight loss and associated DTA endotherm can occur anywhere in the range 450 to 750 °C

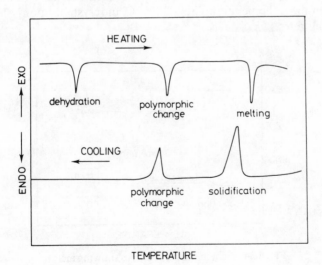

Fig. 4.33 Some schematic reversible and irreversible changes

202

thermic, at some higher temperature. Finally, the sample melts, giving a third endotherm. On cooling, the melt crystallizes, as shown by an exothermic peak, and the polymorphic change also occurs, exothermically, but rehydration does not occur. The diagram shows two reversible and one irreversible process. It should be clear that if a particular process, on heating, is endothermic, then the reverse process, on cooling, must be exothermic.

On studying reversible processes, which are observed on both heating and cooling, it is common to observe *hysteresis*; for instance, the exotherm that appears on cooling may be displaced to occur at lower temperatures than the corresponding endotherm which appears on heating. Ideally, the two processes, should occur at the same temperature but hystereses ranging from a few degrees to several hundred degrees are commonplace. The two reversible changes in Fig. 4.33 are shown with small but definite hystereses.

Hysteresis depends not only on the nature of the material and the structural changes involved—difficult transitions involving, for example, the breaking of strong bonds are likely to exhibit much hysteresis—but also on the experimental conditions, such as the rates of heating and cooling. Hysteresis occurs particularly on cooling at relatively fast rates; in some cases, if the cooling rate is sufficiently fast, the change can be suppressed completely. The change is then effectively irreversible under those particular experimental conditions. An example of this which is of great industrial importance is associated with glass formation, as shown schematically in Fig. 4.34. Beginning with a crystalline substance, e.g. silica, an endotherm appears when the substance melts. On cooling, the liquid does not recrystallize but becomes supercooled; as the temperature drops, so the viscosity of the supercooled liquid increases until eventually it becomes a glass. Hence, the crystallization has been suppressed entirely; in other words, hysteresis is so great that crystallization has not occurred. In the case of SiO_2, the liquid is very viscous even above the melting point, $\sim 1700\,°C$, and crystallization is slow, even at slow cooling rates.

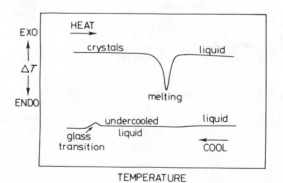

Fig. 4.34 Schematic DTA curves showing melting of crystals on heating and large hysteresis on cooling, which gives rise to glass formation

An important use of DTA and DSC in glass science is to measure the glass transition temperature, T_g. This appears not as a clear peak but rather as a broad anomaly in the baseline of the DTA curve, as shown in Fig. 4.34; T_g represents the temperature at which the glass transforms from a rigid solid to a supercooled, albeit very viscous, liquid. The glass transition is an important property of a glass since it represents the upper temperature limit at which the glass can really be used and also provides a convenient and readily measurable parameter for studying glasses. For glasses which kinetically are very stable, such as silica glass, the glass transition at T_g is the only thermal event observed on DTA since crystallization is usually too sluggish to occur. For other glasses, however, crystallization or devitrification may occur at some temperature above T_g and below the melting point, T_f. Devitrification appears as an exotherm and is followed by an endotherm at a higher temperature that corresponds to the melting of these same crystals. Examples of glasses which readily devitrify are metallic glasses; they may be prepared as thin films by rapidly quenching certain liquid alloy compositions. Other types of important glass-forming materials are amorphous polymers and amorphous chalcogenide semiconductors.

Polymorphic phase transitions may be studied easily and accurately by DTA; since many physical or chemical properties of a particular sample may be modified or changed completely as a consequence of a phase transition, their study is extremely important. Examples are:

(a) Ferroelectric $BaTiO_3$ has a Curie temperature of $\sim 120\,^\circ C$ which may be determined by DTA; substitution of other ions for Ba^{2+} or Ti^{4+} causes the Curie temperature to vary.

(b) In cements, the β-polymorph of Ca_2SiO_4 has superior cementitious properties to the γ-polymorph. The effect of different additives on the transitions can be studied by DTA.

(c) In refractories, transitions such as $\alpha \rightleftharpoons \beta$ quartz or quartz $\rightleftharpoons$ cristobalite have a deleterious effect on silica refractories because volume changes associated with each transition reduce the mechanical strength of the refractory. These transitions, which should be prevented from occurring if possible, may be monitored by DTA.

DTA is a powerful method for the determination of phase diagrams, especially when used in conjunction with other techniques, such as X-ray diffraction for the identification of the crystalline phases present. Its use is illustrated in Fig. 4.35(b) for two compositions in the simple binary eutectic system shown in Fig. 4.35(a). On heating composition A, melting begins to occur at the eutectic temperature, T_2, and gives rise to an endothermic peak. However, this is superposed on a much broader endothermic peak which terminates approximately at temperature T_1, and which is due to the continued melting that occurs over the temperature range T_2 to T_1. For this composition, an estimate of both solidus, T_2, and liquidus, T_1, temperatures may therefore be made. Composition B corresponds to the eutectic composition. On heating, this transforms completely to liquid at the eutectic temperature, T_2, and gives a single, large endotherm, peaking at T_2, on DTA.

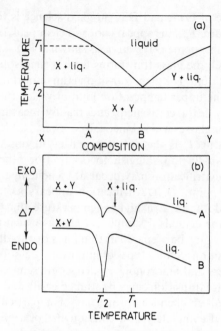

Fig. 4.35 Use of DTA for phase diagram determination: (a) a simple binary eutectic system; (b) schematic DTA traces for two compositions, A and B, on heating

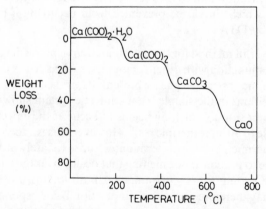

Fig. 4.36 Schematic, stepwise decomposition of calcium oxalate hydrate by TG

Thus, if DTA traces for a range of mixtures between X and Y could be compared, all should give an endotherm at T_2, whose magnitude depended on the degree of melting that occurred at T_2 and hence on the closeness of the sample composition to the eutectic composition, B. In addition, all compositions, apart from B, should give a broad endotherm peaking at some temperature above T_2; this is due to the completion of melting at the liquidus. The temperature of this peak should vary with composition. Polymorphic phase transitions also appear on phase diagrams, at subsolidus temperatures, and these may be determined very readily by DTA, especially if solid solutions form and the transition temperature is composition dependent.

In multistage decomposition processes, TG, either alone or in conjunction with DTA, may be used to separate and determine the individual steps. A good example, the decomposition of calcium oxalate monohydrate, is illustrated in Fig. 4.36, from which it can be seen that decomposition occurs in three stages giving, as intermediates, anhydrous calcium oxalate and calcium carbonate. Many other examples of similar multistage decompositions could be given in hydrates, hydroxides, oxysalts and minerals.

Chapter 5

Crystal Defects, Non-Stoichiometry and Solid Solutions

Crystal defects and non-stoichiometry

Perfect and imperfect crystals

A perfect crystal may be defined as one in which all the atoms are at rest on their correct lattice positions in the crystal structure. Such a perfect crystal can be obtained, hypothetically, only at absolute zero. At all real temperatures, crystals are imperfect. Apart from the fact that atoms are vibrating, which may be regarded as a form of defect, a number of atoms are inevitably misplaced in a real crystal. It is such imperfections or defects in an otherwise perfectly regular atomic array that concern us here. In some crystals, the number of defects present may be very small, $\ll 1$ per cent, as in, for example, high purity diamond or quartz crystals. In other crystals, very high defect concentrations, > 1 per cent, may be present. In the latter, highly defective crystals, the question arises as to whether or not the defects themselves should be regarded as forming a fundamental part of the crystal structure rather than as some imperfection in an otherwise ideal structure.

Crystals are invariably imperfect because the presence of defects up to a certain concentration leads to a reduction of free energy (Fig. 5.1). In order to understand the reasons for this, let us consider the effect on the free energy of a perfect crystal of creating a single defect, say a vacant cation site. This requires a certain amount of energy, ΔH. Creation of this single defect causes a considerable increase in entropy, ΔS, of the crystal, because of the large number of positions which this defect can occupy. Thus, if the crystal contains 1 mole of cations, there are $\sim 10^{23}$ possible positions for the vacancy. The entropy gained by having this choice of positions is called *configurational entropy* and is given by the Boltzmann formula

$$S = k \ln W \tag{5.1}$$

where the probability, W, is proportional to 10^{23}; other, smaller, entropy changes are also present due to the disturbance of the crystal structure in the neighbourhood of the defect. As a result of this considerable increase in entropy, the enthalpy required to form the defect initially is more than offset by the gain in

206

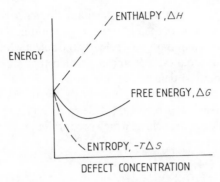

Fig. 5.1 Energy changes on introducing defects into a perfect crystal

entropy of the crystal. Consequently, the free energy, given by

$$\Delta G = \Delta H - T\Delta S \tag{5.2}$$

decreases (Fig. 5.1).

If we go now to the other extreme where, say, 10 per cent of the cation sites are vacant, the change in entropy on introducing yet more defects is small because the crystal is already very disordered in terms of occupied and vacant cation sites. The energy required to create yet more defects may be larger than any subsequent gain in entropy and hence such a high defect concentration would not be stable. In between these two extremes lie most real materials. A minimum in free energy occurs at a certain defect concentration which represents the number of defects present in the crystal under conditions of thermodynamic equilibrium (Fig. 5.1).

Although this is a simplified explanation, it serves to illustrate why crystals are imperfect. It also follows that the equilibrium number of defects present in a crystal increases with increasing temperature; assuming that ΔH and ΔS are independent of temperature, the $-T\Delta S$ term becomes larger and the free energy minimum displaced to higher defect concentrations, with increasing temperature.

For a given crystal or material, curves such as shown in Fig. 5.1 can be drawn

Table 5.1 *Predominant point defects in various crystals*

Crystal	Crystal structure	Predominant intrinsic defect
Alkali halides (not Cs)	Rock salt, NaCl	Schottky
Alkaline earth oxides	Rock salt	Schottky
AgCl, AgBr	Rock salt	Cation Frenkel
Cs halides, TlCl	CsCl	Schottky
BeO	Wurtzite, ZnS	Schottky
Alkaline earth fluorides, CeO_2, ThO_2	Fluorite, CaF_2	Anion Frenkel

for every possible type of crystal defect; the main difference between the different curves will be in the position of their free energy minimum. The type of defect that predominates in a particular material is clearly the defect which is easiest to form, i.e. the defect with the smallest ΔH and for which the free energy minimum is associated with the highest defect concentration. Thus, for NaCl it is easiest to form vacancies (the Schottky defect) and so these are the predominant point defects, whereas in AgCl the reverse is true and interstitial (Frenkel) defects predominate. In Table 5.1 the defects which predominate in a variety of inorganic solids are summarized.

Types of defect

Various schemes have been proposed for the classification of defects, each of which has its uses and none of which is entirely satisfactory. Defects can be broadly divided into two groups: *stoichiometric defects* in which the crystal composition is unchanged on introducing the defects, and *non-stoichiometric defects* which are a consequence of a change in crystal composition. Alternatively, the size and shape of the defect can be used as a basis for classification: *point defects* involve only one atom or site, e.g. vacancies or interstitials, although the atoms immediately surrounding the defect are also somewhat perturbed; *line defects*, i.e. *dislocations*, are effectively point defects in two dimensions but in the third dimension the defect is very extensive or infinite; in *plane defects*, whole layers in a crystal structure can be defective. Sometimes the name *extended defects* is used to include all those which do not come in the category of point defects.

It is convenient here to begin with point defects. These are the 'classical' defects; they were proposed in the 1930s following the work of Schottky, Frenkel, Wagner and others, and it was several decades before direct experimental evidence for their existence was forthcoming.

Schottky defect

The Schottky defect is a stoichiometric defect in ionic crystals. It is a pair of vacant sites, an anion vacancy and a cation vacancy. To compensate for the vacancies, there should be two extra atoms at the surface of the crystal for each Schottky defect. The Schottky defect is the principal point defect in the alkali halides and is shown for NaCl in Fig. 5.2. There are equal numbers of anion and cation vacancies in order to preserve local electroneutrality as much as possible, both inside the crystal and at the surface of the crystal.

The vacancies may be distributed at random in the crystal or may be associated into pairs or larger clusters. The reason why they tend to associate is because vacancies carry an effective charge and therefore, oppositely charged vacancies are attracted to each other. Thus, an anion vacancy in NaCl has a net positive charge of $+1$ because the vacancy is surrounded by six Na^+ ions, each with partially unsatisfied positive charge. Put another way, the anion vacancy has charge $+1$ because, on placing an anion of charge -1 in the vacancy, local

Cℓ Na Cℓ Na Cℓ Na Cℓ Na Cℓ

Na Cℓ Na Cℓ Na Cℓ Na Cℓ Na

Cℓ Na Cℓ Na Cℓ Na Cℓ Na Cℓ

Na Cℓ ☐ Cℓ Na Cℓ Na Cℓ Na

Cℓ Na Cℓ Na Cℓ Na ☐ Na Cℓ

Na Cℓ Na Cℓ Na Cℓ Na Cℓ Na

Cℓ Na Cℓ Na Cℓ Na Cℓ Na Cℓ

Na Cℓ Na Cℓ Na Cℓ Na Cℓ Na

Fig. 5.2 Two-dimensional representation of a Schottky defect with cation and anion vacancies

electroneutrality is restored. Similarly, the cation vacancy has a net charge of -1. In order to dissociate the pairs, energy equivalent to the enthalpy of association, 1.30 eV for NaCl ($\sim 120 \text{ kJ mol}^{-1}$) must be provided.

The number of Schottky defects in a crystal of NaCl is either very small or very large, depending on one's point of view. In NaCl at room temperature, typically only one in 10^{15} of the possible anion and cation sites is vacant, a number which is insignificant in terms of the average crystal structure of NaCl as determined by X-ray diffraction. On the other hand, a grain of salt weighing 1 milligram (and containing approximately 10^{19} atoms) contains $\sim 10^4$ Schottky defects, hardly an insignificant number! The presence of defects, even in small concentrations, often influences greatly the properties of materials. For instance, point defects such as Schottky defects are responsible for the optical and electrical properties of NaCl.

Frenkel defect

This is also a stoichiometric defect and involves an atom displaced off its lattice site into an interstitial site that is normally empty. Silver chloride (which also has the NaCl crystal structure) has predominantly this defect, with silver as the interstitial atom (Fig. 5.3). The nature of the interstitial site is shown in (b). The

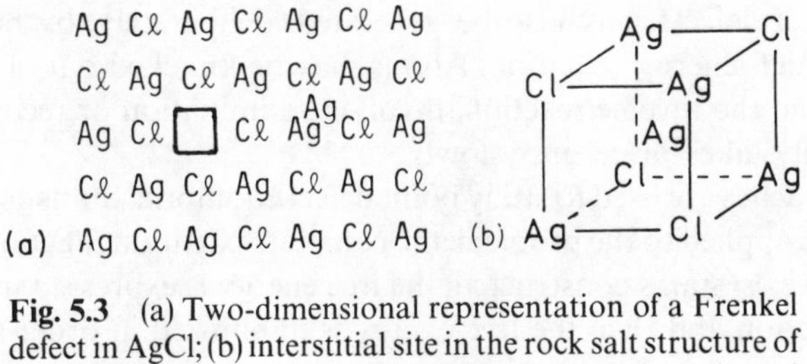

Ag Cℓ Ag Cℓ Ag Cℓ Ag

Cℓ Ag Cℓ Ag Cℓ Ag Cℓ

Ag Cℓ ☐ Cℓ Ag Cℓ Ag

Cℓ Ag Cℓ Ag Cℓ Ag Cℓ

(a) Ag Cℓ Ag Cℓ Ag Cℓ Ag   (b)

Fig. 5.3 (a) Two-dimensional representation of a Frenkel defect in AgCl; (b) interstitial site in the rock salt structure of AgCl showing tetrahedral coordination by both silver and chlorine

interstitial Ag^+ is surrounded tetrahedrally by four Cl^- ions but also, and at the same distance, by four Ag^+ ions. The interstitial Ag^+ ion is in an eight-coordinate site, therefore, with four Ag^+ and four Cl^- nearest neighbours. There is probably some covalent interaction between the interstitial Ag^+ ion and its four Cl^- neighbours which acts to stabilize the defect and give Frenkel defects, in preference to Schottky defects, in AgCl. On the other hand, Na^+, with its 'harder', more cationic character, would not find much comfort in a site which was tetrahedrally surrounded by four other Na^+ ions. Frenkel defects therefore do not occur to any significant extent in NaCl.

Calcium fluoride, CaF_2, also has predominantly Frenkel defects but in this case it is the anion, F^-, which occupies the interstitial site. These interstitial sites (empty cubes) can be seen in Fig. 1.29. Other materials with fluorite and antifluorite structures have similar defects, e.g. ZrO_2 (O^{2-} interstitial) and Na_2O (Na^+ interstitial).

As with Schottky defects, the vacancy and interstitial in a Frenkel defect are oppositely charged and may attract each other to form a pair. These pairs are electrically neutral overall, in both Schottky and Frenkel disorder, but they are dipolar. Pairs can therefore attract each other to form larger aggregates or clusters. Clusters similar to these may act as nuclei for the precipitation of phases of different composition in non-stoichiometric crystals.

Thermodynamics of Schottky and Frenkel defect formation

Both Schottky and Frenkel defects are *intrinsic defects*, i.e. they are present in pure material and a certain minimum number of these defects must be present from thermodynamic considerations. It is possible and, indeed, very common for crystals to have more defects present than corresponds to the thermodynamic equilibrium concentration. This arises because crystals are usually prepared at high temperatures and intrinsically, more defects are present the higher the temperature due to the increasing important of the $T\Delta S$ term in the free energy (Fig. 5.1). On cooling the crystals to room temperature, a small number of the defects may be eliminated, by various mechanisms, but in general and unless the cooling rate is extremely slow, the defects present at high temperature are preserved on cooling and are then present in excess of the equilibrium concentration.

An excess of defects may also be generated, deliberately, by bombarding a crystal with high-energy radiation. Atoms may be knocked out of their normal lattice sites and the reverse reaction, involving elimination or recombination of defects, usually takes place only slowly.

Two approaches are used to study point defect equilibria. Statistical thermodynamics can be applied to the problem; the complete partition function for a model of the defective crystal is constructed, the free energy is expressed in terms of the partition function and then the free energy is minimized in order to obtain the equilibrium condition. This method can also be applied to non-stoichiometric equilibria. Alternatively, the law of mass action can be applied to Schottky and

Frenkel equilibria and the concentration of defects expressed as an exponential function of temperature. Because of the simplicity and ease of application of the law of mass action to stoichiometric crystals, this latter method is used here. The notation used for representing crystal defects is similar to that proposed by Kröger (1974).

Schottky defects

The Schottky equilibria in a crystal of, for example, NaCl can be treated by assuming that a pair of ions are removed from the bulk of the crystal, leaving vacancies, and relocated at the crystal surface. This process may be represented by the following equation;

$$Na^+ + Cl^- + V^s_{Na} + V^s_{Cl} \rightleftharpoons V_{Na} + V_{Cl} + Na^{+,s} + Cl^{-,s} \tag{5.3}$$

where V_{Na}, V_{Cl}, V^s_{Na}, V^s_{Cl}, Na^+, Cl^-, $Na^{+,s}$ and $Cl^{-,s}$ represent cation and anion vacancies, vacant cation and anion surface sites, normally occupied cation and anion sites and occupied cation and anion surface sites, respectively. The equilibrium constant for formation of Schottky defects is given by

$$K = \frac{[V_{Na}][V_{Cl}][Na^{+,s}][Cl^{-,s}]}{[Na^+][Cl^-][V^s_{Na}][V^s_{Cl}]} \tag{5.4}$$

where square brackets represent concentrations of the species involved. The number of surface sites is always constant in a crystal of constant total surface area. Hence the number of Na^+ and Cl^- ions that occupy surface sites is always constant. On formation of Schottky defects, Na^+ and Cl^- ions move out of the crystal to occupy surface sites but, at the same time an equal number of fresh surface sites are created. (In fact, the total surface area of a crystal must increase slightly as Schottky defects are created but this effect can be ignored.) Therefore, $[Na^{+,s}] = [V^s_{Na}]$ and $[Cl^{-,s}] = [V^s_{Cl}]$. Equation (5.4) then reduces to

$$K = \frac{[V_{Na}][V_{Cl}]}{[Na^+][Cl^-]} \tag{5.5}$$

Let N be the total number of sites of each kind. Let N_V be the number of vacancies of each kind and hence the number of Schottky defects. The number of occupied sites of each kind is, therefore, $N - N_V$. Substituting into equation (5.5) gives

$$K = \frac{(N_V)^2}{(N - N_V)^2} \tag{5.6}$$

For small concentrations of defects:

$$N \simeq N - N_V \tag{5.7}$$

and therefore,

$$N_V \simeq N\sqrt{K} \tag{5.8}$$

The equilibrium constant, K, can be expressed as an exponential function of temperature:

$$K \propto \exp\left(-\Delta G/RT\right) \tag{5.9}$$

$$\propto \exp(-\Delta H/RT)\exp(\Delta S/r) \tag{5.10}$$

$$= \text{constant} \times \exp(-\Delta H/RT) \tag{5.11}$$

Therefore,

$$N_V = N \times \text{constant} \times \exp(-\Delta H/2RT) \tag{5.12}$$

where ΔG, ΔH and ΔS are the free energy, enthalpy and entropy of formation of one mole of defects. This equation shows that the equilibrium concentration of Schottky defects increases exponentially with temperature. Also, for a given temperature, the smaller the enthalpy of formation, H, the higher the Schottky defect concentration.

A similar expression can be derived for the concentration of vacancies in a monoatomic crystal of, for example, a metal. The difference is that, because only one type of vacancy is present, equations (5.5) to (5.7) simplify to give

$$N_V = NK \tag{5.13}$$

The factor of 2 therefore drops out of the exponential term in equation (5.12)

Frenkel defects

The Frenkel equilibria in a crystal of, for example, AgCl involve the displacement of Ag^+ ions from their lattice sites into empty interstitial sites and can be represented by

$$Ag^+ + V_i \rightleftharpoons Ag_i^+ + V_{Ag} \tag{5.14}$$

where V_i and Ag_i^+ represent empty and occupied interstitial sites. The equilibrium constant for this is given by

$$K = \frac{[Ag_i^+][V_{Ag}]}{[Ag^+][V_i]} \tag{5.15}$$

Let N be the number of lattice sites that would be occupied in a perfect crystal. Let N_i be the number of occupied interstitial sites, i.e.

$$[V_{Ag}] = [Ag_i^+] = N_i \tag{5.16}$$

and

$$[Ag^+] = N - N_i \tag{5.17}$$

For most regular crystal structures,

$$[V_i] = \alpha N \tag{5.18}$$

i.e. the number of available interstitial sites is simply related to the number of occupied lattice sites. For AgCl, $\alpha = 2$ because there are two tetrahedral interstitial sites for every octahedral site that is occupied by Ag^+ (in the cubic close packed rock salt-like structure of AgCl, there are twice as many tetrahedral sites as octahedral sites). Substituting into equation (5.15) gives:

$$K = \frac{N_i^2}{(N - N_i)(\alpha N)} \simeq \frac{N_i^2}{\alpha N^2} \tag{5.19}$$

Using equation (5.9) for the temperature dependence of the number of Frenkel defects, and substituting equation (5.19) gives:

$$[V_{Ag}] = [Ag_i^+] = N_i = N\sqrt{\alpha}\exp(-\Delta G/2RT) \tag{5.20}$$

$$= \text{constant} \times N\exp(-\Delta H/2RT) \tag{5.21}$$

For Frenkel defects as well as for Schottky defects, a factor of 2 appears in the exponential part of the expression for the number of defects (equations (5.21) and (5.12)). This is because, for both of these, there are two defective sites per defect, i.e. two vacancies per Schottky defect and one vacancy, one interstitial site per Frenkel defect. In each case, therefore, the overall ΔH for defect formation can be regarded as the sum of two component enthalpies.

Experimental values for the number of Frenkel defects in AgCl are given in Fig. 5.4. The method of presentation is the one that is normally used for the Arrhenius (or Boltzmann) equation, i.e. taking logs in equation (5.21) gives:

$$\log_{10}(N_i/N) = \log_{10}(\text{constant}) - (\Delta H/2RT)\log_{10}e \tag{5.22}$$

A plot of $\log_{10}(N_i/N)$ against $1/T$ should be a straight line of slope $-(\Delta H \log_{10}e)/2R$. The experimental data for AgCl fit the Arrhenius equation reasonably well although there is a slight but definite upward departure from

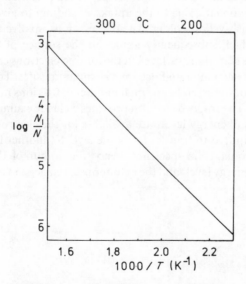

Fig. 5.4 Atomic fraction of Frenkel defects in AgCl as a function of temperature. (Data from Abbink and Martin, *J. Phys. Chem. Solids*, **27**, 205, 1966)

linearity at high temperatures. From Fig. 5.4, and ignoring for the moment the slight curvature at high temperatures, the numbers of vacancies and interstitials increase rapidly with temperature: on extrapolating the data to just below the melting point of AgCl, 456 °C, the concentration of Frenkel defects in equilibrium at ~ 450 °C is estimated as 0.6 per cent, i.e. approximately 1 in 200 of the silver ions move off their octahedral lattice sites to occupy interstitial tetrahedral sites. This defect concentration is 1 to 2 orders of magnitude higher than observed for most ionic crystals, with either Frenkel or Schottky disorder, just below their melting points. The enthalpy of formation of Frenkel defects in AgCl is ~ 1.35 eV (130 kJ mol^{-1}) and of Schottky defects in NaCl $\simeq 2.3$ eV (~ 220 kJ mol^{-1}); these values are fairly typical for ionic crystals.

The departure from linearity at high temperatures in Fig. 5.4 is attributed to the existence of a long-range type of Debye–Hückel attractive force between defects of opposite charge, e.g. between vacancies and interstitials in a Frenkel defect. This attraction reduces somewhat the energy of formation of the defects and so the number of defects increases, especially at high temperatures.

Colour centres

A considerable area of solid state physics is concerned with the study of colour centres in alkali halide crystals. The best known example is the F-centre (from the German *Farbenzentre*), shown in Fig. 5.5; it is an electron trapped on an anion vacancy. F-centres can be prepared by, for instance, heating an alkali metal halide in vapour of an alkali metal. Sodium chloride heated in sodium vapour becomes slightly non-stoichiometric due to the uptake of sodium to give Na$_{1+\delta}$Cl:$\delta \ll 1$, which has a greenish-yellow colour. The process must involve the absorption of sodium atoms, which subsequently ionize on the surface of the crystals. The resulting Na$^+$ ions stay at the surface but the ionized electrons can diffuse into the crystal where they encounter and occupy vacant anion sites. To preserve charge balance throughout the crystals, an equal number of Cl$^-$ ions must find their way out to the surface. The trapped electron provides a classic example of an 'electron in a box'. A series of energy levels are available for the electron within this box and the energy required to transfer from one level to another falls in the visible part of the electromagnetic spectrum; hence the colour of the F-centre. The magnitude of the energy levels and the colour observed depend on the host crystal

Cℓ Na Cℓ Na Cℓ

Na Cℓ Na Cℓ Na

Cℓ Na e Na Cℓ

Na Cℓ Na Cℓ Na

Cℓ Na Cℓ Na Cℓ   **Fig. 5.5** The F-centre, an electron trapped on an anion vacancy

and not on the source of the electron. Thus, NaCl heated in potassium vapour has the same yellowish colour as NaCl heated in sodium vapour, whereas KCl heated in potassium vapour is violet.

Another means of producing F-centres in NaCl is by irradiation. Using one of the normal methods of recording an X-ray diffraction pattern, Chapter 3, powdered NaCl turns a greenish-yellow colour after bombardment with X-rays for half an hour or so. The cause of the colour is again trapped electrons, but in this case they cannot arise from a non-stoichiometric excess of sodium. They probably arise from ionization of some chloride anions within the structure. The F-centre is a single trapped electron which has an unpaired spin and, therefore, an electron paramagnetic moment. A powerful method for studying such colour centres is ESR spectroscopy (Chapter 4), which detects unpaired electrons.

Many other colour centres have been characterized in alkali halide crystals; two of these, the H-centre and V-centre, are shown in Fig. 5.6. Both contain the chloride molecule ion, Cl_2^-, but this occupies one site in the H-centre and two sites in the V-centre; in both cases the axis of the Cl_2^- ion is parallel to the [101] direction. The V-centre occurs on irradiation of NaCl with X-rays. The mechanism of formation presumably involves ionization of a Cl^- ion to give a neutral chlorine atom which then covalently bonds with a neighbouring Cl^- ion.

One method by which defects can be eliminated from crystals is by annihilating each other. For example, if F- and H-centres come together they may cancel each other out, leaving a region of perfect crystal. Other defect centres which have been identified in the alkali halides include:

(a) the F'-centre, which is two electrons trapped on an anion vacancy;
(b) the F_A-centre, which is an F-centre, one of whose six cationic neighbours is a foreign monovalent cation, e.g. K^+ in NaCl;
(c) the M-centre, which is a pair of nearest neighbour F-centres;
(d) the R-centre, which is three nearest neighbour F-centres located on a (111) plane;
(e) ionized or charged cluster centres, such as M^+, R^+ and R^-.

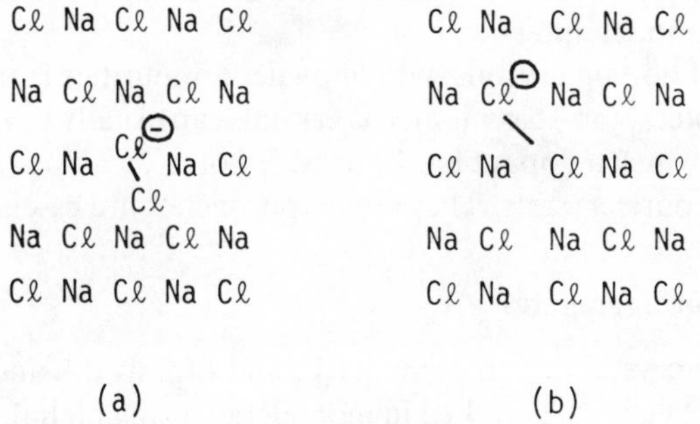

(a)　　　　　　　　(b)

Fig. 5.6 (a) H-centre and (b) V-centre in NaCl

Vacancies and interstitials in non-stoichiometric crystals

Most of the defects described so far are stoichiometric defects, i.e. they are present in the pure (stoichiometric) crystal and do not involve any change in overall composition. Defects can also be non-stoichiometric. Thus, some of the colour centres described in the previous section e.g. the F-centre, involve a change in composition. Non-stoichiometric crystals can also be prepared by other means, especially by doping pure crystals with *aliovalent impurities*, i.e. impurity atoms which have a different valency to those in the host crystal. For instance, NaCl may be doped with $CaCl_2$ to give non-stoichiometric crystals of formula, $Na_{1-2x}Ca_xV_{Na_x}Cl$, where V_{Na} represents a cation vacancy. In these crystals, the cubic close packed chloride arrangement is retained but the Na^+, Ca^{2+} ions and the cation vacancies are distributed over the octahedral cation sites. The overall effect of doping NaCl with Ca^{2+} ions is to increase the number of cation vacancies. Those vacancies that are controlled by the impurity level are termed *extrinsic* defects, in contrast to the thermally created *intrinsic* defects such as Schottky pairs.

For crystals with dilute ($\ll 1$ per cent) defect concentrations, the law of mass action can be applied. From equation (5.5), the equilibrium constant K for the formation of Schottky defects is proportional to the product of anion and cation vacancy concentrations, i.e.

$$K \propto [V_{Na}][V_{Cl}]$$

It is assumed that small aditions of impurities such as Ca^{2+} do not affect the value of K. As the cation vacancy concentration increases with increasing $[Ca^{2+}]$, the anion vacancy concentration must decrease correspondingly.

The practice of doping crystals with aliovalent impurities, coupled with the study of mass transport or electrical conductivity, has proved to be a powerful method for studying point defect equilibria. Mass transport or electrical conductivity in NaCl occurs by migration of vacancies. In reality, an ion adjacent to a vacancy moves into that vacancy, thereby leaving its own site vacant, but this process can effectively be regarded as vacancy migration. Measurements are made of the dependence of conductivity on temperature and defect concentration; from a suitable analysis, various thermodynamic parameters, such as the enthalpies of creation and migration of defects, may be determined. Further details are given in Chapter 7.

The subject of doping crystals with aliovalent impurities is treated more fully later in this chapter. Non-stoichiometric crystals can usually be regarded either as pure crystals to which a dopant has been added or as *solid solutions* based on the structure of the pure crystals. These two approaches are essentially equivalent.

Defect clusters or aggregates

Research into crystal defects is an active and rapidly advancing area of solid state science. As defects are studied in more detail, using high resolution electron microscopy and other techniques, coupled with computer-assisted modelling of defect structures, it is becoming clear that the apparently simple point defects such as vacancies and interstitials are often, in fact, more complex. Instead of single atom defects, larger defect clusters tend to form. Take the example of an interstitial metal atom in a face centred cubic metal. If the assumption is made that creation of defects, such as interstitial atoms, does not perturb the host metal structure then there are two possible sites for the interstitial atom, tetrahedral and octahedral. Recent research shows, however, that interstitial atoms *do* perturb the host structure, especially in the immediate vicinity of the interstitial atom. An example is shown in Fig. 5.7 for platinum metal containing an interstitial platinum atom. Instead of the interstitial platinum atom occupying the octahedral site, it is displaced by about 1 Å off the centre of this site and in the direction of one of the face centre atoms. The platinum atom on this face centre position also suffers a corresponding displacement in the same [100] direction. Thus, the defect involves *two* atoms, both of which are on distorted interstitial sites. This defect complex is known as a *split interstitial* or *dumbell-shaped interstitial*.

A similar split interstitial defect is present in body centred cubic metals such as α-Fe. It involves displacement of a normal lattice atom in the [110] direction i.e. along a face diagonal, together with introduction of the extra iron atom (Fig. 5.8). The 'ideal' site for the interstitial would be in the centre of a cube face, but instead it is displaced off the centre of this site in the direction of one of the corners. (Note that by changing the origin of the cell to the cube body centre it would appear as if the defect involved was displacing a body centred atom in the direction of one of the cube edges. The two descriptions are equivalent).

The exact structure of interstitial defects in alkali halides is not clear. Although Schottky defects predominate, interstitials are also present but in much smaller quantities. Calculations indicate that, in some materials, occupation of undistorted interstitial sites is favoured whereas in others a split interstitial is preferred. However, these results await experimental confirmation.

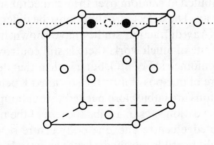

● Interstitial atoms

○ Normally occupied lattice site

□ Octahedral site

Fig. 5.7 Split interstitial defect in a face centred cubic metal

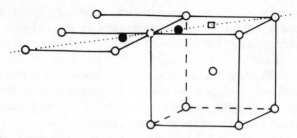

Fig. 5.8   Split interstitial in a body centred cubic metal,
e.g. α-Fe. Symbols as in Fig. 5.7

The presence of vacancies in metals and ionic crystals appears to cause a relaxation of the structure in the immediate environment of the vacancy. In metals, the atoms surrounding the vacancy appear to relax inward by a few per cent, i.e. the vacancy becomes smaller, whereas in ionic crystals the reverse occurs and as a result of an imbalance in electrostatic forces the atoms relax outwards.

Vacancies in ionic crystals are charged and therefore vacancies of opposite charge can attract each other to form clusters. The smallest cluster would be either an anion vacancy/cation vacancy pair or an aliovalent impurity (e.g. Cd^{2+})/cation vacancy pair. These pairs are dipolar, although overall electrically neutral, and so can attract other pairs to form larger clusters.

One of the most studied and best understood defect systems is wüstite, $Fe_{1-x}O:0 \lesssim x \lesssim 0.1$. Stoichiometric FeO has the rock salt structure with Fe^{2+} ions on octahedral sites. Density measurements have shown that the crystal structure of non-stoichiometric $Fe_{1-x}O$ contains a deficiency of iron rather than an excess of oxygen, relative to stoichiometric FeO. Using ideas of point defects, one would anticipate that non-stoichiometric $Fe_{1-x}O$ would have a structure represented by the formula $Fe^{2+}_{1-3x}Fe^{3+}_{2x}V_xO$ in which Fe^{2+}, Fe^{3+} and cation vacancies were distributed at random over the octahedral sites in the cubic close packed oxide ion array. The defect structure must be different to this, however, since neutron and X-ray diffraction studies have shown that Fe^{3+} ions are in *tetrahedral* sites. In spite of much work, there is still controversy over the actual structure of non-stoichiometric $Fe_{1-x}O$, but it is clear that defect clusters must be present. The structure of one possibility, the so-called Koch cluster, is shown in Fig. 5.9. The oxygen ions are cubic close packed throughout the crystals and the cluster involves all the cation sites in a cube the size of the normal f.c.c. rock salt unit cell. The twelve edge centre and one body centre octahedral sites are all empty; four of the eight possible tetrahedral sites contain Fe^{3+} ions (dividing the cube into eight smaller cubes, these tetrahedral sites occupy the body centres of the small cubes). This cluster has a net negative charge of fourteen, since there are thirteen vacant M^{2+} sites $(26-)$ and only four interstitial Fe^{3+} ions $(12+)$. Extra Fe^{3+} ions are distributed over the normal octahedral sites around the clusters to preserve electroneutrality. It has been suggested that these Koch clusters are present in wüstites of different x values. Their number increases with x

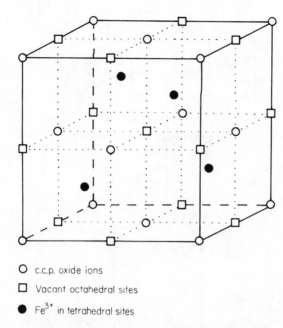

○ c.c.p. oxide ions

□ Vacant octahedral sites

● Fe^{3+} in tetrahedral sites

Fig. 5.9 Koch cluster postulated to exist in wüstite, $Fe_{1-x}O$

and hence the average separation between clusters decreases. Evidence from diffuse neutron scattering indicates that ordering of the clusters into a regularly repeating pattern may occur and this results in a superstructure for wüstite.

Another well-studied defect system is oxygen-rich uranium dioxide. X-ray diffraction is virtually useless for studying this material because information about the oxygen positions is lost in the heavy scattering from uranium. Instead, neutron diffraction has been used. The formula of this non-stoichiometric material is $UO_{2+x}: 0 < x \lesssim 0.25$. Stoichiometric UO_2 has the fluorite structure and non-stoichiometric UO_{2+x} contains interstitial oxide ions, which form part of the cluster shown in Fig. 5.10. In UO_2, the interstitial positions are present at the centres of alternate cubes containing oxygen ions at the corners. In the defect cluster proposed for UO_{2+x}, an interstitial oxygen is displaced off a cube centre interstitial positron in the [110] direction, i.e. towards one of the cube edges. At the same time, the two nearest corner oxygens are displaced into adjacent empty cubes along [111] directions (Fig. 5.10). Thus, in place of a single interstitial atom, the cluster appears to contain three interstitial oxygens and two vacancies.

At this point, it is worth while to emphasize why it is difficult to determine the exact structures of defects in crystals. Diffraction methods (X-ray, neutron, electron), as they are normally used in crystallographic work, yield *average* structures for crystals. For crystals that are pure and relatively free from defects, the averaged structure is usually a close representation of the true structure. For non-stoichiometric and defective crystals, however, the averaged structure may

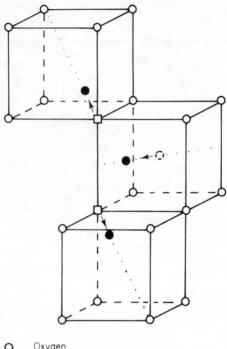

○   Oxygen

◌   Ideal interstitial site for oxygen

●   Interstitial oxygen

□   Vacant lattice site

Fig. 5.10 Interstitial defect cluster in UO_{2+x}. Uranium positions (not shown) are in the centre of every alternate cube

give a very poor and even an incorrect representation of the actual structure in the region of the defect. In order to determine defect structures, one essentially needs techniques that are sensitive to local structure. While study of local structure is a principal application of spectroscopic techniques (Chapter 4), these techniques have been of limited use for studying defects other than point defects. For instance, the sites occupied by impurity atoms can often be determined if the impurity is, in some way, spectroscopically active. However, the spectra are not usually informative about structure extending beyond the immediate coordination environment of a particular atom. What is needed, therefore, are techniques that give information on local structure, but extending over a distance of, say, 10 to 20 Å. Such techniques are very scarce, although there is hope that improved EXAFS methods may be suitable.

Interchanged atoms

In certain crystalline materials, it is common to find that some pairs of atoms or ions have swapped places. This may occur in alloys that contain two or more different elements, each arranged on a specific set of sites. It also occurs in certain ionic structures that contain two or more types of cation, again, each being located on a specific set of sites. If the number of interchanged atoms is large, and especially if it increases significantly as a function of temperature, this takes us into the realm of *order–disorder phenomena*. The limiting situation is reached when sufficient pairs have swapped places that they no longer show any preference for particular sites. The structure is then *disordered* as regards the particular sets of atoms or ions that are involved.

Alloys, by their very nature, involve a distribution of atoms of two or more different metals over one (sometimes more) set of crystallographically equivalent sites. Such alloys are therefore examples of substitutional solid solutions (see later). Alloys can either be disordered, with the atoms distributed at random over the available sites, or ordered, with the different atoms occupying distinct sets of sites. Ordering is generally accompanied by the formation of a supercell which is detected by X-ray diffraction from the presence of extra reflections. The ordered superstructure which is present in β'-brass, CuZn, below $\sim 450\,°C$ is shown in Fig. 5.11. Copper atoms occupy the body centre positions in a cube with zinc atoms at the corners, as in the CsCl structure; the lattice type is therefore primitive. In the disordered alloy of the same composition, β-brass, the copper and zinc atoms are distributed at random over the corner and body centre positions and therefore the lattice is body centred, as in the structure of α-Fe. The main point of interest to this chapter is that it is possible, within the ordered, β'-brass structure, to mix up *some* of the copper and zinc atoms while still retaining the long range order and superstructure. This disordering can be regarded as the introduction of defects into an otherwise perfect structure.

Good examples of interchanged cations in non-metallic oxide structures are provided by materials with the spinel structure. More details are given in Chapters 1 and 8.

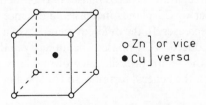

Fig. 5.11 Ordered, primitive cubic unit cell of β'-brass, CuZn

Extended defects—crystallographic shear structures

For a long time it was known that certain transition metal oxides could be prepared with an apparent wide range of non-stoichiometry, e.g. WO_{3-x}, MoO_{3-x}, TiO_{2-x}. Following the work of Magneli, it was recognized that in some of these systems, instead of continuous solid solution formation, a series of closely related phases with very similar formulae and structures existed. In oxygen-deficient rutile, a homologous series of phases was prepared of formula Ti_nO_{2n-1} with $n = 3 \ldots 10$. Thus, $Ti_8O_{15}(TiO_{1.875})$ and $Ti_9O_{17}(TiO_{1.889})$ are each homogeneous, physically separate phases. The structural principal underlying these phases was worked out by Magneli, Wadsley and others and the term *crystallographic shear* was coined for the type of 'defect' involved. In the oxygen-deficient rutiles, regions of normal rutile structure occur. These are separated from each other by crystallographic shear planes (CS planes) which are thin lamellae of rather different structure and composition. All of the oxygen deficiency is concentrated within these CS planes. With increased reduction, the variation in stoichiometry is accommodated by increasing the number of CS planes and decreasing the thickness of the blocks of rutile structure between adjacent CS planes.

For the purpose of understanding CS structures, it is useful to consider schematically how they might form on reduction of WO_3 (Fig. 5.12); the CS planes in WO_3 are easier to draw than those in rutile. The structure of WO_3 may be regarded as a three-dimensional framework of corner-sharing octahedra (a). The first step in the reduction of WO_3 involves the formation of vacant oxygen sites together with the reduction of W^{6+} ions to W^{5+}. The vacant oxygen sites are not distributed at random, however, but are located on certain planes within the crystal (b). Such a structure would, of course, be unstable and so a partial collapse of the structure occurs to eliminate the layer of vacancies and form the CS planes (c). As a result of this condensation, octahedra within the CS planes share some edges whereas in unreduced regions of WO_3, the corresponding linkages are by corner-sharing only. The CS plane is outlined in heavy lines in (c) and runs obliquely through the structure.

CS planes may occur at random through the structure or may be spaced at regular intervals. If they are randomly spaced, they may be regarded as planar defects in the crystal structure and are known as *Wadsley defects*. If they are regularly spaced, then separate, ordered phases may be identified. With increased reduction, the number of CS planes increases and hence the spacing between adjacent CS planes decreases. In the different members of a homologous series of phases, such as the reduced WO_3 phases, the integrity and composition of each member is given by the separation distance of regularly repeating CS planes. Thus, as n decreases in the general formula, so the spacing between adjacent CS planes decreases.

So far, we have considered homologous series of phases in which the structure and orientation of the CS planes is the same in each member of the series. CS planes may occur in different orientations, however, in which case their detailed

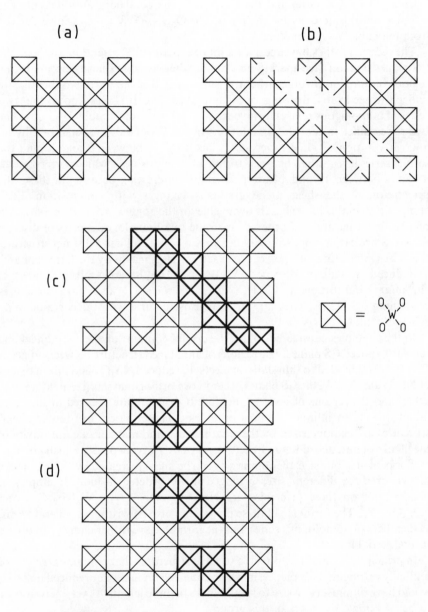

Fig. 5.12 Formation of CS planes in WO_3 and related structures. Each crossed square represents a chain of WO_6 octahedra linked by sharing corners. The parent, unreduced part of the structure is the same as that of cubic ReO_3

structure also changes. This is shown in (d); the CS plane is in a different orientation to that in (c) and the linkages of the octahedra are different. The example in (d) is actually for the phase Mo_nO_{3n-1}:$n = 8$, which is also derived from the cubic WO_3 structure.

The reduced rutiles have received a lot of experimental attention. In them, the CS planes contain face-sharing octahedra, whereas edge-sharing occurs in the regions of unreduced rutile structure. Since the CS planes in reduced rutile are difficult to see from a drawing, one is not given here. It is now known that two series of homologous phases form, both with the general formula Ti_nO_{2n-1}. The two series differ in that the orientations of the CS planes are separated by an angular rotation of $11.53°$. One series has n values between 3 and 10 and the second series has n values between 16 and 36. Compositions which correspond to n values between 10 and 14 are particularly interesting because, in these, the orientation of the shear plane gradually changes with composition. Each composition in this range has its own value for the angular orientation of the CS planes; the latter are well ordered, as in the two homologous series of discrete phases. Since each composition in the range $n = 10$ to 14 has a distinct structure, this raises interesting questions as to exactly what is meant by the term 'a phase'.

Reduced vanadium dioxides, V_nO_{2n-1}:$4 \leqslant n \leqslant 8$, and mixed oxides of chromium and titanium, $Cr_2Ti_{n-2}O_{2n-1}$: $6 \leqslant n \leqslant 11$, also have structures containing CS planes. They appear to be isostructural with the corresponding reduced rutiles.

In the examples considered so far, the crystal contains one set of parallel and regularly spaced CS planes. The regions of unreduced structure between adjacent CS planes are limited to thin slabs or sheets. In reduced Nb_2O_5 and mixed oxides of Nb, Ti and Nb, W, the CS planes occur in *two* orthogonal sets (i.e. at 90° to each other) and the regions of perfect (unreduced) structure are reduced in size from infinite sheets to infinite columns or blocks. These 'block' or 'double shear' structures are characterized by the length, width and manner of connectivity of the blocks of unreduced ReO_3 structure. As well as having phases which are built of blocks of only one size, the complexity can be much increased by having blocks of two or three different sizes arranged in an ordered fashion. Examples of phases built on these principles are $Nb_{25}O_{62}$, $Nb_{47}O_{116}$, $W_4Nb_{26}O_{77}$ and $Nb_{65}O_{161}F_3$. The formulae of these phases can also be written in general terms, as members of homologous series, but the formulae are rather clumsy, involving several variables.

In principle, it should be possible to have structures that contain *three* sets of mutually orthogonal CS planes, in which cases the regions of unreduced material would have diminished in size to small blocks of finite length. As yet, there appear to be no known members of this group.

Structures that contain CS planes are normally studied by X-ray diffraction and high-resolution electron microscopy. Single crystal X-ray diffraction is, of course, by far the most powerful method for solving crystal structures and in this sense electron microscopy usually plays a minor role; its use is limited to the determination of unit cells and space groups for very small crystals and to studying defects such as stacking faults and dislocations. However, with the technique of direct lattice imaging, electron microscopy has found great application for structural studies of CS phases. In favourable cases, an image of the projected structure at about 3 Å resolution can be obtained. This usually takes the form of fringes or lines which correspond to the more strongly diffracting heavy metal atoms. The separation of the fringes can be measured and, in a 'perfect' crystal, should be absolutely regular. Whenever a CS plane is imaged, irregularity in the fringe spacing occurs because, as a result of the condensation to form a shear plane, the metal atom separation across the CS plane is reduced. A pair of fringes that is more closely spaced than normal therefore indicates a CS plane. By counting the number of normal fringes in each block, the n value of the phase in the homologous series can usually be determined. If the crystal structure of one of the members of the series has been worked out in detail using X-ray methods, it is a relatively simple matter to deduce the structures of the remainder from the electron microscope results.

For studying defects in CS structures, the electron microscope is indispensable. Thus Wadsley defects (random CS planes) can be recognized immediately; heterogeneities within supposedly single crystals can be detected, e.g. if a crystal is zoned and has a slight variation in composition with position or if the crystal is composed of intergrowths of two or more phases.

Stacking faults

Stacking faults occur commonly in materials that have layered structures, especially those which also exhibit polytypism. Stacking faults are examples of two-dimensional or plane defects, as also are CS planes. A metal which exhibits both polytypism and stacking faults is cobalt. It can be prepared in two main forms (polytypes) in which the arrangement of the metal atoms is either cubic close packing (...ABCABC...) or hexagonal close packing (...ABABAB...). In these two polytypes, the structures are the same in two dimensions, i,e, within the layers, and differ only in the third dimension, i.e. in the sequence of layers. *Stacking disorder* occurs when the normal stacking sequence is interrupted at irregular intervals by the presence of 'wrong' layers, e.g. schematically ...ABABABA*BCA*BABA.... The letters in italics correspond to layers that either are completely wrong (*C*) or do not have their normal neighbouring layers (*A* and *B*) on either side. Graphite is another element which exhibits polytypism (usually h.c.p. but sometimes c.c.p. of carbon atoms) and stacking disorder (mixed h.c.p. and c.c.p.).

Subgrain boundaries and antiphase domains (boundaries)

One type of imperfection in so-called single crystals is the presence of a domain or mosaic texture. Within the domains, which are typically $\sim 10,000$ Å in size, the structure is relatively perfect, but at the interface between domains there is a structural mismatch (Fig. 5.13). This mismatch may be very small and involve a

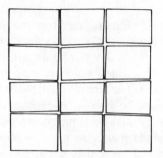

Fig. 5.13 Domain texture in
a single crystal

difference in angular orientation between the domains that is several orders less than 1°. The interfaces between grains are called *subgrain boundaries* and can be treated in terms of dislocation theory (see later).

Subgrain boundaries involve a difference in the relative angular orientation between two parts of essentially the same crystal. Another type of boundary, called an *antiphase boundary*, involves a relative lateral displacement of two parts of the same crystal. This is shown schematically for a two-dimensional crystal, AB, in Fig. 5.14. Across the antiphase boundary, like atoms face each other and the ...ABAB... sequence (in the horizontal rows) is reversed. The term arises because, if the A and B atoms are regarded as the positive and negative parts of a wave, then a phase change of π occurs at the boundary.

The occurrence of antiphase domains in metals has been known for about thirty years. By dark field imaging in the electron microscope, the boundaries can be seen as fringes. If the boundaries are regularly spaced, unusual diffraction effects occur: a superlattice is associated with the ordering of domains but because, at the boundaries, an expansion of the structure occurs due to repulsions

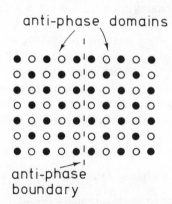

Fig. 5.14 Antiphase domains and boundaries in an ordered crystal AB: A, open circles; B, closed circles

between like atoms, satellite spots appear on either side of the supercell positions in the reciprocal lattice. This has been observed in alloys such as CuAu and in some silicate minerals such as plagioclase feldspars.

Dislocations and mechanical properties of solids

Dislocations are an extremely important class of crystal defect. They are responsible for the relative weakness of pure metals and in certain cases (after work-hardening) for just the opposite effect of extra hardness. The mechanism of crystal growth from either solution of vapour appears to involve dislocations. Reactions of solids often occur at active surface sites where dislocations emerge from the crystal.

Dislocations are stoichiometric line defects. Their existence was postulated long before direct experimental evidence of their occurrence was obtained. There were several types of observation which indicated that defects other than point defects must be present in crystals:

(a) Metals are generally much softer than expected. Calculations of the shearing stress of metals gave values of $\sim 10^6$ p.s.i. whereas experimental values for many metals are as low as $\sim 10^2$ p.s.i. This indicated that there must be some kind of weak link in their structures which allows metals to cleave so easily.

(b) Many well-formed crystals were seen under the microscope or even with the naked eye to have spirals on their surfaces which clearly provided a mechanism for crystal growth. Such spirals could not occur in perfect crystals, however.

(c) The malleable and ductile properties of metals were difficult to explain without invoking dislocations. Thus, ribbons of magnesium metal can be stretched out to several times their original length, almost like chewing gum, without rupture.

(d) The process of work-hardening of metals was difficult to explain without invoking dislocations.

Dislocations can be one of two extreme types, edge or screw dislocations, or can have any degree of intermediate character.

Edge dislocations

An *edge dislocation* is shown schematically in Fig. 5.15 and constitutes an extra half-plane of atoms, i.e. a plane of atoms that goes only part of the way through a crystal structure. Planes of atoms within the crystal structure are shown in projection as lines. These lines are parallel except in the region where the extra half-plane terminates. The centre of the distorted region is a line that passes right through the crystal, perpendicular to the paper, and approximates to the end of the extra half-plane; this is the *line* of the dislocation. Outside this stressed region, the crystal is essentially normal; the top half in the drawing must be slightly wider than the bottom half in order to accommodate the extra half-plane.

Fig. 5.15 Edge dislocation in projection

In order to understand the effect of dislocations on the mechanical properties of crystals, consider the effect of applying a shearing stress to a crystal that contains an edge dislocation, Fig. 5.16. The top half of the crystal is being pushed to the right and the bottom half to the left. Comparing (a) and (b), the extra half-plane which terminates at 2 in (a) can effectively move simply by breaking the bond 3–6 and forming a new linkage 2–6. Thus, with a minimum of effort, the half-plane has moved one unit of distance in the direction of the applied stress. If this process continues, the extra half-plane eventually arrives at the surface of the crystal as in (c). All that is now needed is an easy means of generating half-planes

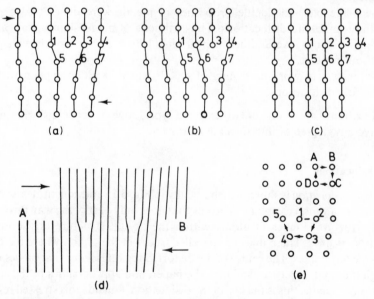

Fig. 5.16 Migration of an edge dislocation under the action of a shearing stress

and by a process of repetition the crystal will eventually completely shear. In (d), it is assumed that half-planes are generated at the left-hand end of the crystal (mechanisms of generation are discussed later). With each one that is generated, another one, equal and opposite in orientation and sign, is left behind. At the left-hand side, five half-planes in the bottom part of the crystal have accumulated at the surface. Balancing these in the top part of the crystal, three half-planes have arrived at the right-hand end and two more (positive dislocations with the symbol $\perp$) are still inside the crystal and in the process of moving out.

The easy motion of dislocations has an analogy in one's experience of trying to rearrange a carpet, preferably a large one. The method of lifting up one end and tugging needs a great deal more effort than making a ruck in one end of the carpet and gliding this through to the other end.

The process of movement of dislocations is called *slip* and the pile-up of half-planes at opposite ends gives ledges or *slip steps*. The line AB in (d) represents the projection of the plane over which the dislocation moves and is called the *slip plane*.

Dislocations are characterized by a vector, the *Burgers vector*, **b**. To find the magnitude and direction of **b**, it is necessary to make an imaginary atom-to-atom circuit around the dislocation (e). In normal regions of the crystal a circuit such as ABCDA, involving one unit of translation in each direction, is a closed loop and the starting point and finishing point are the same, A. However, the circuit 12345 which passes round the dislocation is not a closed circuit because 1 and 5 do not coincide. The magnitude of the Burgers vector is given by the distance 1–5 and its direction by the direction 1–5 (or 5–1). For an edge dislocation, **b** is perpendicular to the line of the dislocation and parallel to the direction of motion of the line of the dislocation under the action of an applied stress. It is also parallel to the direction of shear.

Screw dislocations

The *screw dislocation* is a little more difficult to visualize and is shown in Fig. 5.17. In (b), the line SS' represents the line of the screw dislocation. In front of this line the crystal has undergone slip but behind the line it has not. The effect of continued application of a shearing stress, arrowed in (b), is such that the slip step gradually extends across the whole face (side faces) of the crystal at the same time as the line SS' moves towards the back face (c). To find the Burgers vector of a screw dislocation, consider the circuit 12345 (a) which passes around the dislocation. The magnitude and direction of the distance 1–5 defines **b**. For a screw dislocation, the Burgers vector is parallel to the line of the dislocation (SS') and perpendicular to the direction of motion of this line. This contrasts with the edge dislocation and it can be seen that edge and screw dislocations are effectively at 90° to each other. For both edge and screw dislocations, **b** is parallel to the direction of shear or slip. The origin of the term 'screw' is easy to see by considering the atoms 54321 in (a). They lie on a spiral which passes right through the crystal and emerges with opposite hand or sign at S'. As in the case of an edge

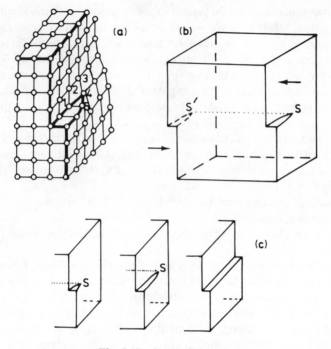

Fig. 5.17 Screw dislocation

dislocation, it is necessary to break only a few bonds in order for a screw dislocation to move. Thus in (a), the bond between atoms 2 and 5 has just broken and 2 has joined up with 1; the bond between 3 and 4 will be the next to break. Note that it is useful to consider bonds as breaking and forming although in practice it is not nearly as clear-cut as this; in metals and ionic crystals the bonds are certainly not covalent.

Dislocation loops

The process of generation of dislocations is complicated but seems always to involve the formation of *dislocation loops*. Consider the crystal shown in Fig. 5.18. On looking at the right-hand face it is obvious that a screw dislocation emerges near to point S, where the slip step terminates. However, this dislocation does not

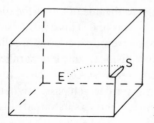

Fig. 5.18 A quarter dislocation loop

reach the left-hand face because there is no corresponding slip step on it. Looking now at the front face, in isolation, it is clear that there is an extra half-plane and a positive edge dislocation in the top part of the crystal which emerges, say, at E. However, once again this dislocation does not extend to the appropriate point on the opposite face, i.e. the back face. If a dislocation enters a crystal it must reappear somewhere because, by their nature, dislocations cannot just terminate inside a crystal. What happens in this case is that the two dislocations change direction inside the crystal and meet up to form a *quarter dislocation loop*. Thus, the same dislocation is pure edge at one end, pure screw at the other, and in between has a whole range of intermediate properties. It is difficult without the aid of a three-dimensional model to picture the structural distortions which must occur around that part of the dislocation which has a mixed edge and screw character.

The origin of dislocation loops is by no means well understood. One source, discussed later, is by means of clustering of vacancies on to a plane, followed by an inward collapse of the structure. This generates loops which are entirely inside the crystal. For purpose of illustration, an alternative source is shown in Fig. 5.19. To start a loop, all that is needed is a nick in one edge of the crystal; the displacement of a few atoms is then sufficient to create a small quarter loop (a). Once created, the loops can expand very easily (b). Usually loops do not expand symmetrically because edge dislocations move more easily and more rapidly than corresponding screw dislocations. The result is shown in (c); the edge component has sped across to the opposite, left-hand face and the slip process for the front few layers of the crystal is complete. The dislocation which remains is mainly or entirely screw in character and this continues to move slowly to the back of the crystal, thus terminating the slip process for this dislocation.

As a consequence of the very easy generation of dislocation loops, the mechanical strength of materials, especially metals, is greatly reduced. This is obviously very serious for metals which are used in construction. The impact resistance of metals to high stresses is easy to measure because the experiments can be carried out rapidly. However, much more difficult to assess is the resistance over a long period of time to much smaller stresses. These small stresses

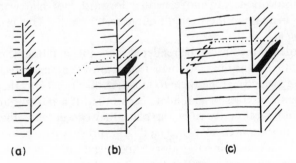

(a)　　　　　　(b)　　　　　　(c)

Fig. 5.19 Generation and motion of a dislocation loop

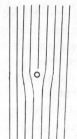

Fig. 5.20 Locking of an edge dislocation at an impurity atom

may be sufficient to generate and move dislocations only very slowly but the results are cumulative and, generally, non-reversible, so that one day catastrophic failure of the metal may occur, perhaps for no apparent reason.

As well as dislocations being a great weakener of materials they can also have the reverse effect and greatly increase the strength or hardness. One process that occurs is the 'locking' of dislocations at certain impurity atoms (Fig. 5.20), e.g. interstitial carbon atoms dissolved in iron. The dislocation moves freely until it encounters the impurity atom and then effectively becomes trapped by the impurity atom(s).

A very important strengthening process is that which occurs in the work-hardening of metals. On hammering a metal, an enormous number of disloc-ations are generated which in a polycrystalline material are in a large number of orientations. These start to move through the crystal but sooner or later, depending on the metal and its crystal structure, movement stops. Grain boundaries provide an effective means of stopping dislocations inside the grains. This is because, as a dislocation passes out of a grain, the surface of the grain becomes deformed and the resulting stress imposed upon neighbouring grains may act to prevent the dislocation from ever reaching the surface. Because the area around a dislocation is stressed, two dislocations may repel each other if they get too close. Thus, once the leading dislocations become trapped at grain boundaries or against dislocations arriving from other directions, the succeeding dislocations pile up. A log-jam of dislocations then forms rapidly and the individual dislocations can move neither forwards nor backwards, so that a considerable increase in the strength of the metal results. This process is called *strain-hardening*.

Strain-hardened metals can be rendered malleable and ductile once again by high-temperature annealing. At high temperatures, atoms have considerable thermal energy which enables them to move. Dislocations may therefore be able to reorganize themselves or annihilate each other. If a positive and a negative edge dislocation meet on the same slip plane they cancel each other out, leaving behind a strain-free area in the crystal. The process can be very rapid; e.g. platinum crucibles used in laboratory experiments may be softened in a few minutes by placing them at, say, 1200 °C.

Observation of dislocations

Optical and electron microscopy are by far the most important methods for the study of dislocations. Using the technique of lattice imaging, dislocations may be seen directly in electron micrographs of thin crystals. An older procedure is to etch the surface of the material under study and view the resulting etch patterns by optical microscopy.

Much of the early work on dislocations (in the 1950s) was done by metallurgists using crystals of LiF and the technique of etching followed by examination with optical microscopy. Metals are difficult to study by this technique because it is not usually possible to prepare metals that are free from dislocations. LiF is an excellent material to study, however, because it can be prepared relatively free from dislocations, in single crystal form and, on the application of moderate stresses, dislocations can be introduced in a controlled manner. The dislocations are observed, indirectly, by etching the LiF crystals in H_2O_2 solution, in which LiF is normally only very sparingly soluble. At the sites of emergent dislocations on the crystal surface, the crystal is in an abnormally stressed condition, the atoms or ions are not located in deep free energy minima and this region of the crystal dissolves rapidly in H_2O_2. This results in the formation of deep, pointed etch pits, in the form of inverted pyramids. The method is rather destructive because, on an atomic scale, enormous holes are created just so that one tiny dislocation can be observed. However, it clearly demonstrates the effect of dislocations on the reactivity, in this case solubility, of crystals. If the pits are etched to a size of $10 \, \mu m$ at their base (i.e. at the surface of the crystal) the dislocations must not be closer than $10 \, \mu m$ (i.e. $10^5 \, \text{Å}$) in order for the etch pits to be seen in isolation. During slip, dislocations may be much closer than this, in which case a continuous linear etch pit may form. The pits occur in lines or sets of lines because of the repeated generation of dislocations at certain surface sites.

With this combined approach of etching followed by microscopic examination, the number and distribution of dislocations in a crystal can be assessed. This also gives an indirect, yet sensitive method of studying the movement of dislocations. Etch pits have pointed bottoms only as long as there is an emergent dislocation at the bottom of the pit. If the dislocation moves away and etching is continued, the pit continues to grow, but only laterally; flat-bottomed pits result. In favourable cases, it is therefore possible to study the migration of dislocations under the action of an applied stress by periodically releasing the stress, re-etching the crystal and observing the dislocation network from the pattern of etch pits; new dislocation sites can be distinguished from old ones by the shape of the pits. One interesting result is that, in crystals that have been studied in this way, edge dislocations appear to move much more rapidly than do screw dislocations.

An overall picture of the presence of dislocations may be obtained by viewing the crystal between crossed polars in the optical microscope. LiF is cubic and therefore dark in crossed polars. The presence of dislocations caused by a compressive stress distorts the structure from cubic symmetry in the vicinity of the dislocations and this shows up as *stress birefringence*. The dislocations in LiF are concentrated on the planes {110} and {100} and this gives a criss-cross or tartan appearance to the stress birefringence pattern.

Dislocations and crystal structure

For a given crystal structure there is usually a preferred plane or set of planes on which dislocations can occur and also preferred directions for dislocation motion. The energy, E, that is required to move a dislocation by one unit of translation is proportional to the square of the magnitude of the Burgers vector, **b**, i.e.

$$E \propto |\mathbf{b}|^2 \qquad (5.23)$$

Thus, the dislocations that generally are most important in a particular material are those that have the smallest **b** value.

In metals, the direction of motion of dislocations is usually parallel to one of the directions of close packing in the metal crystal structure. In Fig. 5.21(a), two rows of spheres are shown and each represents a layer of atoms in projection; the rows are close packed because the individual atoms of each row are touching neighbouring atoms in the same row. Under the action of a shearing stress and with the aid of a dislocaton (not shown) the top row shears over the bottom row by one unit. After shear, (b), atom 1′ occupies the position that was previously occupied by atom 2′ in (a), and so on. The distance 1′–2′ corresponds to the unit of translation for slip and for dislocation motion. It is equivalent to the magnitude of the Burgers vector, i.e. $|\mathbf{b}'| = d$, where d is the diameter of the spheres. Therefore $E_b \propto d^2$.

An example of dislocation motion in a non-close packed direction is shown in (c) and (d). The distance 1′–2′ is now longer than in (a) and (b); by a simple geometrical construction, it can be shown that $|\mathbf{b}''| = d\sqrt{2}$. Therefore, $E_b'' \propto 2d^2$.

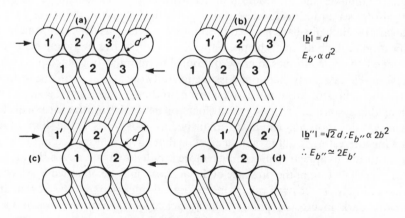

Fig. 5.21 Burgers vector in (a), (b) close packed and (c), (d) non-close packed directions in a crystal

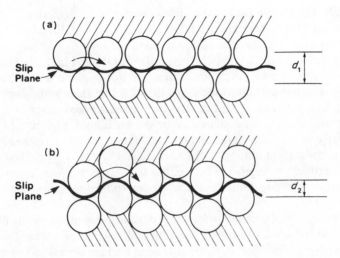

Fig. 5.22 Slip occurs more easily if the slip plane is a close packed plane, as in (a)

Consequently, motion of dislocations in a close packed direction requires only half the energy needed for motion in the particular non-close packed direction shown in (c).

It has been shown that the preferred direction of motion of dislocations is parallel to close packed directions. It can also be shown that the preferred plane of shear or slip is usually a close packed plane. This is because the distance of separation between two close packed planes, d_1, is greater than the corresponding distance between two non-close packed planes, d_2 (Fig. 5.22). During shear, the atoms to either side of the slip plane move, relatively, in opposite directions and the distortion of the structure for intermediate or saddle positions is much less in (a) than in (b). Alternatively, the energy barrier to be surmounted by an atom in moving from one position to an equivalent one (represented by the curved arrows) is much greater for example (b).

Mechanical properties of metals

Metals which are face centred cubic—Cu, Ag, Au, Pt, Pb, Ni, Al, etc.—are generally more malleable and ductile than either hexagonal close packed—Ti, Zr, Be—or body centred cubic metals—W, V, Mo, Cr, Fe—although there are notable exceptions much as Mg (h.c.p.) and Nb(b.c.c.) which are malleable and ductile. Many factors influence malleability and ductility. In part, malleability and ductility depend on the numbers of close packed planes and directions possessed by a structure.

Face centred cubic metals have four different sets of close packed planes which are perpendicular to the body diagonals of the cubic unit cell (Fig. 1.16). Each close packed layer possesses three close packed directions, X–X', Y–Y', Z–Z'

236

(Fig. 1.12). These close packed directions are the face diagonals of the cube and have indices [110]; there are a total of six close packed directions.

A hexagonal close packed structure contains only one set of close packed layers, which is parallel to the basal plane of the unit cell (Fig. 1.17). There are also only three close packed directions; these are in the plane of the close packed layers. It is rather difficult to show in a drawing that there is only one set of c.p. layers in hexagonal close packing and the reader is asked either to take this on trust or to make a three-dimensional model for himself to verify it.

A body centred cubic unit cell does not contain any close packed layers. The atomic coordination number in b.c.c. is eight whereas close packed structures have a coordination number of twelve. The b.c.c. structure does, however, contain four close packed directions; these correspond to the four body diagonals of the cube, i.e. $\langle 111 \rangle$.

The behaviour of a metal under stress depends very much on the direction of the applied stress relative to the direction and orientation of the slip directions and slip planes. The effect of tensile stress on a single crystal rod of magnesium metal, whose crystallographic orientation is such that the basal plane of the hexagonal unit cell is at 45° to the rod axis, is shown in Fig. 5.23. For stresses below ~ 100 p.s.i., the crystal undergoes *elastic deformation*, i.e. no permanent elongation occurs. Above the *yield point*, ~ 100 p.s.i., *plastic flow* begins to occur and the crystal suffers an irreversible elongation. Magnesium metal is quite

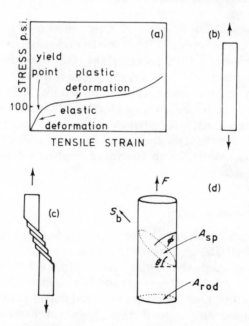

Fig. 5.23 (a) Tensile stress–strain curve for single crystal Mg; (b), (c) tensile stress resulting in slip and elongation; (d) definition of the resolved shear stress

remarkable in that it can be stretched out to an elongation of up to several times its original length. How this can occur is shown schematically in (c). Slip occurs on a massive scale and the resulting slip steps are so large that they can be observed directly with an optical microscope. If the passage of a dislocation gives rise to a slip step that is, say, 2 Å high (or wide), then in order that such features are visible in the microscope the slip steps must have dimensions of at least 2 μm. At least 10,000 dislocations must therefore have passed on the slip plane for every slip step that can be seen.

If the basal planes of magnesium are oriented at other than 45° to the rod axis, it is found experimentally that a larger applied stress is needed in order for plastic flow to occur. This is because the important factor is the magnitude of the applied shear stress when resolved parallel to the basal plane. This is shown as follows (Fig. 5.23d).

For an applied force, F, the stress S_A, on the slip plane is given by

$$S_A = F/A_{sp} = F(\cos \theta/A_{rod}) \tag{5.24}$$

where A_{rod} is the cross-sectional area of the rod, A_{sp} is the area of the slip plane and θ is the angle between A_{rod} and A_{sp}. The stress S_A on the slip plane has a component S_B parallel to the direction of slip in the slip plane, given by

$$S_B = S_A \cos \phi = (F/A_{rod}) \cos \theta \cos \phi \tag{5.25}$$

where ϕ is the angle between the slip direction and the stress axis. S_B is the resolved shear stress and the value which is needed to cause plastic flow is the *critical resolved shear stress*. From the interrelation that exists between θ and ϕ, the maximum value of S_B occurs when $\theta = \phi = 45°$, i.e.

$$S_B = \tfrac{1}{2}(F/A_{rod}) = \tfrac{1}{2}S \quad \text{for } \theta = \phi = 45° \tag{5.26}$$

where S is the stress applied to the crystal. It follows from equation (5.25) that when the slip plane is either perpendicular or parallel to the direction of applied stress, the resolved shear stress is zero and therefore slip cannot occur.

The ability of face centred cubic metals to undergo severe plastic deformation in contrast to hexagonal and especially body centred cubic metals can now be explained qualitatively. Face centred cubic metals have four close packed planes and six close packed directions which are suitable for slip. For any given applied stress, at least one and usually more of these slip planes and directions is suitably oriented, relative to the direction of applied stress, such that slip can occur. In contrast, hexagonal metals slip easily only when in a certain range of orientations relative to the applied stress. The difference between these groups of metals shows up markedly in the mechanical properties of polycrystalline pieces of metal. Polycrystalline hexagonal metals almost certainly contain grains whose slip planes are either parallel or perpendicular to the direction of applied stress. The amount of plastic deformation that is possible in these is therefore more limited. On the other hand, face centred cubic metals are malleable and ductile whether in the form of single crystals or polycrystalline pieces.

The critical resolved shear stresses for face centred cubic metals are usually small, e.g. in pounds per square inch: Cu 92, Ag 54, Au 132 and Al 148. Values for the hexagonal metals fall into two groups. One group is similar to values for the face centred cubic metals, e.g. Zn 26, Cd 82, Mg 63. The other group has much higher values, e.g. Be 5,700, Ti 16,000 (slip in titanium occurs more easily on a non-basal plane, but one which has a close packed direction; it has a critical resolved shear stress of 7,100 p.s.i.) The reason for the much higher values of Be, Ti (and Zr) probably has something to do with the fact that the unit cell dimensions of these metals show that the structures are somewhat compressed in the c direction (perpendicular to the basal plane). An ideal hexagonal unit cell has a $c:a$ ratio of 1.632; values of $c:a$ for Zn, Cd and Mg are 1.856, 1.886 and 1.624, respectively, and are similar to or greater than the ideal value. Values for Be, Ti and Zr are 1.586, 1.588 and 1.590, respectively, and are all consistently *less* than the ideal c/a ratio. Thus, it would seem that this decrease in distance between adjacent basal planes makes it much more difficult for basal slip to occur.

The critical resolved shear stress for body centred metals is also high, e.g. α-Fe $\simeq 4000$ p.s.i., because, although they possess four close packed directions, they do not have any closed packed planes.

Dislocations, vacancies and stacking faults

Dislocations are usually of most interest to people concerned with the mechanical properties of materials. There is a close relation, however, between dislocations and other defects such as point defects (vacancies) and plane defects (stacking faults), as shown by the following examples.

The process of *climb* is a special mechanism of dislocation motion which involves vacancies, as shown in Fig. 5.24. Suppose that near to the line of an edge dislocation, which is in fact an extra half-plane of atoms, there is a vacancy. If an atom can move from the dislocation line into this vacancy, then effectively, the extra half-plane has become one atom shorter at that point. If this process is repeated along the length of the dislocation line, then the half-plane effectively is beginning to climb out of the crystal.

An intimate structural relation exists between dislocations and stacking faults. Two adjacent layers of a face centred cubic metal are shown in projection in Fig. 5.25 (a). The upper layer (white circles) has a zigzag row of atoms missing.

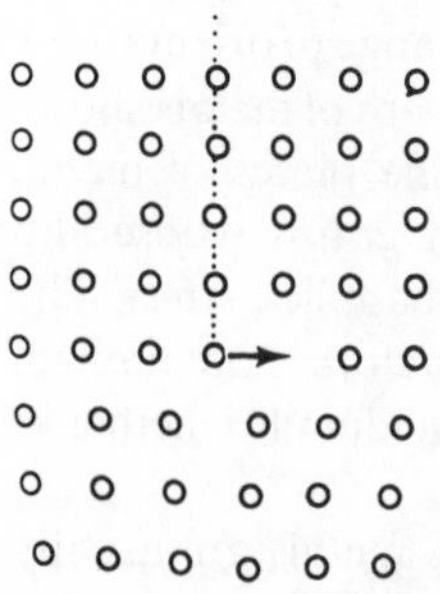

Fig. 5.24 Climb of an edge dislocation by vacancy migration

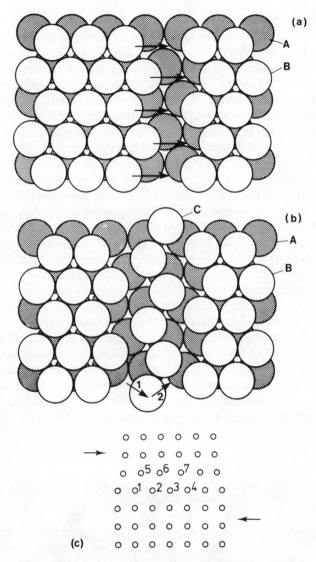

Fig. 5.25 (a) An edge dislocation and (b) a partial dislocation in a face centred cubic metal; (c) the edge dislocation in projection at atom 3

This may alternatively be regarded as an extra row of atoms or negative edge dislocation in the lower layer (shaded circles), immediately below the row of missing atoms in the upper layer. Looked at from this latter point of view, the presence of the edge dislocation in the lower layer effectively forces apart the atoms in the upper layer in the vicinity of the dislocation line, to the extent that a row of atoms in it appears to be missing. This same effect is shown from a different angle in (c). The row containing atoms 1–4 corresponds to the lower layer in (a)

240

and that containing atoms 5–7 corresponds to the upper layer. The negative edge dislocation is effectively represented by atom 3 in (c). Motion of the edge dislocation is shown arrowed in (a) and (c) and effectively involves one row of white circles in the upper layer moving to occupy the row of missing atoms. This movement to the right by one unit effectively causes the negative edge dislocation to move to the left by one unit. The stacking sequence of the layers is unchanged by passage of the dislocation, since atoms in the upper layer occupy B sites before and after slip. The Burgers vector is given by the direction of the arrows and has magnitude approximately equal to one atomic diameter.

The atomic motions arrowed in (a) are relatively difficult to carry out because atoms in the upper layer virtually have to climb over the top of atoms in the layer below in order to follow the direction indicated by the arrows. A much easier alternative route is shown in (b) in which the migration is divided into two smaller steps (arrows 1 and 2). For each of these, the B atoms traverse a low pass between two A atoms; after the first pass (arrow 1), the atoms find themselves in a C position; a single row of atoms in C positions is shown in (b). This process is repeated (arrow 2) and after the second pass the atoms enter the adjacent set of B positions. In this manner a dislocation is divided into two *partial dislocations*. The formation of partial dislocations can be regarded from two viewpoints. In terms of the energy barriers to be surmounted, two small passes are easier than one big hump, even though the combined distance is longer. In terms of the Burgers vector for the two alternatives, the longer, divided pathway is favoured, provided the distance is not too much longer. Thus, if the direct jump is across a distance b, the energy E_1 is given by $E_1 \propto |b|^2$. For each partial jump, the distance involved is $b/\sqrt{3}$, because the direction is at 30° to the horizontal. Thus for the two partial jumps the combined energy E_2 is proportional to $(b/\sqrt{3})^2 + (b/\sqrt{3})^2$, i.e.

$$E_2 \propto \tfrac{2}{3}|b|^2$$

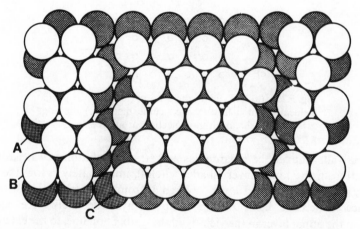

Fig. 5.26 Separation of two partial dislocations to give a stacking fault

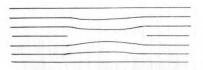

Fig. 5.27 Collapse of the structure around a cluster of vacancies, thereby generating a dislocation loop

Therefore,

$$E_2 < E_1$$

Separation of two partial dislocations may occur as in Fig. 7.26, due to their mutual repulsion. In between the two partial dislocations the stacking sequence is 'wrong' because all the white atoms occupy C positions; we now have the makings of a *stacking fault*. If the two partial dislocations can be encouraged to go to the edges of the crystal the whole layer will be faulty. Whether or not this can happen in practice is unknown but, nevertheless, a clear relationship exists between dislocations, partial dislocations and stacking faults.

One suggested way by which dislocations are generated involves the clustering of vacancies onto a certain plane in a structure followed by an inward collapse of the structure, thereby generating a dislocation loop. This mechanism would be important only at high temperatures where vacancies both form readily and are relatively mobile. If, in a pure metal, vacancies cluster onto a certain plane in the structure such that every atomic site within, say, a certain radius is empty, then a disc-shaped hole would be present in the structure. The two sides of the disc would then cave in, as shown in cross-section in Fig. 5.27, and it can be seen that the two opposite ends have the appearance of edge dislocations of opposite sign. This therefore gives a means of generating dislocations that has nothing to do with mechanical stresses.

Dislocations and grain boundaries

An elegant relationship exists between dislocations and grain boundaries such that the interface between two grains in a polycrystalline material may be regarded as a dislocation network, provided the angular difference in orientation of the grains is not too large. In Fig. 5.28 are shown six positive edge dislocations at different heights in a crystal. With each one, the top half of the crystal becomes slightly wider than the bottom half and the effect of having several such dislocations in a similar orientation and position is to introduce an angular misorientation between the two halves of the crystal. Thus the surface perpendicular to the paper on which these dislocations terminate can be regarded as an interface or boundary between the left-hand and right-hand grains, although the continuity in the structures of the crystals across this boundary is generally excellent. This type of interface is known as a *low angle grain boundary*. The mosaic or domain texture of single crystals (Fig. 5.13) can often be described in terms of low angle grain boundaries.

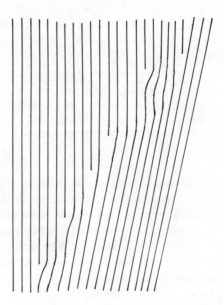

Fig. 5.28 Array of edge dislocations which constitutes a low angle grain boundary

The angular misorientation of two grains can be calculated readily. Suppose that in a piece of crystal which is say 1 μm (i.e. 10,000 Å) wide there are five positive edge dislocations forming a low angle grain boundary; therefore one end of the crystal will be five atomic planes, say 10 Å, wider than the other end. The angular misorientation of the grains, $\theta = \tan^{-1}(10/10,000) = 0.057°$. Low angle grain boundaries can sometimes be revealed by appropriate etching of the surface, although for the pits to be seen in isolation in the optical microscope they must be at least 1 to 2 μm in size and separated by the same distance. The dislocation model for grain boundaries is useful for angular misorientations of up to several degrees. For misorientations > 10 to $20°$, however, the model is probably no longer meaningful because the edge dislocations would have to be spaced too closely and would lose their separate identity.

Solid solutions

Solid solutions are very common in crystalline materials. A solid solution is basically a crystalline phase that can have variable composition. Simple solid solution series are one of two types: in *substitutional* solid solutions, the atom or ion that is being introduced directly replaces an atom or ion of the same charge in the parent structure; in *interstitial* solid solutions, the introduced species occupies a site that is normally empty in the crystal structure and no ions or atoms are left out. Starting with these two basic types, a considerable variety of more complex solid solution mechanisms may be derived, by having both substitution and interstitial formation occurring together and/or by introducing ions of different charge to those in the host structure. Let us first look in some detail at the simple solid solutions.

Substitutional solid solutions

An example of a substitutional solid solution is the series of oxides formed on reacting together Al_2O_3 and Cr_2O_3 at high temperatures. Both of these end-member phases have the corundum crystal structure (approximately hexagonal close packed oxide ions with Al^{3+}, Cr^{3+} ions occupying two-thirds of the available octahedral sites) and the solid solution may be formulated as $(Al_{2-x}Cr_x)O_3$: $0 \leqslant x \leqslant 2$. At intermediate values of x, Al^{3+} and Cr^{3+} ions are distributed at random over those octahedral sites that are normally occupied by Al^{3+} in Al_2O_3. Thus, while any particular site must contain either a Cr^{3+} or an Al^{3+} ion, the probability that it is one or the other is related to the composition x. When the structure is considered as a whole and the occupancy of all the sites is averaged out, it is useful to think of each site as being occupied by an 'average cation' whose properties, atomic number, size, etc., are intermediate between those of Al^{3+} and Cr^{3+}.

For a range of simple substitutional solid solutions to form, there are certain requirements that must be met. First, the ions that replace each other must have the same charge. If this were not the case, then other structural changes, involving, e.g. vacancies or interstitials, would be required to maintain electroneutrality. Many examples in which these latter effects occur are known, but such cases are more complex than the simple substitutional solid solutions with which we are currently concerned.

Second, the ions that are replacing each other must be fairly similar in size. From a review of the experimental results on metal alloy formation, it has been suggested that a difference of 15 per cent in the radii of the metal atoms that replace each other is the most that can be tolerated if a substantial range of substitutional solid solutions is to form. For solid solutions in non-metallic systems, the limiting difference in size that is acceptable appears to be somewhat larger than 15 per cent, although it is very difficult to quantify this. To a certain extent, this is because it is difficult to quantify the sizes of the ions themselves (see Chapter 2 for a discussion of ionic radii), but also because solid solution formation is very temperature dependent. Thus, in oxide systems, for example, extensive solid solutions often form at high temperatures whereas at lower temperatures, these may be much more restricted or practically non-existent.

There is a simple thermodynamic explanation for such variations with temperature. For a solid solution to form and be stable, it must have a lower free energy than the corresponding mixture of phases without any solid solution formation. Free energy may be separated into its enthalpy and entropy components according to equation (5.2); let us consider the effect on each component of solid solution formation. The entropy term always favours solid solution formation. This is because the entropy of a solid solution, in which say two types of cation are distributed at random over a set of sites in the crystal structure, is much greater than the entropy of the corresponding mixture of two essentially stoichiometric component phases. The enthalpy term may or may not favour solid solution formation. If the enthalpy of formation is negative, then solid solution formation is favoured by both enthalpy and entropy terms and the solid solutions form and are stable at all temperatures. If, however, the enthalpy of formation is positive, then the enthalpy and entropy effects are opposed; in such cases, the enthalpy term is likely to dominate the free energy at low temperatures and extensive solid solutions do not form. With increasing temperature, the $T\Delta S$ term increases in magnitude until such a temperature that it more than compensates for the positive enthalpy. Consequently, solid solutions are favoured by high temperatures.

Let us now see some examples of ions which can and cannot replace each other in solid solution formation. We need to use a consistent set of radii in order to make comparisons; using Shannon and Prewitt radii for octahedral coordination, based on an oxide ion radius of 1.26 Å, monovalent ions have radii (in Å): Li, 0.88; Na, 1.16; Ag, 1.29; K, 1.52; Rb, 1.63; Cs, 1.84. The radii of K/Rb and Rb/Cs pairs are both within 15 per cent of each other and it is common to get solid solutions between say, pairs of corresponding Rb/Cs salts. However, Na/K salts also sometimes form solid solutions with each other, especially at high temperatures (e.g. KCl and NaCl at 600 °C) and the K^+ ion is ~ 30 per cent larger than Na^+. The difference in size of Li and K appears to be too large, however, and these ions do not generally replace each other in solid solutions. The Ag^+ ion is similar in size to Na^+ and solid solution between pairs of corresponding Na/Ag salts is common.

Turning now to divalent ions, with octahedral radii: Mg, 0.86; Ca, 1.14; Sr, 1.30; Ba, 1.50; Mn, 0.96 (high spin); Fe, 0.91 (high spin); Co, 0.88 (high spin); Ni, 0.84; Cu, 0.87; Zn, 0.89; Cd, 1.09. It is common to find solid solutions in which the divalent transition ions, Mn...Zn, substitute for each other and we can see that their radii are not too different. Mg also commonly forms solid solutions with them and is of similar size; Ca does not, however, but it is some 20–30 per cent larger than these transition metal ions. Similar considerations apply to trivalent ions; similar-sized Al, Ga, Fe and Cr (0.67–0.76 Å) commonly substitute for each other, as do many of the trivalent lanthanides (0.99–1.20 Å).

In summary, then, ions of very similar size (e.g. Zr, 0.86; Hf, 0.85) substitute for each other easily and extensive solid solutions form which are stable at all temperatures; the enthalpy of mixing of such similar-sized ions is likely to be small and the driving force for solid solution formation is the increased entropy. With ions that differ in size by 15–20 per cent, solid solutions may form, especially at high temperatures where the entropy of solid solution formation is able to offset the significant positive enthalpy term. With ions that differ in size by more than ~ 30 per cent, solid solutions are not expected to form to any significant extent.

In considering whether or not solid solutions form, an important factor is the crystal structure of the two end members. In systems that exhibit complete ranges of solid solution, it is clearly essential that the two end members be isostructural. The reverse is not necessarily true, however, and just because two phases are isostructural, it does not follow that they will form solid solutions with each other; e.g. LiF and CaO both have the rock salt structure but they do not form solid solutions.

While complete ranges of solid solutions form in favourable cases, as for example between Al_2O_3 and Cr_2O_3, it is far more common to have only partial or limited ranges of solid solutions. In such cases, it is no longer necessary that the end members be isostructural. For example, the silicates, Mg_2SiO_4 (forsterite, an olivine) and Zn_2SiO_4 (willemite) can each dissolve about 20 per cent of the other one in solid solution formation, even though their crystal structures are quite different. In the forsterite solid solutions, whose formula may be written

$$Mg_{2-x}Zn_xSiO_4$$

some of the octahedrally coordinated magnesium ions are replaced by zinc. In the willemite solid solutions on the other hand, of formula

$$Zn_{2-x}Mg_xSiO_4$$

magnesium ions substitute partially for the tetrahedrally coordinated zinc ions. Such solid solutions are possible because Mg and Zn are of similar size and are happy in both tetrahedral and octahedral coordination.

Some ions, especially transition metal ions, have strong preference for a particular kind of site symmetry or coordination. An example is Cr^{3+} which is almost always found in octahedral sites, whereas similar-sized Al^{3+} can occupy either tetrahedral or octahedral sites. Such site preferences can preclude the formation of certain solid solutions. For example, $LiCrO_2$ contains octahedral Cr^{3+} and forms an extensive range of solid solutions, of formula

$$LiCr_{1-x}Al_xO_2$$

in which Al partially replaces Cr on the octahedral sites. The converse solid solution does not form, however, since in $LiAlO_2$ the coordination of Al is tetrahedral and Cr has a strong dislike for tetrahedral coordination.

Many other types of atom or ion may replace each other to form substitutional solid solutions. Silicates and germanates are often isostructural and form solid solutions with each other by $Si^{4+} \rightleftharpoons Ge^{4+}$ replacement. The lanthanide elements, because of their similarity in size, are notoriously good at forming solid solutions with each other in, say, their oxides. Indeed, one cause of the great difficulty experienced by the early chemists in trying to separate the lanthanides was this very easy solid solution formation. Anions may also replace each other in substitutional solid solutions, e.g. AgCl–AgBr solid solutions, but these are not nearly as common as the solid solutions formed by cation substitution, probably because there are not many pairs of anions that have similar size and coordination/bonding requirements. Many alloys are nothing more than substitutional solid solutions, e.g. in brass, copper and zinc atoms replace each other over a wide range of compositions, with general formula $Cu_{1-x}Zn_x$.

Interstitial solid solutions

Many metals form interstitial solid solutions in which small atoms, e.g. hydrogen, carbon, boron, nitrogen, etc., enter empty interstitial sites within the host structure of the metal. Palladium metal is well known for its ability to 'occlude' enormous volumes of hydrogen gas and the product hydride is an interstitial solid solution of formula PdH_x: $0 \leqslant x \lesssim 0.7$, in which hydrogen atoms occupy interstitial sites within the face centred cubic palladium metal structure. There is still uncertainty as to whether hydrogen is in octahedral or tetrahedral holes and it appears that the sites occupied may depend on the composition x.

Possibly the technologically most important interstitial solid solution is that of carbon in the octahedral sites of face centred cubic γ-Fe. This solid solution is the starting point for the manufacture of steel. It is useful to consider why carbon dissolves in γ-Fe but not in the body centred cubic α polymorph, since this illustrates structural aspects of interstitial solid solution formation.

Iron exists in three polymorphic forms: body centred cubic α, stable below 910 °C: face centred cubic γ, stable between 910 and 1400 °C; and body centred cubic (again!) δ, stable between 1400 °C and the melting point 1534 °C. γ-Iron can dissolve appreciable amounts of carbon, up to 2.06 wt%, in solid solution formation, whereas the α and δ forms dissolve very much less carbon, up to a maximum of 0.02 and 0.1 wt%, respectively.

There is a simple explanation for the very different solubilities of carbon in the γ and α polymorphs of iron. Although the face centred cubic γ structure is more densely packed than the body centred cubic α structure, the interstitial holes (suitable for occupation by carbon) are larger although much less numerous in γ-Fe. Unit cells of the two forms are shown in Fig. 5.29 together with the octahedral sites that are available for occupation by carbon. These sites are at the cube face centres in α-Fe and at the body centre in γ-Fe. The iron–carbon distances and, hence, the sizes of the interstitial sites are considerably larger in γ-Fe than in α-Fe.

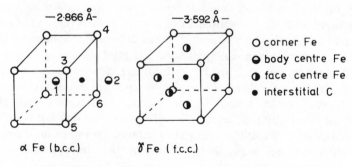

Fig. 5.29 Interstitial sites for carbon in α-Fe and γ-Fe

In α-Fe, these sites are also grossly distorted, as shown by the following calculation. The cubic unit cell edge in α-Fe is 2.866 Å. Consequently, two Fe–C distances, involving Fe atoms 1 and 2 in the centres of adjacent cells (Fig. 5.29a) would be $a/2 = 1.433$ Å. The other four Fe–C distances, involving Fe atoms 3–6 at the corners of the cell face, would be equal to $a\sqrt{2}/2 = 2.03$ Å. In γ-Fe, by contrast, (b), the octahedral sites are undistorted and the Fe–C distance is equal to $a/2 = 1.796$ Å. This is the value calculated from the room temperature a value for γ-Fe. The bond length at 900 °C is likely to be a few percent greater which is then comparable to expected Fe–C bond distances, as judged by values in the range 1.89–2.15 Å in for instance, Fe_3C. The two shortest distances of 1.433 Å calculated in α-Fe are prohibitively short, therefore, and make this an unattractive site for an interstitial carbon atom.

More complex solid solution mechanisms

Consider now what happens when cation substitution occurs but the two cations are of different charge. There are four possibilities, summarized in Fig. 5.30. A similar scheme is possible for anion substitution but is not considered further because anion substitution occurs rather infrequently in solid solutions.

1. Creating action vacancies

If the replaceable cation of the host structure has a lower charge than that of the cation which is replacing it, additional changes are needed to preserve electroneutrality. One way is to create cation vacancies by leaving out more cations of the host structure. For example, NaCl is able to dissolve a small amount of $CaCl_2$ and the mechanism of solid solution formation involves the replacement of two Na^+ ions by one Ca^{2+} ion; one Na^+ site therefore becomes vacant. The formula of this solid solution may be written

$$Na_{1-2x}Ca_xV_xCl$$

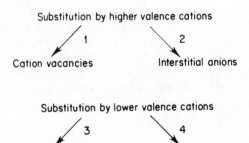

Fig. 5.30 Various complex solid solution mechanisms involving the substitution of aliovalent cations

in which V represents a cation vacancy. It is found experimentally that at, e.g. 600 °C, $0 < x < 0.15$. In such solid solutions, the Ca^{2+} ions, the cation vacancies and the host Na^+ ions are all disordered over the octahedral sites of the rock salt structure. In practice, there may be some local ordering of the defects since the substituted Ca^{2+} ion, with an effective excess site charge of $+1$, will be attracted to Na^+ vacancies with their effective site charges of -1, in much the same way that Schottky defects tend to associate into pairs.

There are countless other examples of similar solid solutions. Size constraints, as discussed above for simple substitutional solid solutions, again apply, but it is difficult to make quantitative rules and predictions as to what solid solutions may or may not form. Other charge combinations for the substituting ions also occur. Thus, divalent ions may be substituted by trivalent ones in e.g. spinel, $MgAl_2O_4$. In this, Mg^{2+} ions on tetrahedral sites are replaced by Al^{3+} ions, in the number ratio, 3:2. Consequently, the solid solution formula may be written

$$Mg_{1-3x}Al_{2+2x}O_4$$

in which x tetrahedral site vacancies are also created.

Many transition metal compounds are non-stoichiometric and may exist over a range of compositions because the transition metal ion is present in more than one oxidation state. This can give rise to a range of solid solutions as in e.g. wüstite, $Fe_{1-x}O$, which was discussed earlier in this chapter and which contains a mixture of Fe^{2+}, Fe^{3+} ions and cation vacancies. A similar example is provided by nickel oxide, NiO, which when heated in oxygen, picks up oxygen to give a solid solution of formula, $Ni_{1-x}O$. This contains a mixture of Ni^{2+}, Ni^{3+} ions and cation vacancies disordered over the octahedral sites of the rock salt structure, with a general formula

$$Ni^{2+}_{1-3x}Ni^{3+}_{2x}O$$

2. Creating interstitial anions

The other mechanism by which a cation of higher charge may substitute for one of lower charge is to create, at the same time, interstitial anions. This mechanism is not nearly as common as the one discussed above involving cation vacancies, mainly because most structures do not have interstitial sites that are large enough to accommodate extra anions. It does appear to be favoured by the fluorite structure in certain cases, however. For example, calcium fluoride can dissolve small amounts of yttrium fluoride. The total number of cations remains constant with Ca^{2+}, Y^{3+} ions disordered over the calcium sites. In order to retain charge balance, fluoride interstitials are created to give the solid solution formula

$$Ca_{1-x}Y_xF_{2+x}$$

These extra fluoride ions occupy large sites which are surrounded by eight other fluoride ions at the corners of a cube (Fig. 1.25).

Another fluorite-related structure capable of containing extra interstitial anions is UO_2. On oxidation it forms a solid solution phase, UO_{2+x}, which contains interstitial oxide ions and the formula of which may be given as

$$U^{4+}_{1-x}U^{6+}_x O_{2+x}$$

Careful structural studies have shown that the defect structure of this material is in fact rather complex and involves defect clusters such as shown in Fig. 5.10, rather than randomly distributed interstitial oxide ions.

3. Creating anion vacancies

If the replaceable ion of the host structure has a higher charge than that of the replacing cation, charge balance may be maintained by creating either anion vacancies or interstitial cations. The most well-known examples of anion vacancies occur again with the fluorite structure and in particular with oxides such as zirconia, ZrO_2. For instance, anion vacancies occur in cubic, lime-stabilized zirconia of formula

$$Zr_{1-x}Ca_x O_{2-x}$$

Experimentally, $0.1 < x < 0.2$. In these solid solutions, the total number of cations remains constant and Ca^{2+}, Zr^{4+} ions are disordered over the eight-coordinate Zr sites in the fluorite structure. In order to retain charge balance, oxide ion vacancies are created. These materials are very important in modern technology, both as engineering ceramics and as oxide ion conducting solid electrolytes (Chapter 7).

4. Creating interstitial cations

The alternative mechanism to the preceding one is to create interstitial cations at the same time that a cation of lower charge substitutes for one of higher charge. This is a common substitution mechanism and occurs provided the host structure has suitably sized interstitial sites to accommodate the extra cations. Good examples, although rather complex structurally, are the various 'stuffed silica' phases. These are aluminosilicates in which the structure of one of the three polymorphs of silica—quartz, tridymite or cristobalite—may be modified by partial replacement of Si^{4+} by Al^{3+} and, at the same time, alkali metal cations enter normally empty interstitial holes in the silica framework.

Stuffed quartz structures have formulae such as $Li_x(Si_{1-x}Al_x)O_2$ for $0 < x \lesssim 0.5$; special compositions exist at $x = 0.5$ ($LiAlSiO_4$, eucryptite) and $x = 0.33$ ($LiAlSi_2O_6$, spodumene). β-spodumene has the unusual property of a very small, perhaps even slightly negative, coefficient of thermal expansion; ceramics containing β-spodumene as a major constituent are therefore dimensionally stable and resistant to thermal shock. As such, they find many high-temperature applications (Chapter 7). The interstitial holes in the quartz structure are too small to accommodate cations larger than Li^+. Tridymite and cristobalite have lower densities than quartz with larger interstices in their structures. Stuffed tridymite and stuffed cristobalite solid solutions, similar to the stuffed quartz solid solutions, form but in these the interstitial or stuffing cations are Na^+ and K^+.

A variety of other complex solid solution mechanisms occur, two of which are given next.

Double substitution

In such processes, two substitutions take place simultaneously. For example, in synthetic olivines, Mg^{2+} may be replaced by Fe^{2+} at the same time as Si^{4+} is replaced by Ge^{4+} to give solid solutions

$$(Mg_{2-x}Fe_x)(Si_{1-y}Ge_y)O_4.$$

Silver bromide and sodium chloride form a complete range of solid solutions in which both anions and cations replace each other,

$$(Ag_{1-x}Na_x)(Br_{1-y}Cl_y):0 < x, y < 1.$$

The substituting ions may also be of different charge, providing that overall electroneutrality prevails; e g. in the plagioclase feldspars a complete range of solid solutions forms between anorthite, $CaAl_2Si_2O_8$, and albite, $NaAlSi_3O_8$. Their formulae may be written

$$(Ca_{1-x}Na_x)(Al_{2-x}Si_{2+x})O_8:0 < x < 1$$

and the two substitutions $Na \rightleftharpoons Ca$ and $Si \rightleftharpoons Al$ must occur simultaneously and to the same extent.

Double substitutional processes occur in *sialons*, which are solid solutions in the system Si—Al—O—N and based on the Si_3N_4 parent structure. β-Silicon nitride is built of SiN_4 tetrahedra linked at their corners to form a 3-D network. Each nitrogen is in planar coordination and forms the corner of three SiN_4 tetrahedra. In the sialon solid solutions, Si^{4+} is partly replaced by Al^{3+} and N^{3-} is partly replaced by O^{2-}. In this way charge balance is retained. The structural units in the solid solutions are $(Si, Al)(O, N)_4$ tetrahedra and the solid solution mechanism may be written as

$$(Si_{3-x}Al_x)(N_{4-x}O_x)$$

Silicon nitride is potentially a very useful high-temperature ceramic. The discovery of sialon and its derivatives by Jack and coworkers at Newcastle has opened up a new field of crystal chemistry and increased the possible applications of nitrogen-based ceramics.

Further comments on the requirements for solid solution formation

The factors that govern whether or not solid solutions, especially the more complex ones, form are understood only qualitatively. For a given system, it is not usually possible to predict whether solid solutions will form or, if they do form, what is their compositional extent and their mechanism of formation. Instead, this has to be determined experimentally. If we are restricted to solid solutions that exist under equilibrium conditions (Fig. 5.1) and are represented on the appropriate phase diagram, then solid solutions form only if they have lower free energy than any other phase or assemblage of phases with the same overall composition. Under non-equilibrium conditions, however, and by using 'chemie douce' or other preparative techniques, it is often possible to prepare solid solutions that are much more extensive than those existing, if at all, under equilibrium conditions. A simple example is provided by the β-aluminas, $Na_2O \sim 8Al_2O_3$. Part or all of the Na^+ ions may be ion exchanged for a variety of other monovalent ions, including Li^+, K^+, Ag^+ and Cs^+, even though most of these ion exchanged materials are not thermodynamically stable (Chapter 7).

The limitations on the relative sizes of ions that either substitute directly for each other or enter interstitial sites have been referred to above. In the more complex solid solution mechanisms, the charges of the ions that take part are often different to each other. Clearly charge balance must be preserved overall, but within this limitation and provided the size requirements are met, there is often much scope for introducing ions of different charge. A rather extreme example of this is Li_2TiO_3. This has, at high temperatures, a rock salt structure in which Li^+, Ti^{4+} ions are disordered over the octahedral sites in a cubic close packed oxide ion array. It can form two series of solid solutions with either excess Li_2O or excess TiO_2 and of respective formulae

$$\text{(a)} \quad Li_{2+4x}Ti_{1-x}O_3: \quad 0 < x \lesssim 0.08$$
$$\text{(b)} \quad Li_{2-4x}Ti_{1+x}O_3: \quad 0 < x \lesssim 0.19.$$

Both of these involve the interchange of mono- and tetravalent ions with the creation of either interstitial Li^+ ions (a) or Li^+ vacancies (b) to maintain electroneutrality. The large difference in charge of Li^+ and Ti^{4+} obviously does not prevent solid solution formation. Part of the reason why solid solutions form appears to be that both Li^+ and Ti^{4+} are able to occupy similar-sized octahedral sites with metal–oxygen distances in the range 1.9 to 2.2 Å.

There are many other examples in which ions of similar size but very different charge can replace each other in solid solution formation. The ilmenite-like phase $LiNbO_3$, an important optoelectronic material, forms a limited range of solid solution by $5Li^+ \rightleftharpoons Nb^{5+}$ substitution on octahedral sites. Na^+ is of a similar size to Zr^{4+} and these ions replace each other on octahedral sites in the solid solution series, $Na_{5-4x}Zr_{1+x}P_3O_{12}: 0.04 < x < 0.15$.

Experimental methods for studying solid solutions

X-ray powder diffraction

There are two main ways in which powder diffraction may be used to study solid solutions. One is as a simple fingerprint method in which qualitative phase analysis is carried out. The objective is to determine the crystalline phases that are present in a sample without necessarily measuring the patterns very accurately. The second way is to measure the powder pattern accurately in order to obtain information about the composition of the solid solution. Usually, the unit cell undergoes a small contraction or expansion as the composition varies across a solid solution series and once a calibration graph of d-spacing or cell volume against composition has been drawn, the compositions of solid solutions may be obtained from an accurate measurement of their unit cell parameters or the d-spacings of certain lines in the powder X-ray pattern.

The use of the qualitative fingerprint method may be shown by reference to the phase diagram of $MgAl_2O_4$–Al_2O_3 (Fig. 5.31). The spinel solid solutions are much more extensive at 1800 °C than at 1000 °C, as shown by the *solvus*, which is the curve limiting the maximum compositional extent of the solid solutions. Let us do some imaginary experiments on a sample of composition 65 mol% Al_2O_3, 35 mol% MgO. According to the phase diagram, such a composition, under equilibrium conditions, would give a single phase, homogeneous spinel solid solution of the same composition (i.e. 65:35) above about 1550 °C. Below 1550 °C, two phases would be present, essentially stoichiometric Al_2O_3 and a spinel solid solution containing less than 65% Al_2O_3; e.g. at 1200 °C, the spinel solid solution composition is given by the position of the solvus at 1200 °C and is about 55% Al_2O_3. The relative amounts of the two phases, spinel solid solution and corundum (i.e. the phase composition), are given by a level rule calculation (Chapter 6) and it can be seen that with decreasing temperature the proportion of alumina present increases gradually. Powder X-ray diffraction may be used to

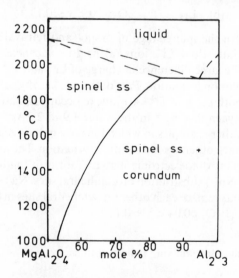

Fig. 5.31 Phase diagram for part of the system $MgAl_2O_4$–Al_2O_3 showing solid solutions of the spinel phase. (Data from Roy, Roy and Osborn, *J. Amer. Ceram. Soc.*, **36**, 149, 1953)

detect the presence or absence of alumina in samples that have been subjected to various heat treatment schedules, i.e. alumina should be present in samples heated at e.g. 1500 °C but not in samples heated at 1600 °C. The method may therefore be used to determine phase diagrams, i.e. to determine whether or not solid solutions form and, if so, their compositional extent, perhaps as a function of temperature. This is practicable only if the solid solutions that exist at the temperature of the experiment are preserved to room temperature on rapid quenching. In many cases, however, precipitation from supersaturated solid solution occurs during cooling.

Information about the composition of solid solutions may be obtained if the d-spacings of the powder lines can be measured accurately. Methods for doing this are discussed in Chapter 3. Usually, the unit cell expands if a small ion is being replaced by a larger one, and vice versa. From Bragg's law and the d-spacing formulae, an increase in the unit cell parameters leads to an increase in the d-sapcings of the powder lines; the whole pattern shifts to lower values of 2θ, therefore, although all the lines do not necessarily move by the same amount. In non-cubic crystals, the expansion or contraction of the unit cell with changing composition may not be the same for all three axes and sometimes one axis may expand while the other contract (or vice versa).

According to *Vegard's law*, unit cell parameters should change linearly with composition. In practice, Vegard's law is often obeyed only approximately and accurate measurements reveal departures from linearity. Vegard's law is not really a law but rather a generalization that applies to solid solutions formed by random substitution or distribution of ions. It assumes implicitly that the changes in unit cell parameters with composition are governed purely by the relative sizes of the atoms or ions that are 'active' in the solid solution mechanism, e.g. the ions that replace each other in a simple substitutional mechanism.

Density measurements

The mechanism of solid solution formation may sometimes be inferred by a combination of density and unit cell volume measurements for a range of compositions. Broadly speaking, an interstitial mechanism leads to an increase in density because extra atoms or ions are added to the unit cell, whereas a mechanism involving vacancy creation may lead to a decrease in density.

As an example, consider the stabilized zirconia solid solutions formed between ZrO_2 and CaO over the compositional range ~ 10 to $25\%CaO$. Two simple mechanisms could be postulated for the solid solutions: (a) the total number of oxide ions remains constant and, therefore, interstitial Ca^{2+} ions are created according to the formula $(Zr_{1-x}Ca_{2x})O_2$; (b) the total number of cations remains constant and, therefore, O^{2-} vacancies are created according to the formula $(Zr_{1-x}Ca_x)O_{2-x}$. In mechanism (a), two calcium ions replace one zirconium and the formula unit decreases in mass by 11 g as x varies, hypothetically, from 0 to 1. In (b), one zirconium and one oxygen is replaced by one calcium with a decrease in mass of the formula unit by 67 g as x varies from 0 to 1. Assuming that the unit

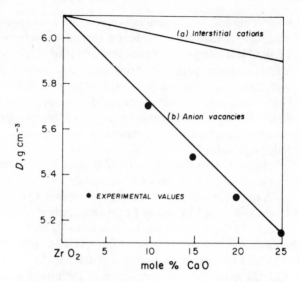

Fig. 5.32 Density data for cubic CaO-stabilized zirconia solid solutions for samples quenched from
1600 °C. (Data from Diness and Roy, *Solid State
Commun.*, **3**, 123, 1965)

cell volume does not change with composition (this is not strictly true), mechanism (b) would lead to a larger decrease in density with increasing x than would mechanism (a).

Experimental results (Fig. 5.32) confirm that mechanism (b) is operative, at least for samples heated at 1600 °C. It is possible, in theory at least, to propose alternative and more complex mechanisms than (a) and (b); e.g. the total number of zirconium ions remains constant, in which case both interstitial Ca^{2+} and O^{2-} ions are needed. Usually, however, simple mechanisms operate and there is no need to invoke more far-fetched possibilities.

Density data for the CaF_2–YF_3 solid solutions described earlier are given in Fig. 5.33. These data clearly show that a model based on interstitial fluoride ions fits the data rather than a model based on cation vacancies.

Density measurements do not, of course, give any atomistic details of the vacancies or interstitials involved, but only a bulk mechanism. Other techniques, such as diffuse neutron scattering, are needed to probe the defect structure, and as more systems are studied in detail it is becoming increasingly apparent that often, simple point defects such as vacancies and interstitials do not occur. Instead, defect clusters form by a relaxation of the crystal structure in the immediate vicinity of the point defect.

Densities can be measured by several simple techniques. The volume of a few grams of material may be measured by displacement of liquid from a specific gravity bottle whose volume is known accurately. From the difference in weight of the bottle filled with displacement liquid, e.g. CCl_4, and the bottle containing the solid topped up with liquid, the volume of the solid may be calculated if the

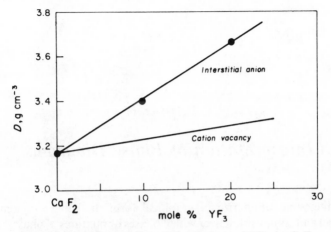

Fig. 5.33 Density data for solid solutions of YF$_3$ in CaF$_2$ (From Kingery, Bowen and Uhlmann, *Introduction to Ceramics*, Wiley, New York, 1976)

density of the displacement liquid is known. In the float-sink method, a few crystals of the material are suspended in liquids of a range of densities until a liquid is found in which the crystals neither sink nor float. The density of the crystals then equals that of the liquid. A variant of this method is to use a density gradient column which is a column of liquid of gradually increasing density. Crystals are dropped in at the top and sink until their density equals that of the liquid. The crystal density is then obtained from a calibration curve of height in the column against density. In all of the above methods, it is important that air bubbles are not trapped on the surface of the crystals, otherwise anomalously low density values may be obtained.

A good method for measuring the density of larger samples (10 to 100 g) is by gas displacement pycnometry. In this, the sample is compressed in a gas-filled chamber by a piston until a certain pressure, e.g. 2 atmospheres, is reached. The volume of the sample is obtained from the difference in position of the piston when the sample is in the chamber compared to when the chamber is empty of solid and contains only gas compressed to the same pressure.

Changes in other properties—thermal activity and DTA

Many materials undergo abrupt changes in structure or property on heating and, if the material forms a solid solution, the temperature of the change usually varies with composition. The changes can usually be studied readily by DTA since most phase transitions have an appreciable enthalpy of transition. This provides a very sensitive way of studying solid solutions because often a transition temperature varies over tens or hundreds of degrees as the composition changes, e.g. addition of carbon to iron causes the temperature of the $\alpha \rightleftharpoons \gamma$ transition to drop rapidly from 910 to 723 °C with addition of only 0.02 wt% carbon.

Chapter 6

Interpretation of Phase Diagrams

Phase diagrams summarize in graphical form the ranges of temperature (or pressure) and composition over which phases or mixtures of phases are stable under conditions of thermodynamic equilibrium. They therefore contain information on the compositions of compounds and solid solution series, phase transitions and melting temperatures.

The fundamental rule upon which phase diagrams are based is the *phase rule* of W. J. Gibbs. Its derivation is not given here, nor are detailed thermodynamic considerations of the free energies of phases, solid solutions and mixtures. Instead, we concentrate on the interpretation and practical aspects of phase diagrams. Also, we are concerned primarily with solid–solid and solid–liquid equilibria, which contrasts with the emphasis on liquid–gas equilibria found in most chemistry textbooks.

The phase rule, phases, components and degrees of freedom

The phase rule is given by the equation

$$P + F = C + 2 \tag{6.1}$$

where P is the number of phases present in equilibrium, C is the number of components needed to describe the system and F is the number of degrees of freedom or independent variables taken from temperature, pressure and composition of the phases present. Each term is now explained more fully.

The number of *phases* in a sample at equilibrium is the number of physically distinct and mechanically separable (in principle) portions, each phase being itself homogeneous. The distinction between different *crystalline* phases is usually clear. For example, the differences between chalk, $CaCO_3$, and sand, SiO_2, are obvious. The distinction between crystalline phases made from the same components but of different composition is also usually clear. Thus, the magnesium silicate minerals enstatite, $MgSiO_3$, and forsterite, Mg_2SiO_4, are different phases. They have very different composition, structure and properties. With solids it is also possible to get different crystalline phases having the *same* chemical composition. This is known as *polymorphism*. For example, two polymorphs of Ca_2SiO_4 can be prepared at room temperature, the stable γ form

256

and the metastable β form, but these have quite distinct physical and chemical properties and crystal structures.

One complicating but very important factor in classifying solid phases is the occurrence of solid solutions (Chapter 5): a solid solution is a single phase that has variable composition. For example, α–Al_2O_3 and Cr_2O_3 have the same crystal structure (corundum) and form a continuous range of solid solutions at high temperature. Any mixture of Al_2O_3 and Cr_2O_3 can react at high temperature to form a single, homogeneous phase whose composition may be altered without changing the integrity or homogeneity of the single phase.

Sometimes, as with crystallographic shear structures (Chapter 5), it may be difficult to decide exactly what constitutes a separate phase. This is because a minute change in composition can lead to a different arrangement of defects in a structure. In the oxygen-deficient tungsten oxides, WO_{3-x}, what was previously thought to be a range of homogeneous solid solutions is now known to be a large number of phases that are very close in composition and similar, but distinct, in structure. Some of these phases have formulae belonging to the homologous series W_nO_{3n-1}. Thus $W_{20}O_{59}$ and $W_{19}O_{56}$ are physically distinct phases. With the majority of solid compounds however, such complexities do not occur and there is no problem in deciding exactly what constitutes a phase.

In the *liquid* state, the number of possible, separate homogeneous phases that can exist is much more limited than in the solid state. This is because single-phase liquid solutions form much more readily and over wider compositional ranges than do single-phase solid solutions. Take the Na_2O–SiO_2 system for instance. In the liquid state at high temperatures, Na_2O and SiO_2 are completely miscible to give a single, liquid, sodium silicate phase. In the crystalline state, however, the number of phases is quite large. Crystalline sodium silicate phases form at five different compositions and at least one of these shows polymorphism.

In the *gaseous* state, the maximum number of possible phases appears always to be 1; there are no known cases of immiscibility between two gases, if the effects of gravity are ignored.

The number of *components* of a system is the number of constituents that can undergo independent variation in the different phases; alternatively, it is the *minimum* number of constituents needed in order to describe completely the compositions of the phases present. This is best understood with the aid of examples:

(a) All of the crystalline calcium silicates can be considered to be built from CaO and SiO_2 in varying proportions. CaO–SiO_2 is therefore a *two-component system* even though there are three elements present, Ca, Si and O. Compositions between CaO and SiO_2 can be regarded as forming a *binary* (i.e. two-component) *join* in the ternary system Ca–Si–O (Fig. 6.1).

(b) The system MgO is a *unary* (*one-component*) system, at least up to the melting point 2700 °C, because the composition of MgO is always fixed.

(c) The composition 'FeO' is part of the two-component system, iron–oxygen, because wüstite is actually a non-stoichiometric, iron-deficient phase, $Fe_{1-x}O$, caused by having some Fe^{3+} present (Chapter 5). The bulk

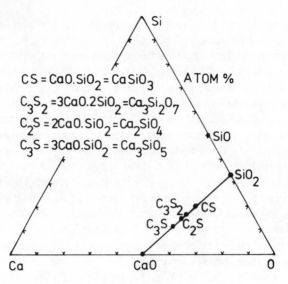

Fig. 6.1 Binary join CaO–SiO_2 in the ternary system Ca–Si–O. Note the method used for the labelling of phases, C=CaO, etc. This type of abbreviation is widely used in oxide chemistry

composition 'FeO' in fact contains a mixture of two phases at equilibrium: $Fe_{1-x}O$ and Fe metal.

The number of *degrees of freedom* of a system is the number of independently variable factors taken from temperature, pressure and composition of phases, i.e. it is the number of these variables that must be specified in order that the system be completely defined. Again, let us see some examples:

(a) A system that consists of boiling water, i.e. water and steam in equilibrium, does not have a composition variable since both water and steam contain molecules of the same fixed formula, H_2O. To define the system it is necessary to specify only the steam pressure because then the temperature of boiling is automatically fixed (or vice versa). Application of the phase rule to this system gives:

$$P + F = C + 2; \quad C = 1 \text{ (i.e. } H_2O), P = 2 \text{ (vapour and liquid)}$$

and so $F = 1$ (either temperature or pressure but not both).

At sea level, water boils at $100\,°C$ but at the high altitude of Mexico City atmospheric pressure is only $580\,mmHg$ and water in equilibrium with steam at this pressure boils at $92\,°C$. The system water–steam is therefore *univariant* because only one degree of freedom, either P or T, is needed to describe completely the sytem at equilibrium. It should be emphasized that the relative amounts of water and steam are not given by the phase rule. As long as there is sufficient steam present to maintain the equilibrium pressure, the volume of vapour is unimportant.

(b) A solid solution in the system Al_2O_3–Cr_2O_3 has one composition variable because the Al_2O_3:Cr_2O_3 ratio can be varied and the same homogeneous phase obtained. The temperature of these single-phase solid solutions can also be varied. Two degrees of freedom are therefore needed in order to fully characterize a certain solid solution, namely its composition and temperature.

In many solid systems with high melting temperatures, the vapour pressure of the solid phases and even that of the liquid phase is negligible in comparison with atmospheric pressure. The vapour phase is effectively non-existent, therefore, and need not be regarded as a possible variable for work at atmospheric pressure. Such systems are called *condensed systems* and the phase rule is modified accordingly to give the *condensed phase rule*,

$$P + F = C + 1 \tag{6.2}$$

In using phase diagrams, it is important to define what is meant by *equilibrium*. The equilibrium state is always that which has the lowest free energy. It may be thought of as lying at the bottom of a free energy well (Fig. 6.2). The problem in determining whether equilibrium has been reached is that other free energy minima may exist but not be as deep as the equilibrium well. There may be a considerable energy barrier involved in moving from a *metastable* to the *stable* state and under many conditions this barrier may be prohibitively high. Such an example is the metastability of diamond relative to graphite at room temperature. The energy barrier or activation energy for the diamond → graphite reaction is so high that, once formed, diamond is kinetically stable although thermodynamically metastable.

The thermodynamic meaning of the term *unstable* is also shown in Fig. 6.2. If a ball is perched on a hill-top, the slightest movement is sufficient to cause it to start rolling down one side or the other. In the same way, there is no activation energy

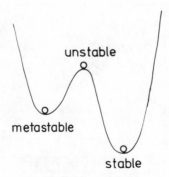

Fig. 6.2 Schematic diagram showing stable, unstable and metastable conditions

involved in changing from a thermodynamically unstable to either a stable or a metastable state. Examples of unstable equilibrium are difficult to find (because of their instability!) but one can point to their would-be occurrence. Inside a region of liquid immiscibility (Chapter 8) exists an area bounded by a dome called a *spinodal*. Within the spinodal, a homogeneous liquid would be unstable and would spontaneously separate into two liquids by the process known as *spinodal decomposition*.

One-component systems

The independent variables in a one-component system are limited to temperature and pressure because the composition is fixed. From the phase rule, $P + F = C + 2 = 3$. The system is *bivariant* ($F = 2$) if one phase is present, *univariant* ($F = 1$) if two are present and *invariant* ($F = 0$) if three are present. Schematic phase relations are given in Fig. 6.3 for a one-component system in which the axes are the independent variables, pressure and temperature. Possible phases are two crystalline modifications or polymorphs (sometimes called allotropes), X and Y, liquid and vapour. Each of these phases occupies an area or *field* on the diagram since $F = 2$ when $P = 1$ (both pressure and temperature are needed to describe a point in one of these fields). Each of these single-phase regions is separated from the neighbouring single-phase regions by univariant curves ($P = 2$ and so $F = 1$). On these curves, if one variable, say pressure, is fixed, then the other, temperature, is automatically fixed. The univariant curves on the diagram represent the following equilibria:

(a) BE—transition curve for polymorphs X and Y; it gives the change of transition temperature with pressure.
(b) FC—change of melting point of polymorph Y with pressure.
(c) AB, BC—sublimation curves for X and Y, respectively.
(d) CD—vapour pressure curve for the liquid.

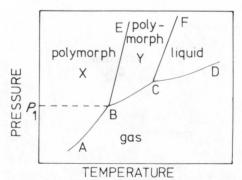

Fig. 6.3 Schematic pressure versus temperature phase diagram of a one-component system

From Fig. 6.3, crystals of polymorph X can never melt directly under equilibrium conditions because the fields of X and liquid never meet on the diagram. On heating, crystals of X can either sublime at a pressure below P_1 or transform to polymorph Y at pressures above P_1. They cannot melt directly. Also present in Fig. 6.3 are two invariant points B and C for which $P = 3$ and $F = 0$. The three phases that coexist at point B are: polymorph X, polymorph Y and vapour. Points B and C are also called *triple points*.

The system H₂O

This important one-component system, shown in Fig. 6.4, gives examples of solid–solid and solid–liquid transitions. Ice I is the polymorph that is stable at atmospheric pressure; several high-pressure polymorphs are also known—ice II to ice VI. At first sight, there is little similarity between the diagram for a schematic one-component system (Fig. 6.3) and that for water (Fig. 6.4), but this is mainly because of the location of the univariant curve XY that separates the fields of ice I and water. It is well known that ice I has the unusual property of being less dense than liquid water at 0 °C. The effect of pressure on the ice I-water transition temperature can be understood from Le Chatelier's principle which states: 'When a constraint is applied to a system in equilibrium the system adjusts itself so as to nullify the effects of this constraint.' The melting of ice I is accompanied by a decrease in volume; increased pressure makes melting easier and so melting temperature decrease with increased pressure, in the direction YX. The water system (Fig. 6.4) is also more complex than Fig. 6.3 since additional invariant points exist which correspond to three solid phases in equilibrium (e.g. point Z). The rest of the diagram should be self-explanatory; thus the curves YXABC give the variation of melting point with pressure for some of the different ice polymorphs. Liquid–vapour eqilibria are omitted from Fig. 6.4 because, with the pressure scale used, these equilibria lie very close to the temperature axis and in the high temperature corner.

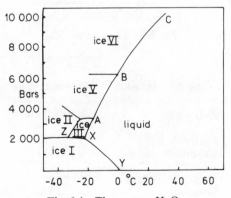

Fig. 6.4 The system H₂O

The system SiO$_2$

Silica is the main component of many ceramic materials as well as being the most common oxide, apart from H$_2$O, in the earth's crust. The polymorphism of SiO$_2$ is complex with major, first-order phase changes such as quartz–tridymite and minor changes such as α(low)–β(high) quartz. The polymorphism at atmospheric pressure can be summarized by the following sequence of reactions on heating:

$$\alpha\text{-quartz} \xrightarrow{573\,^{\circ}\text{C}} \beta\text{-quartz} \xrightarrow{870\,^{\circ}\text{C}} \beta\text{-tridymite} \xrightarrow{1470\,^{\circ}\text{C}} \beta\text{-}$$

$$\text{cristobalite} \xrightarrow{1710\,^{\circ}\text{C}} \text{liquid}$$

With increasing pressure, two main changes are observed (Fig. 6.5); first, the contraction of the field of tridymite and its eventual disappearance, at ~ 900 atm; second, the disappearance of the field of cristobalite at ~ 1600 atm. Above 1600 atm, quartz is the only stable crystalline polymorph and exists up to much higher pressures. The disappearance of tridymite and cristobalite with increasing pressure can be correlated with the lesser density of these phases relative to that of quartz, Table 6.1; the effect of pressure generally is to produce polymorphs that have a higher density and therefore smaller volume. Above 20,000 to 40,000 atm

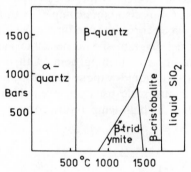

Fig. 6.5 The system SiO$_2$

Table 6.1 *Densities of* SiO$_2$ *polymorphs*

Polymorph	Density (g cm^{-3})
Tridymite	2.298
Cristobalite	2.334
Quartz	2.647
Coesite	2.90
Stishovite	4.28

(depending on temperature), quartz transforms to another polymorph, coesite, and above 90,000 to 120,000 atm yet another polymorph, stishovite, is the stable polymorph of SiO_2. These pressures are well off the scale of Fig. 6.5. Again, it may be noted that the densities of these polymorphs, Table 6.1, are much higher those of the other forms.

It may be noted that there are many metastable polymorphs of SiO_2 which are absent from Fig. 6.5; e.g. it is very easy to undercool cirstobalite and to observe a reversible α(low)–β(high) transformation at $\sim 270\,^\circ C$. However, at these temperatures, cristobalite is metastable relative to quartz and so this transformation is omitted from Fig. 6.5.

Condensed one-component systems

For most systems and applications of interest in solid state chemistry, the condensed phase rule is applicable, pressure is not a variable and the vapour phase is not important. The phase diagram for a condensed, one-component system then reduces to a line since temperature is the only degree of freedom. It is not normal practice to represent such a line phase diagram in graphical form, unless it forms part of, say, a binary system. For instance, the condensed phase diagram at 1 atmosphere pressure for SiO_2 would simply be a line showing the polymorphic changes that occur with changing temperature. In such cases it is easier to represent the changes as a 'flow diagram', as indicated above for SiO_2.

Two-component condensed systems

Two-component or binary systems have three independent variables: pressure, temperature and composition. In most systems of interest in the general sphere of solid state chemistry, the vapour pressure remains low for large variations in temperature and so, for work at atmospheric pressure, the vapour phase and the pressure variable need not be considered. In almost all of what follows, the condensed phase rule $P + F = C + 1$ is used. In binary systems under these conditions an invariant point occurs when three phases coexist in equilibrium: a univariant curve for two phases and a bivariant condition for one phase. Conventionally, temperature is the vertical scale and composition the horizontal one in binary phase diagrams.

A simple eutectic system

The simplest possible type of two-component condensed system is the simple eutectic system shown in Fig. 6.6(a). In the solid state there are no intermediate compounds or solid solutions but only a mixture of the end-member crystalline phases, A and B. In the liquid state, at high temperatures, a complete range of single-phase, liquid solutions occurs. At intermediate temperatures, regions of partial melting appear on the diagram. These regions contain a mixture of a crystalline phase and a liquid of different composition to the crystalline phase.

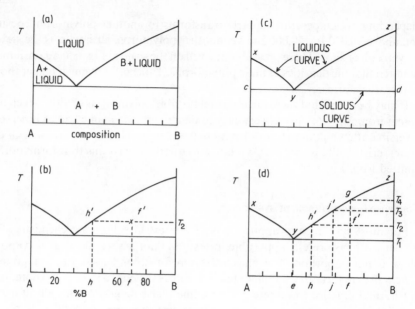

Fig. 6.6 Simple eutectic binary system

The phase diagram shows several regions or areas which contain either one or two phases. These areas are separated from each other by solid curves or lines. The area 'liquid' at high temperatures is single phase and bivariant. Every point within this area represents a different state for the liquid, i.e. a different temperature and composition. In this region, $P = 1$ and $F = 2$.

The other three areas shown each contain two phases—A + B, A + liquid and B + liquid—and are univariant since $P = 2$ and $F = 1$. Let us consider one of these in detail, the region B + liquid. Let a mixture of B and liquid have overall composition f and temperature T_2 (Fig. 6.6b), i.e. this mixture is represented in temperature and overall composition by point f' in (b). In this region, there is only one degree of freedom, either temperature or composition, but not both, and we have decided to fix the temperature at T_2.

In order to determine the compositions of the two phases, B and liquid, construction lines, dashed in (b), are drawn. First, an *isotherm* is drawn at the temperature of interest, T_2. This is the horizontal dashed line that terminates on the *liquidus* at point h'. This point, h', represents the liquid that is present in the mixture of B + liquid at temperature T_2. The composition of this liquid is given by drawing a vertical line or *isopleth* which intersects the composition (horizontal) axis at h, at which point the composition of the liquid may be simply read off the composition scale; here, h is at 43% B, 57% A. The other phase that is present in the mixture is B, whose composition is fixed, as pure B, in this example.

An important distinction to be made here is between different meanings of the word 'composition'. It has at least three meanings:

(a) The composition of a particular phase. In the above example, the liquid phase has composition h, 43% B, 57% A.

(b) The relative amounts of the different phases present in a mixture. This may be referred to as the *phase composition*. In the above example, B and liquid are present in the ratio ~ 1:1 (see later for an explanation of the lever rule used to determine phase compositions).

(c) The overall composition of a mixture, in terms of the components and irrespective of the phases present. This may be termed the *component composition*. In the above example, the component composition of mixture f is 30%A, 70%B.

Since there is no universally adopted convention over the use of the word 'composition', one can only say, be careful!

In the sense to which the phase rule applies, composition may be regarded as a degree of freedom only when it refers to the actual compositions of the phases involved, category (a). The relative amounts of the different phases in a mixture, category (b), do not constitute a degree of freedom. Thus, along the isotherm passing through h' and f' at temperature T_2, the relative amounts of the phases B and liquid vary, but the compositions of the two individual phases do not vary. The component composition, category (c), is included in the phase rule, not as a degree of freedom but as the number of components.

The *liquidus* curve, xyz in (c), gives the highest temperature at which crystals can exist as a function of overall composition. Liquids that cross the liquidus curve between points x and y on cooling enter the two-phase region: A + liquid. For these compositions, A is said to be the *primary phase*, because it is the first phase to crystallize on cooling. Similarly, for liquids between y and z, B is the primary phase. The line cyd is the *solidus* and gives the lowest temperature at which liquids can exist over this composition range, i.e. above the solidus, melting commences. In summary, above the liquidus mixtures are completely molten, below the solidus they are completely solid, and in between partial melting occurs.

Point y is an *invariant point* at which three phases coexist: A, B and liquid. It is a *eutectic* and its temperature is the lowest temperature at which a composition (here 70%A, 30%B) can completely liquid. It is also a minimum point on the liquidus curve, xyz. In simple eutectic systems such as this one, the eutectic and solidus temperatures are the same.

In order to determine the relative amounts of two phases in a mixture (the phase composition, category (b) above), the *lever rule* is used. This is the same as the 'principle of moments' which operates when children are balancing on a see-saw. The pivot point or fulcrum of the see-saw is equivalent to the overall composition of a mixture. The two phases in the mixture are equivalent to the two children on the see-saw. In the same way that children of different sizes sit at different distances from the centre of the see-saw to achieve balance, so the compositions and amounts of two phases in a mixture are interrelated. For the see-saw, the balance condition is as shown in Fig. 6.7(a) and is given by

$$m_1*(\text{distance } xy) = m_2*(\text{distance } yz) \qquad (6.3)$$

This may be reorganized to give, for instance, the ratio m_1/m_2 or the ratio $m_1/(m_1 + m_2)$.

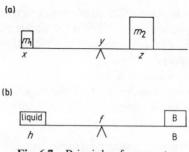

Fig. 6.7 Principle of moments

Returning now to the lever rule and mixtures of phases, the liquid in Fig. 6.6(b) has composition h on the A–B axis, the crystals are pure B and the overall composition is f. This is reproduced schematically in Fig. 6.7(b). Applying the principle of moments,

$$\text{(amount of liquid)} * \text{(distance } hf) = \text{(amount of B)} * \text{(distance } fB)$$
$$(6.4)$$

If we consider the amounts of B and liquid as a fraction of the total, such that

$$\text{(amount of B)} + \text{(amount of liquid)} = 1 \qquad (6.5)$$

then equation (6.4) may be rearranged, to give

$$\frac{\text{(amount of B)}}{\text{(amount of liquid)}} = \frac{hf}{fB} = \frac{\text{(amount of B)}}{1 - \text{(amount of B)}}$$

i.e. $\text{(amount of B)} * Bf = hf - [hf * \text{(amount of B)}]$

therefore, $\qquad \text{(amount of B)} = hf/Bh \qquad (6.6)$

and $\qquad \text{(amount of liquid)} = Bf/Bh \qquad (6.7)$

The lever rule may be used to determine how the relative amounts of phases in a mixture change with temperature. Thus at temperature T_2, (Fig. 6.6b), the amount of liquid present in composition f is given by Bf/Bh, i.e. 0.53. At a higher temperature T_3, (Fig. 6.6d) the amount of liquid is given by Bf/Bj, i.e. 0.71. At a lower temperature, just above the solidus T_1, the amount of liquid is given by Bf/Be, i.e. 0.43. Clearly, therefore, the effect of raising the temperature above T_1 is to cause an increase in the degree of melting from 43% at T_1 to 71% at T_3. The limit is reached at T_4 where the fraction of liquid is 1 and melting is complete. As the degree of melting increases with increasing temperature, so the composition of the liquid phase must change accordingly: since crystals of B disappear into the liquid phase on melting, the liquid must become richer in B. Thus the first liquid that appears on heating, at temperature T_1, has composition e, i.e. 30% B, 70% A. As melting continues, the liquid follows the liquidus curve $yh'j'g$ until, when melting is complete at T_4, the liquid has composition f, i.e. 70% B, 30% A. On cooling the liquid of composition f, the reverse process should be observed under equilibrium conditions. At T_4, crystals of B begin to form and with falling temperature the liquid composition moves from g to y as more crystals of B precipitate.

The *eutectic reaction* which occurs on cooling through temperature T_1 gives a good example of the use of the lever rule. Just above T_1, the fraction of B present in composition f is given by fe/Be and is roughly 0.57. Just below T_1 the fraction of B is fA/BA and is roughly 0.70. Thus, the residual liquid, of composition e, has crystallized to a mixture of A and B, i.e. the quantity of B present has increased even further and crystals of A are formed for the first time. A solid mixture of A and B of eutectic composition e undergoes complete melting at temperature T_1 and conversely, on cooling, a homogeneous liquid of this composition completely crystallizes to a mixture of A and B at T_1.

The reactions described above are those that should occur under equilibrium conditions. This usually means that slow rates of heating and, especially, cooling are necessary. Rapid cooling rates often lead to different results, especially in systems which have more complicated phase diagrams. However, the equilibrium diagram can often be very useful in rationalizing these non-equilibrium results (see later).

The liquidus curve xyz may be regarded in various ways. As well as giving the maximum temperature at which crystals can exist, it is also a *saturation solubility curve*. Thus, curve yz could be regarded as giving the solubility limit with temperature for crystals of B dissolved in liquid. Above yz a homogeneous solution occurs but below this curve undissolved crystals of B are present. On cooling, precipitation of crystals B would therefore occur below the temperatures of curve yz; otherwise a metastable undercooled, supersaturated solution would be present.

Another interpretation of the liquidus is that it shows the effect of soluble impurities on the melting points of pure compounds. If a specimen of B is held at temperature T_4 and a small amount of A is added, then some liquid of composition g will form. As the amount of added A increases, so the amount of liquid g increases until, when sufficient *flux*, A, has been added to bring the overall composition to g, the solid phase disappears and the sample will be all liquid. Therefore a small amount of soluble impurity A has lowered the melting point of B from T_5 to T_4. A familiar practical example of this is the addition of salt to icy roads. In the binary system $H_2O-NaCl$, addition of NaCl lowers the melting point of ice below $0\,°C$; the system contains a low temperature eutectic at $\simeq -21\,°C$.

Binary systems with compounds

Three types of binary system with a compound AB are shown in Figs 6.8 and 6.9. A stoichiometric binary compound such as AB is represented on the phase diagrams by a vertical line. This shows the range of temperatures over which

268

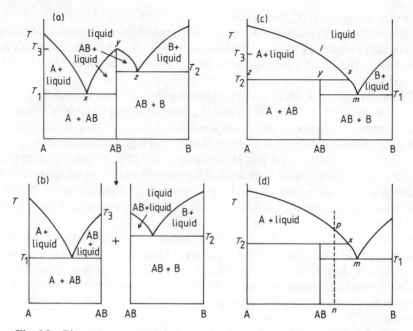

Fig. 6.8 Binary systems showing a compound AB melting congruently (a) and incongruently (c), (d)

compound AB is stable. Compound AB melts *congruently* in Fig. 6.8(a) because it changes directly from solid AB to liquid of the same composition at temperature T_3. Figure 6.8(a) may be conveniently divided into two parts, given by the composition ranges A–AB and AB–B as shown in (b); each part may be treated as a simple eutectic system in exactly the same manner as Fig. 6.6. Although the horizontal lines at T_1 and T_2, corresponding to the two eutectic temperatures, meet the vertical line representing crystalline AB, no changes would be observed at T_1 and T_2 on heating pure AB. This is because these horizontal lines should peter out as composition AB is approached; composition AB, by itself, is a one-component system and only when another component, A or B, is added are changes observed at T_1 or T_2.

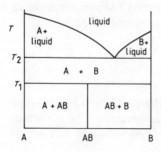

Fig. 6.9 Binary system showing compound AB with an upp limit of stability

In Fig. 6.8(c), compound AB melts *incongruently* at T_2 to give a mixture of crystals A and liquid of composition x. The relative amounts of liquid and crystals A just above T_2 are given by the lever rule; fraction liquid $= yz/xz$. On further heating, crystals of A gradually dissolve as the liquid becomes richer in A and moves along the liquidus curve in the direction xl. At T_3, the liquid composition has reached l and the last A crystals should disappear.

Point x is an invariant point at which three phases coexist: A, AB and liquid. It is a *peritectic* point because the composition of the liquid cannot be represented by positive quantities of the two coexisting solid phases, i.e. composition x does not lie between A and AB, as is the case for a eutectic point (Fig. 6.8a). Another characteristic feature of a peritectic is that it is not a minimum point on the liquidus, as is a eutectic.

Phase AB has a primary phase field. It is the first phase to crystallize on cooling liquids in the composition range x–m. However, the composition of AB is separated from its primary phase field. This is different to the case for congruently melting AB in Fig. 6.8(a), where the composition AB lies *within* the range xyz over which AB is the primary phase.

The behaviour of liquid of composition, n, Fig. 6.8(d), on cooling is worth describing. At point p, crystals of A start to precipitate; more A crystals form as the temperature drops and the liquid composition moves from p to x. At T_2, the *peritectic reaction* liquid (x) + A → liquid (x) + AB occurs. Thus, the crystalline phase changes from A to AB and the amount of liquid present must diminish. From the level rule, just above T_2, the mixture is ~ 85 per cent liquid and just below T_2 only ~ 50 per cent liquid. Therefore, *all* of phase A has reacted with *some* of the liquid to give AB. On further cooling from T_2 to T_1, more AB crystallizes as the liquid composition moves from x to m; finally, at T_1, the residual liquid of composition m crystallizes to a mixture of AB and B. Just above T_1, the mixture is $\sim 40\%$ liquid and 60% AB and just below T_1, $\sim 20\%$ B and 80% AB.

The behaviour on cooling of any liquid of composition between A and AB is similar but with one important difference. At T_2, the peritectic reaction for these compositions involves *some* of A reacting with *all* the liquid to give AB. Below T_2, a mixture of A and AB coexists and no further changes occur on cooling.

In systems that contain incongruently melting compounds, such as Fig. 6.8(c), it is very easy to get non-equilibrium products on cooling. This is because the peritectic reaction that *should* occur between A and liquid on cooling is slow, especially if the crystals of A are much more dense than the liquid and have settled to the bottom of the sample. What commonly happens in practice is that the crystals of A which have formed are effectively lost to the system and there is not time for much peritectic reaction to occur at T_2. The liquid of composition x then effectively begins crystallizing again below T_2, but this time crystals of AB form; at the eutectic temperature T_1, the residual liquid m crystallizes to a mixture of AB and B, as usual. Thus, it is quite common to obtain a mixture of *three* crystalline phases on cooling: A, AB and B, at least one of which should be absent under equilibrium conditions.

Another common type of non-equilibrium assemblage occurs when the intermediate, incongruently melting compound AB fails completely to crystallize on cooling. If this happened on cooling liquids in Fig. 6.8(c), the crystalline products would be A and B with no AB. Hence the peritectic reaction A + liquid → AB would have been suppressed entirely.

Sometimes, compounds decompose before their melting point is reached, as shown for AB in Fig. 6.9. Compound AB has an *upper limit of stability* and at temperature T_1 disproportionates into a mixture of crystalline A and B; at higher temperatures the system is simple eutectic in character.

There are also many examples of systems which contain compounds with a *lower limit of stability*, i.e. below a certain temperature, compound AB decomposes into a mixture of A and B. The behaviour of AB at higher temperatures can then be any of the three types described above.

The CaO–SiO$_2$ diagram

A phase diagram which contains most of the binary features discussed above and which is one of the most important diagrams in silicate technology is that of the system CaO–SiO$_2$, part of which is shown in Fig. 6.10. The compound Ca$_2$SiO$_4$ ($\equiv$ 2CaO·SiO$_2$ $\equiv$ 33.3%SiO$_2$·66.7%CaO) melts congruently at 2130 °C. The compound Ca$_3$SiO$_5$ melts incongruently at 2150 °C to CaO and liquid. It also has a lower limit of stability at 1250 °C, decomposing to give a mixture of CaO and Ca$_2$SiO$_4$ at lower temperatures. The diagram has one peritectic, P and one eutectic, E.

The CaO–SiO$_2$ phase diagram is important to cement manufacture. The key ingredient in rapid-hardening Portland cement is Ca$_3$SiO$_5$, but from the phase

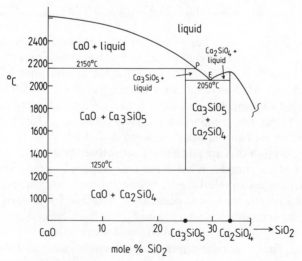

Fig. 6.10 Partial diagram for lime-rich compositions in the system CaO–SiO$_2$

diagram, it is unstable below 1250 °C. In order to obtain it, therefore, the reactants are fired in a kiln at 1400–1500 °C and cooled rapidly in a blast of air. With such rapid cooling rates, there is no time for the Ca_3SiO_5 to decompose and it is quenched to room temperature where it is kinetically stable.

Binary systems with solid solutions

The simplest form of solid solution system is one that shows complete miscibility in both solid and liquid states (Fig. 6.11). The melting point of one end member, A, is depressed by addition of the other end member, B, and vice versa, that of B is increased by the addition of A. The liquidus and solidus are both smooth curves which meet only at the end-member compositions A and B. At low temperatures, a single-phase solid solution exists and is bivariant ($C = 2$, $P = 1$, and so $F = 2$). At high temperatures, a single phase liquid solution exists and is similarly bivariant. At intermediate temperatures, a two-phase region of solid solution + liquid exists. Within this two-phase region, the compositions of the two phases in equilibrium are found by drawing isotherms or *tie-lines* at the temperature of interest, e.g. T_1. The intersection of the tie-line and the solidus gives the composition of the solid solution, a, and the intersection of the tie-line and the liquidus gives the liquid composition, b.

On cooling liquids in a system such as this, the crystallization pathways are complicated, Fig. 6.11(b). A liquid of bulk composition b begins to crystallize a solid solution of composition a at temperature T_1. At a lower temperature, T_2, and at equilibrium, the amount of solid solution present increases but also its composition changes to a'. The fraction of solid solution a' is given by the lever rule and is equal to $bb'/a'b'$, i.e. the equilibrium mixture is approximately one-third solid solution and two-thirds liquid at T_2. Crystallization is therefore a complex process because with decreasing temperature, the composition of the solid solution has to change continuously in order to maintain equilibrium. With falling temperature, both crystals and liquid become progressively richer in B but

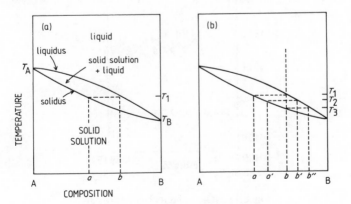

Fig. 6.11 Binary system with a complete range of solid solutions

272

the quantities of the two phases change in accord with the lever rule; the overall composition must obviously always be b. Finally, at temperature T_3, the solid solution composition reaches the bulk composition b and the last remaining liquid, of composition b'', disappears.

In systems with phase diagrams such as this, metastable or non-equilibrium products are often produced by a process of *fractional crystallization*. This occurs unless cooling takes place sufficiently slowly so that equilibrium is reached at each temperature. The crystals that form first on cooling liquid b have composition a. If these crystals do not have time to re-equilibrate with the liquid on further cooling they are effectively lost from the system. Each new crystal that precipitates will be a little bit richer in B and the result is that crystals form which have composition ranging from a to somewhere between b and B. In practice, the crystals that precipitate during cooling are often '*cored*'. The central part that formed first may have composition a and on moving out radially from the centre the crystal becomes increasingly rich in B.

Coring occurs in many rocks and metals. The plagioclase feldspars, which are solid solutions of anorthite, $CaAl_2Si_2O_8$, and albite, $NaAlSi_3O_8$, have the simple phase diagram shown in Fig. 6.12. Igneous rocks contain plagioclase feldspars and form by slow cooling of liquids. Such feldspar liquids and crystals are notoriously slow to equilibrate and although the cooling of melts in nature may have been very slow, it is common nevertheless to find rocks in which fractional crystallization has occurred. In these, the plagioclase crystals have calcium-rich centres and sodium-rich outer regions.

Coring may occur in metals during the manufacture of bars and ingots. The molten metal is poured into moulds (or 'sand cast' in moulds of sand) and allowed

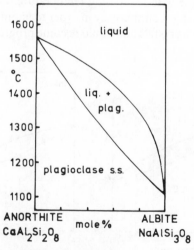

Fig. 6.12 The plagioclase feldspar system, anorthite–albite

to cool. If the metal composition is part of a solid solution then coring may occur as the melt freezes. Coring is usually deleterious to the properties of the metal and has to be eliminated. This can be done by subsequently heating the bars to just below the solidus temperature at which homogenization of the metal, with the elimination of coring, occurs rapidly.

It has been noted that the equilibrium subsolidus assemblage is obtained if cooling rates are sufficiently slow, but that with somewhat faster cooling fractional crystallization may occur. This is a property of the phase diagram and can occur with any kind of material, whether it be rocks, metals, or synthetic inorganic or organic materials. At still faster cooling rates, other types of pathway may be followed.

Sometimes, if the liquid is cooled rapidly to room temperature, there may not be time for any crystallization to occur and a glass forms. Glass formation is common in inorganic materials such as silicates. Recently there has been much scientific and technological interest in glassy semiconductors and metals, materials with unusual electrical and mechanical properties.

The simplest type of solid solution phase diagram is that shown in Fig. 6.11. Other relatively simple types of phase diagram are possible that show complete solubility in both the solid and liquid states but have either a thermal minimum or a thermal maximum in the liquidus and solidus curves (Fig. 6.13). These thermal maxima and minima are called *indifferent points* because they are not true invariant points. For an invariant point, three phases are needed in equilibrium ($F = 0, P = C + 1 = 3$), but this condition can never exist in solid solution systems such as these because there are never more than two phases present, i.e. solid solution and liquid solution. The liquidus and solidus are continuous through the thermal maximum or minimum and do not show a discontinuity such as is observed for peritectics and eutectics.

Complete solid solubility, such as shown in Figs. 6.11 to 6.13, occurs only when the cations or anions that replace each other are similar in size, e.g. Al^{3+} and Cr^{3+}. It is far more common to have phase diagrams in which the crystalline phases have

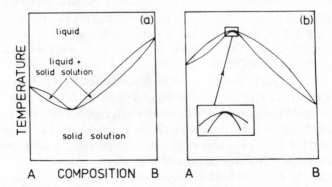

Fig. 6.13 Binary solid solution systems with (a) thermal maxima and (b) thermal minima in liquidus and solidus curves

274

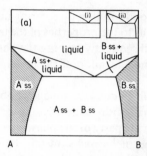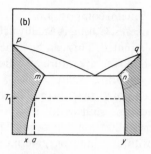

Fig. 6.14 Simple eutectic system showing partial solid solubility of the end members

only partial solubility in each other. The simplest possible case, shown in Fig. 6.14, is a straightforward extension of the simple eutectic system, shown in inset (1). Crystals of B dissolve in crystals of A to form a solid solution (shaded) whose maximum extent depends on temperature and is given by the curves xmp (b). The extent of solid solution is a maximum at the solidus temperature, point m. At lower temperatures, the range of solid solutions contracts, with the solubility limit given by the curve mx. At the other end of the diagram, crystalline B is able to partially dissolve A (shaded region, B s.s., at the right-hand end of the diagram). The extent of the B s.s. region varies with temperature, as given by the curve, ynq; again, the maximum extent is at the solidus, point n.

In the two-phase region (A s.s. + B s.s.), the composition of the A solid solution is given by the intersection of the tie-line (dashed) at the temperature of interest, T_1, and the curve mx which limits the extent of the A s.s. field. This composition corresponds to point a on the composition axis. The composition of the B solid solution in this two-phase region is given similarly by the intersection of the tie-line and curve ny.

In many phase diagrams, the solid phases are shown as line phases, i.e. as being stoichiometric and without a range of homogeneity or solid solution formation (inset i). In practice, however, the phases may have a slightly variable composition, as in inset (ii). An example of a real system similar to Fig. 6.14 is forsterite (Mg_2SiO_4)-willemite (Zn_2SiO_4), shown in Fig. 6.15; the solid solutions in this system were discussed in Chapter 5.

Another type of simple binary system with partial solid solubility is shown in Fig. 6.16. This rather strange looking diagram can be derived from a simple system showing complete solubility (inset a). First, suppose that an *immiscibility dome* exists within the solid solutions which has an *upper consolute temperature* as shown in inset (b). Above the upper consolute temperature, a single-phase solid solution exists, but below it a mixture of two phases exists.

Second, let the dome expand to higher temperatures until it intersects the melting curves. The result is shown in inset (c) and on an expanded scale as the main diagram in Fig. 6.16.

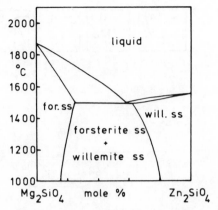

Fig. 6.15 The system Mg_2SiO_4–Zn_2SiO_4. (Data from E.R. Segnit and A.E. Holland, *J. Amer. Ceram. Soc.*, **48**, 412, 1965)

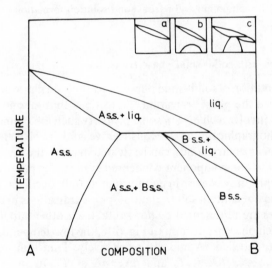

Fig. 6.16 Binary system with partial solid solution formation

A more complex phase diagram containing an incongruently melting phase that forms a range of limited solid solutions is shown in Fig. 6.17. The progressive introduction of solid solutions is shown in insets (a), (b), (c) and (d), again working on the principle that the maximum extent of solid solution occurs at the solidus (temperatures T_1 and T_2).

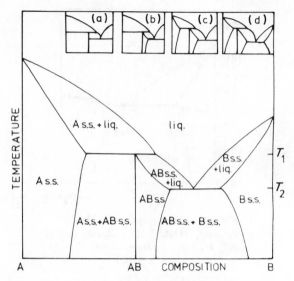

Fig. 6.17 Binary system with incongruently melting compound and partial solid solution formation

Binary systems with solid–solid phase transitions

The representation of solid–solid phase transitions on phase diagrams depends on the nature of the phase transition. Transitions that thermodynamically are first order and involve a change in some property such as volume or enthalpy, or which, crystallographically, are reconstructive and involve the breaking and forming of many primary bonds, can be treated in much the same way as melting phenomena. In a one-component condensed system, e.g. pure A or pure B, two solid polymorphs may coexist in equilibrium at only one fixed point. In binary systems that do not contain solid solutions, phase transitions in either of the end-member phases are represented by horizontal (i.e. isothermal) lines, there being one line for each phase transition. In Fig. 6.18, the low-temperature polymorphs of A and B are labelled αA and αB, respectively. Transition temperatures are αB$\rightleftharpoons$$\beta$B at T_1, αA$\rightleftharpoons$$\beta$A at T_2 and βA$\rightleftharpoons$$\delta$A at T_4. In the absence of solid solutions, these transition temperatures are the same for the pure phase as for the phase mixed with other phase (s).

In systems that exhibit complete solid solubility, as well as phase transitions, three types of phase diagram are possible (Fig. 6.19). For the end-member phases, A and B, the transitions occur at a fixed temperature, as indeed they must according to the phase rule ($C = 1$, $P = 2$, and so $F = 0$). However, for the intermediate compositions, two phases can coexist over a range of temperatures or bulk compositions because there is now one degree of freedom ($C = 2$, $P = 2$, and so $F = 1$). Thus, two-phase regions containing two solid phases, e.g. ($\alpha + \beta$), are generally observed. The treatment of the $\alpha \rightleftharpoons \beta$ change in Fig. 6.19 (a) is

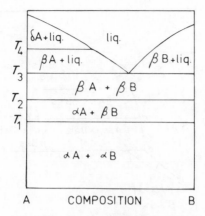

Fig. 6.18 Simple eutectic system with solid–solid polymorphic phase transitions

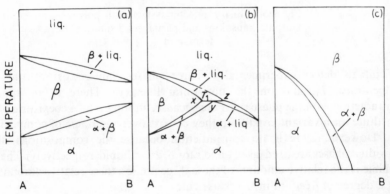

Fig. 6.19 Binary solid solution systems with polymorphic phase transitions

exactly the same as for the melting of β solid solutions, as discussed in detail for Fig. 6.11. The melting relations in Fig. 6.19(a) are the same as in Fig. 6.11 but in fact could be any of the three types given in Figs 6.11 and 6.13. In Fig. 6.19(a), both end members A and B and the entire range of solid solutions show both α and β polymorphs.

In Fig. 6.19(b), the $\alpha \rightleftharpoons \beta$ phase transition curves intersect the solidus curve because the $\alpha \rightleftharpoons \beta$ transition in B and in solid solutions rich in B now occurs, hypothetically, above the melting point of B and the B-rich solid solutions. The nature of the intersection of the three one-phase fields (α, β and liquid) and the three two-phase regions at T_1 is typical of solid solution systems. Always, two one-phase areas such as α and β must be separated from each other by a two-phase area ($\alpha + \beta$), although in practice the width of the two-phase regions may be

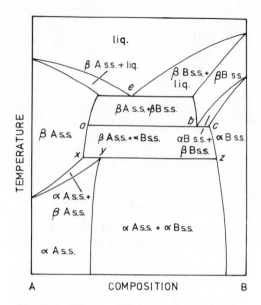

Fig. 6.20 Binary eutectic system with polymorphic transitions and partial solid solution formation

difficult to detect experimentally. Three phases can coexist at only one temperature, T_1, and on the horizontal line *xyz*. There is an apparent contradiction with the phase rule here because, although three coexisting phases constitute an invariant condition, they coexist over the *range* of compositions *xyz*. However, there is no contradiction because the compositions of the individual phases are fixed (as *x*, *y* and *z* for β, α and liquid, respectively). The only variable is the relative amounts of these three phases; these relative amounts are not a degree of freedom in the phase rule.

In Fig. 6.19 (c), the temperature of the $\alpha \rightleftharpoons \beta$ transition decreases increasingly rapidly as the solid solutions become more rich in B. For pure B, the α polymorph does not exist at any real temperature.

A typical binary system that has both phase transitions and partial solid solubility is shown in Fig. 6.20. The $\alpha \rightleftharpoons \beta$ transition occurs at one fixed temperature in pure A and pure B but, in the solid solutions, two-phase regions of $(\alpha A + \beta A)$ and $(\alpha B + \beta B)$ solid solution exist.

We have already seen that there are analogies between melting behaviour and phase transition behaviour in solid solution systems. A further analogy exists between the eutectic, *e*, and the *eutectoid*, *b*, in Fig. 6.20. The line *a–b–c* represents an invariant condition over which three phases coexist, βA s.s. (composition *a*), βB s.s. (composition *b*) and αB s.s. (composition *c*). The *eutectic reaction* at *e* on cooling is 1 liquid → 2 solids (βA s.s. + βB s.s.). The *eutectoid reaction* at *b* on cooling is 1 solid (βB s.s.) → 2 solids (βA s.s. + αB s.s.). Thus, both the eutectic and eutectoid reactions are disproportionation reactions.

Point y is a *peritectoid*. The line $x–y–z$ represents an invariant condition over which three phases coexist, αA s.s. (composition y), βA s.s. (composition x) and αB s.s. (composition z). On heating, the reaction 1 solid (αA s.s.) → 2 solids (βA s.s. + αB s.s.) occurs. This is analogous to the melting of an incongruent compound at a peritectic temperature, which involves the reaction 1 solid → 1 solid + 1 liquid.

Iron and steel making

A eutectoid reaction in the system Fe–C is of great importance is steel making. The iron-rich part of this phase diagram is shown in Fig. 6.21. The changes that occur on thermal cycling of Fe–C alloys can be studied and understood using it.

Iron exists in three polymorphic forms: body centred cubic α, stable below 910 °C; face centred cubic γ, stable between 910 and 1400 °C; and body centred cubic (again) δ, stable between 1400 °C and the melting point 1534 °C. γ-Iron can dissolve appreciable amounts of carbon, up to 2.06 wt%, in solid solution formation, whereas the α and δ forms dissolve very much less carbon, up to a maximum of 0.02 and 0.1 wt%, respectively.

Most carbon steels contain less than 1 wt% carbon, viz. 0.2 to 0.3 per cent for use as structural members. On cooling from the melt, and in the temperature range

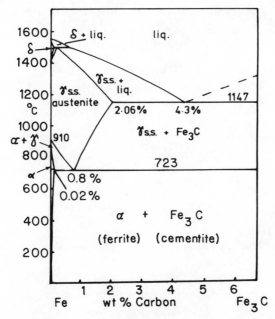

Fig. 6.21 The Fe–C diagram. The diagram is not a true equilibrium one because cementite, Fe_3C, is metastable thermodynamically. However, it is kinetically very stable and appears on the phase diagram as though it were thermodynamically stable

800 to 1400 °C, these steels form a solid solution of carbon in γ-Fe called *austenite*. However, the austenite solid solutions are unstable at lower temperatures (< 723 °C) because from Fig. 6.20, when the structure changes from that of γ-Fe to α-Fe, exsolution or precipitation of the carbide phase, Fe_3C, must occur. In practice this decomposition starts at the boundaries of the austenite grains; the ferrite (α-Fe) and cementite (Fe_3C) crystals grow side by side to give a lamellar texture known as *pearlite*.

If the steel is cooled quickly, there is not time for decomposition to ferrite and cementite to occur; instead *martensite* forms. Martensite has a deformed austenite structure in which the carbon atoms are retained in solid solution. It is possible to release these carbon atoms, as cementite, by tempering, i.e. reheating, to give a fine scale pearlite texture.

The hardness of steel depends very much on the cooling conditions and/or tempering treatment. Martensitic steels are hard largely, it seems, because of the stressed state of the martensite crystals which prevents easy motion of dislocations. In steels with pearlite texture, the hardness depends on the size, amount and distribution of the very hard cementite grains; a finer texture with a large number of closely spaced grains gives harder steel. If the steel is cooled slowly or held just below 723 °C, the decomposition is slow, giving a coarse pearlite texture. A finer texture is obtained by using a faster cooling rate to produce martensite which is then subsequently tempered at a low temperature, e.g. 200 °C.

Chapter 7

Electrical Properties

Survey of electrical properties and materials

The electrical properties of materials depend on whether the materials are electrical conductors or insulators and, if the former, on whether the current carriers are electrons or ions. A wealth of electrical phenomena occur and find a wide variety of applications. Here, we are concerned with the chemical aspects of electrical properties, which primarily involve the relation between crystal structure and properties.

Electronic conductivity occurs to varying extents and by different mechanisms in a wide variety of materials. It is responsible for the characteristic electrical properties of metals, superconductors and semiconductors. *Metallic conductivity* has the following characteristics.

(a) A significant proportion of the outer shell or valence electrons are effectively free to move throughout the structure and are completely delocalized.

(b) Collisions between these electrons and *phonons* (lattice vibrations) occur and are responsible for the residual resistance to current flow and for the heat losses when current passes through a metal.

(c) Metallic bonding and conductivity are usually treated in terms of band theory.

(d) Metallic conductivity is not confined to elemental metals and alloys, but occurs in a variety of inorganic solids, such as certain oxides and sulphides. In these, the metal atoms form a sublattice with overlapping valence orbitals. It also occurs in certain conjugated organic systems, such as doped polyacetylene and polyaniline.

The closely related phenomenon of *superconductivity* has the characteristics:

(a) Valence shell electrons again appear to be delocalized.

(b) The electrons move cooperatively, perhaps as pairs.

(c) No electron–phonon collisions occur and consequently, there is no heat loss and no resistance to current flow.

(d) Until 1986, superconductivity was confined to very low temperatures, $< 23\,K$, but has now been found in 'ceramic superconductors' such as $YBa_2Cu_3O_7$, at temperatures as high as $91\,K$.

Semiconductivity has the following characteristics:

(a) It is associated with a limited degree of electronic conduction.
(b) It is intermediate between metallic conduction, in which a significant number of outer shell electrons are free to move, and insulating behaviour, in which all of the valence electrons are either tightly bound to individual atoms or are localized in bonds between atoms.
(c) It is very common and occurs in, e.g., most transition metal compounds (apart from the metallic ones), Si, Ge and related materials and some organic solids, e.g. anthracene.
(d) It either may be regarded as an electron hopping process, or may be described using band theory, depending on the particular material.
(e) The number of electrons contributing to semiconductivity depends on temperature and impurity level, in contrast to metallic conductivity.
(f) The ability to dope semiconductors and modify their properties gives rise to many applications.

Ionic conductivity occurs in materials known variously as *solid electrolytes, superionic conductors* or *fast ion conductors*.

(a) These solid materials usually have a rigid framework structure, but within which one set of ions forms a mobile sublattice.
(b) In favourable cases, conductivity values as high as $1 \, ohm^{-1} \, cm^{-1}$, similar to that of strong liquid electrolytes, occur.
(c) Solid electrolytes are intermediate between typical ionic solids, in which none of the ions can move off their lattice sites, and liquid electrolytes, in which all the ions are mobile.
(d) The structural requirements for high ionic conductivity are that (i) there must be empty sites available for ions to hop into, and (ii) the energy barrier that ions have to overcome in order to hop between sites must be small.
(e) High ionic conduction is not common but modest amounts of ionic conduction are common, especially in non-stoichiometric or doped materials.

Typical conductivities for a range of materials, both ionic and electronic are given in Table 7.1. Conductivities are usually temperature dependent and for all materials except metals and superconductors, increase with temperature.

Table 7.1 *Typical values of electrical conductivity*

	Material	$\sigma(ohm^{-1} cm^{-1})$
Ionic conduction	Ionic crystals	$< 10^{-18} - 10^{-4}$
	Solid electrolytes	$10^{-3} - 10^{1}$
	Strong (liquid) electrolytes	$10^{-3} - 10^{1}$
Electronic conduction	Metals	$10^{-1} - 10^{5}$
	Semiconductors	$10^{-5} - 10^{2}$
	Insulators	$< 10^{-12}$

Dielectric materials are characterized by the complete absence of electrical conductivity, both electronic and ionic. Their bonding may be strongly ionic, as in MgO, Al_2O_3; strongly covalent, as in diamond, C; or strongly polar covalent, as in SiO_2.

Ferroelectricity has the following characteristics:

(a) It is intermediate between solid electrolyte behaviour, in which ions migrate over long distances, and dielectric behaviour, in which there is no mean displacement of ions from their regular lattice sites.

(b) It is associated with limited atomic displacements of, e.g. 0.1 Å, leading to a net polarization and creation of a dipole moment.

(c) It is limited to a small group of materials. A well-studied case is $BaTiO_3$, perovskite, in which Ti atoms are displaced towards one corner of the TiO_6 octahedra; this happens because Ti is slightly too small for a regular octahedral site.

(d) It is related to *pyroelectricity* and *piezoelectricity*. In pyroelectricity, ionic displacements occur spontaneously and vary with temperature; in piezoelectricity, they are induced by an applied pressure; and in ferroelectricity, they may be induced by an applied electric field.

Metallic conductivity: organic metals

The description of bonding and conductivity in metals in terms of band theory was outlined in Chapter 2. It was also shown how, in transition metal-containing solids, orbital overlap can give rise to bands of energy levels for the *d* electrons; provided certain conditions are met regarding the strength of overlap and the number of *d* electrons available, then metallic conduction may occur. A good example is the oxide TiO.

The possibility of preparing electrically conducting polymers or 'organic metals' is a tantalizing one. Such materials would combine the mechanical properties of polymers–flexibility and ease of fabrication as thin films–with the high electrical conductivity normally reserved for metals. A great deal of research has been carried out in the last few years on such materials but because of problems of atmospheric instability and degradation, none have been commercially exploited as yet. There are two main categories of 'organic metal': conjugated systems and charge transfer complexes.

Conjugated systems

Doped polyacetylene

Organic solids are usually electrical insulators. Electrons cannot move freely within molecules or from one molecule to another in a crystal. Exceptions are conjugated systems that contain a skeleton of alternate double and single carbon-to-carbon bonds, as in graphite. Polymers such as polyethylene are insulators because although the polymer precursor, ethylene, contains a $\overset{\displaystyle}{C}=\overset{\displaystyle}{C}$ double

bond, polyethylene itself is saturated and contains only single $-\overset{|}{\underset{|}{C}}-\overset{|}{\underset{|}{C}}-$ bonds (Fig. 7.1a).

A conjugated long-chain polymer with the potential for electrical conductivity is polyacetylene. The acetylene precursor contains a $-C\equiv C-$ triple bond whereas polyacetylene contains alternate single and double bonds (b). In fact, polyacetylene has a modest electrical conductivity, in the range 10^{-9} ohm^{-1} cm^{-1} (cis form) to 10^{-5} ohm^{-1} cm^{-1} (trans form), which is comparable to that of semiconductors such as silicon.

These conductivity values are quite low because the π electron system is not completely delocalized in polyacetylene. It has a band gap of 1.9 eV. A significant discovery, by MacDiarmid, Heeger and coworkers (1977), was that on doping polyacetylene with suitable inorganic compounds, its conductivity increases dramatically. With dopants such as: (a) Br_2, SbF_5, WF_6 and H_2SO_4, all of which may act as electron acceptors to give e.g. $(CH)_n^{\delta+} Br^{\delta-}$; and (b) alkali metals, which may act as electron donors, conductivities as high as 10^3 ohm^{-1} cm^{-1} in trans-polyacetylene have been obtained. This is similar to the conductivity of many metals and indeed, such materials have been termed 'synthetic metals'. The

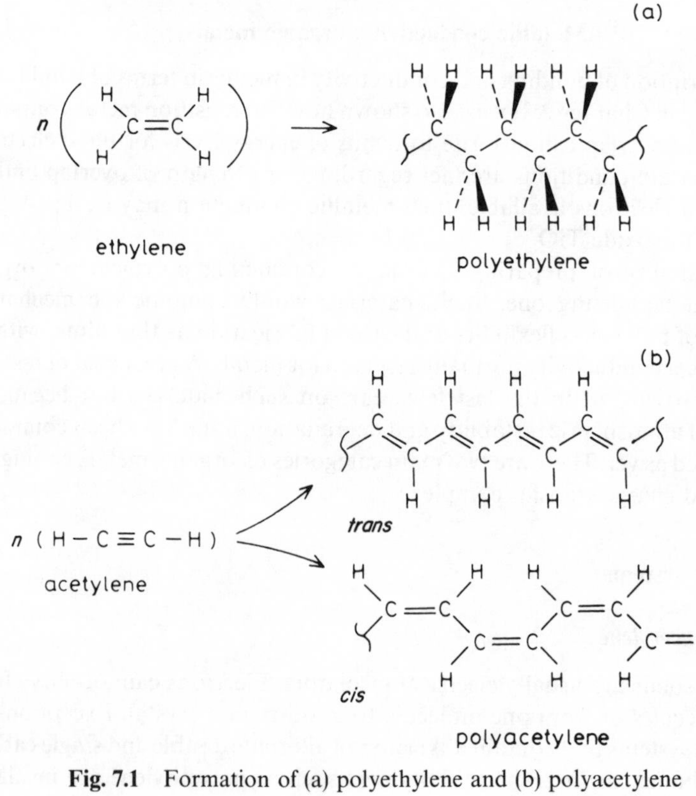

Fig. 7.1 Formation of (a) polyethylene and (b) polyacetylene

conductivity increases extremely rapidly as the dopant is added and a semiconductor-to-metal transition occurs at about 1 to 5 mol% added dopant. Polyacetylene is prepared by the catalytic polymerization of acetylene in the absence of oxygen. A Ziegler–Natta catalyst may be used which is a mixture of $Al(CH_2CH_3)_3$ and $Ti(OC_4H_9)_4$. In one method, acetylene is bubbled through a solution of the catalyst and a solid polyacetylene precipitate forms. In another method, acetylene gas is introduced into a glass tube whose inner surface is coated with a thin layer of catalyst; a layer of polyacetylene forms on the surface of the catalyst. It is usually desired to prepare the *trans* form because it has higher conductivity. It may be prepared directly, at 100 °C, or by heating the *cis* form, which changes rapidly to the *trans* form on heating to ~ 150 °C. Doping may be achieved simply be exposing polyacetylene to gaseous or liquid dopant.

The electronic structure of polyacetylene films is unclear, especially the mechanism of electron transfer between polyacetylene chains. The films have a complex morphology in which chains of polyacetylene appear to fold up to form plates and the plates overlap to form fibres. More work is needed in order to better characterize and control the film texture and to relate specific textural features to the desirable, high electronic conductivity. Conducting polyacetylene has a variety of potential applications. Heavily doped materials could be used in place of metals for certain electrical applications. Undoped or lightly doped materials could be used in place of semiconductors. Thus, *pn diode junctions* could be fabricated in which two films of polyacetylene are brought into contact, one doped with an electron acceptor and p-type, the other doped with an electron donor and n-type, (see next section). Such devices would be easy to make and, with their large surface areas, would have potential application in solar energy conversion. One problem that remains to be solved is the sensitivity of polyacetylene to oxygen. Perhaps substituted or modified polyacetylenes will be made in the future which retain the high conductivity but are not liable to atmospheric attack.

The above discussion is concerned with electronic conductivity. Recently, MacDiarmid and coworkers have shown that ionic conductivity is also possible and that certain doped polyacetylenes may be used as reversible electrodes in new types of battery. The doping is carried out electrochemically. In one arrangement, a polyacetylene film is dipped into a liquid electrolyte composed of $LiClO_4$ dissolved in propylene carbonate. A lithium metal electrode also dips into the electrolyte. On charging the cell at 1.0 V, at room temperature, perchlorate ions from the electrolyte enter the polyacetylene electrode to give $(CH)_y^+(ClO_4)_y^-$, where y is in the range 0 to 0.06. At the same time, Li^+ ions are deposited at the lithium electrode in order to preserve charge balance. The perchlorate ions enter the polyacetylene structure reversibly, i.e. they subsequently diffuse out of the polymer and back into the electrolyte on allowing the cell to discharge. The polyacetylene therefore behaves as a mixed ionic-electronic conductor.

The possibility of using polymeric electrodes in batteries, especially in solid-state batteries, is very attractive for at least two reasons: (a) the polymers are very light (compared to lead in lead/acid batteries): (b) polymers have a flexible

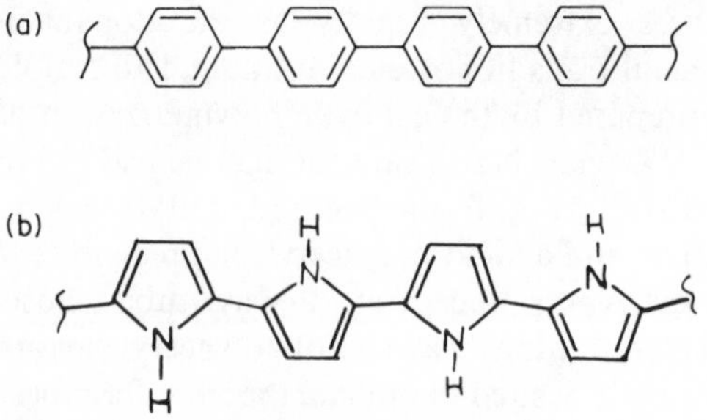

Fig. 7.2 (a) Polyparaphenylene and (b) polypyrrole

structure and this should reduce problems of contact resistances at the electrode–solid electrolyte interface.

Polyparaphenylene and polypyrrole

Another long–chain polymer with considerable potential is polyparaphenylene (Fig. 7.2a). It consists of a long chain of benzene rings and is also a semiconductor. It has been less extensively studied than polyacetylene but has been doped successfully to give much increased electronic conductivity. For instance, on doping with $FeCl_3$, a product of approximate composition $(C_6H_4(FeCl_3)_{0.16})_x$ has been obtained with a conductivity of $0.3\,ohm^{-1}\,cm^{-1}$ at room temperature.

Pyrrole is a heterocyclic molecule with a five-membered ring C_4H_5N. It can be polymerized to give a long-chain structure which effectively has alternate double and single bonds, giving a delocalized π electron system (b). Polypyrrole itself has a low conductivity but can be oxidized by perchlorate to give p-type conductivity as high as $10^2\,ohm^{-1}\,cm^{-1}$. It also has the advantages of being stable in air and able to withstand temperatures of up to $250\,°C$.

Organic charge transfer complexes

There has been much interest for several years in two-component organic systems in which one component is a π electron donor and the other an electron acceptor. Some of these behave as highly conducting synthetic metals and a few are superconducting at very low temperature. The tantalizing possibility of achieving superconductivity at a higher temperature, e.g. at $100\,K$ or higher, provides added impetus to research on related new materials.

Examples of strong π electron acceptors, shown in Fig. 7.3, are (a) tetracyanoquinodimethane (TCNQ) and (b) chloranil. Some strong π electron donors are (c) paraphenylenediamine (PD), tetramethylparaphenylenediamine (TMPD) and (d) tetrathiofulvalene (TTF). In crystalline complexes of π donors and acceptors, the molecules form stacks in which, often, donor and acceptor alternate. The π

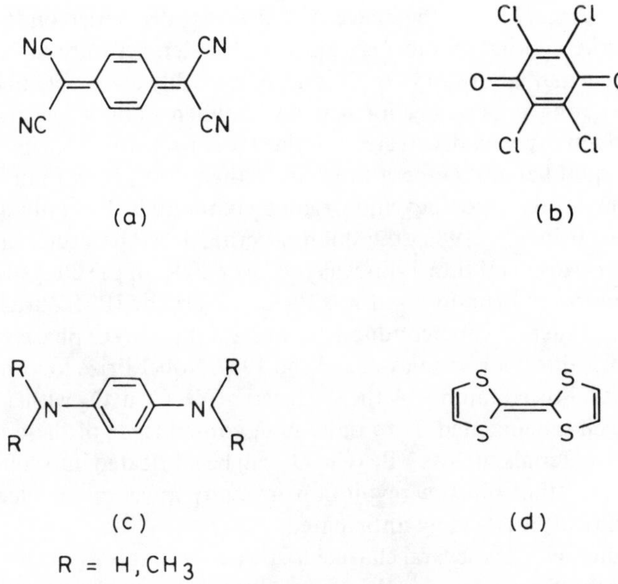

(a)

(b)

(c)

(d)

$R = H, CH_3$

Fig. 7.3 (a) Tetracyanoquinodimethane (TCNQ), (b) chloroanil, (c) p-phenylenediamines and (d) tetrathiofulvalene

electron systems on alternate molecules overlap and permit transfer of charge in the stack direction. For example, TMPD and chloranil form mixed stacks which can be represented as:

$$\ldots(TMPD)^+(chloranil)^-(TMPD)^+(chloranil)^-(TMPD)^+\ldots$$

Complexes are also possible in which only one component is an aromatic system. For instance, TCNQ forms charge transfer complexes with alkali metals, e.g. $(TCNQ)_3Cs_2$, and TTF forms complexes with halogens, e.g. $(TTF)_2Br$.

Although many organic charge transfer complexes are semiconductors, some, such as TTF–TCNQ and NMP–TCNQ (NMP = N-methyl phenazine), have very high conductivity, $\sim 10^3\,ohm^{-1}\,cm^{-1}$. The complex $(TTF)_2Br$ is superconducting at low temperature (below 4.2 K) and high pressure (at 25 kbar).

Superconductivity

Ceramic superconductors

One of the most exciting scientific breakthroughs in recent years has been the discovery in 1986–87 of superconductivity at relatively high temperatures in the ceramic oxides, $La_{2-x}Ba_xCuO_{4-x}$ and $YBa_2Cu_3O_7$. Prior to this discovery, superconductivity was confined to very low temperatures, within a few degrees of absolute zero. Now, with the material $YBa_2Cu_3O_7$ in particular, superconductivity occurs at temperatures ranging from absolute zero to above the boiling point of liquid nitrogen, 77 K. In the space of a few months, superconductivity has changed from being what for most people was a low-temperature curiosity, out of the reach of normal laboratory conditions, to a readily accessible phenomenon with many exciting possibilities for new developments and applications.

Superconductivity was discovered by the Dutch scientist Onnes in 1911, in mercury at liquid helium temperatures. Since then it has been found in a large variety of metals, alloys, oxides and organic compounds, but only at very low temperatures. Prior to 1986, the highest critical temperature, T_c, for the superconductor-to-metal transition was just over 20 K, in the alloys Nb_3Ge and Nb_3Sn. An enormous leap forward was the discovery, by IBM Zurich scientists Bednorz and Muller, of superconductivity in a La–Sr–Cu–O phase with a T_c of about 36 K; for this, they were awarded the 1987 Nobel Prize for Physics. This discovery was followed rapidly by the synthesis of $YBa_2Cu_3O_7$ with a T_c of 91 K, the highest well-documented T_c to date. As a consequence of these discoveries and provided materials such as $YBa_2Cu_3O_7$ can be fabricated into suitable forms such as wires or thin films, a revolution in many areas of the electrical and electronics industries is widely anticipated.

Superconductivity has several characteristic features. It is associated with zero resistance to the flow of electrical current, implying that the current could, in principle, flow indefinitely. The electrical resistance of a small pellet of $YBa_2Cu_3O_7$ as a function of temperature is shown in Fig. 7.4. Below 90 K in the superconducting state, the resistance is zero. Above about 92 K, the material is metallic and the resistance gradually rises with increasing temperature. This is a characteristic feature of metallic behaviour and is associated with electron–phonon collisions, which increase with rising temperature.

Prior to the discovery of the 'high T_c ceramic superconductors', as materials such as $YBa_2Cu_3O_7$ are known, the generally accepted 'BCS' theory of superconductivity (due to Bardeen, Cooper and Schrieffer), involved a loose association of electrons into pairs, known as 'Cooper pairs'. These pairs were supposed to move cooperatively through the lattice in such a way that electron–phonon collisions were avoided. In support of this theory, it is known that the free

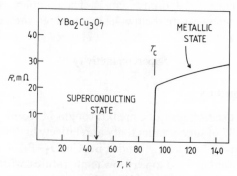

Fig. 7.4   Electrical resistance of $YBa_2Cu_3O_7$
as a function of temperature

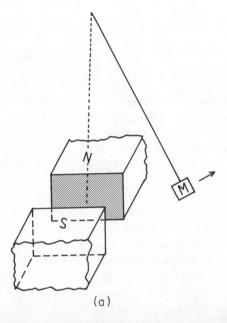

Fig. 7.5 (a) The Meissner effect showing a superconducting material, M, being repelled from a magnetic field, (b) levitation of a sample of $YBa_2Cu_3O_7$ above a magnet. Photograph provided by R. Treviño and C. Piña, UNAM, Mexico.

energy of the superconducting state is lower than that of the corresponding metallic state in the same material; a low-energy 'band gap' is present and absorption of radiation in the microwave region may cause promotion of electrons from the superconducting to the metallic state. Additional evidence for a reduced free energy in the superconducting state comes from heat capacity measurements. These show the superconducting state to have a lower entropy and hence to be more ordered than the metallic state.

In traditional superconductors, the 'correlation length' of the Cooper pairs is large, of the order of hundreds or thousands of angstroms. This is quite different from the situation in chemical bond formation, which involves electron pairs on an angstrom scale. In the new ceramic superconductors, however, the BCS theory leads to correlation lengths of the order of a few angstroms. The tantalizing possibility exists that the superconducting state may be associated with the occurrence of more closely coupled itinerant electron pairs and in some way, with bond formation in solids. It should be emphasized, however, that the current level of understanding of superconductivity in the new ceramic materials is limited and that much more theoretical work remains to be done.

Superconducting materials exhibit 'perfect diamagnetism' and expel a magnetic field, provided the field is below the critical field strength, H_c. This is demonstrated convincingly by the Meissner effect, shown schematically in Fig. 7.5. When in the superconducting state, prepared by cooling in, e.g. liquid nitrogen, a sample, M, of $YBa_2Cu_3O_7$ suspended by a thread swings out from between the jaws of a magnet. As the sample warms, it loses its superconductivity and swings back into the magnetic field.

Structure of $YBa_2Cu_3O_7$

The crystal structure of $YBa_2Cu_3O_7$ is fascinating. Superficially, it may be regarded as a perovskite with 2/9 of the oxygens missing (compare the stoichiometry of perovskite, $BaTiO_3$, with that of $YBa_2Cu_3O_7$). The unit cell is shown in Fig. 7.6 and comprises three perovskite-related units stacked in the c direction. Two of the 'units', comprising the top and bottom thirds of the unit cell, have the net stoichiometry '$BaCuO_{2.5}$' and have two of the twelve edge centre oxygen sites vacant. This gives barium a coordination number of ten, as shown for the upper barium in Fig. 7.6. The middle unit has the effective stoichiometry '$YCuO_2$' and is characterized by four missing edge centre oxygens. Consequently, yttrium has a coordination number of eight.

Two kinds of copper site are present. In perovskite these sites would be octahedral but, because of the oxygen deficiency, one is five-coordinate (Cu(2), square pyramidal arrangement) and the other is four-coordinate (Cu(1), square planar). The square planar units containing Cu(1) link at their corners to form chains parallel to y, Fig. 7.7(a). The square pyramidal units containing Cu(2) also form chains but parallel to both y (a) and x (b); these therefore link up to form sheets in the xy plane. Pairs of adjacent square pyramidal sheets are linked, via the square planar chains, to form symmetrical complex triple layers. The repeat unit of this complex layer is indicated in the bottom right corner of Fig. 7.6. Sections through the layers are shown in Fig. 7.7, in the yz plane (a) and the xz plane (b).

The oxidation state of the copper atoms in $YBa_2Cu_3O_7$ is unusual. If we assume that Y, Ba and O have their usual oxidation states of $+3$, $+2$ and -2,

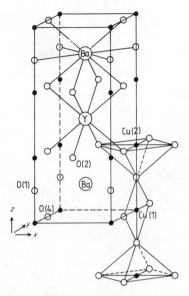

Fig. 7.6 Unit cell of $YBa_2Cu_3O_7$. Orthorhombic, $a = 3.817$, $b = 3.882$, $c = 11.671$ Å. Atomic coordinates: Ba at 1/2, 1/2, 0.18 and 1/2, 1/2, 0.82; Y at 1/2, 1/2, 1/2; Cu(1) at 0, 0, 0; Cu(2) at 0, 0, 0.35 and 0, 0, 0.65; O(1) at 0, 0, 0.16 and 0, 0, 0.84; O(2) at 1/2, 0, 0.38 and 1/2, 0, 0.62; O(3) at 0, 1/2, 0.38 and 0, 1/2, 0.62; O(4) at 0, 1/2, 0

respectively, then for charge balance, the copper must have an average of $+2.33$. This may be rationalized in terms of a $+2$ state for Cu(2), of which there are two in the unit cell, and a $+3$ state for Cu(1), of which there is one per cell.

The key to the superconducting behaviour in $YBa_2Cu_3O_7$ appears to be the copper atoms, their oxidation states and the manner in which they link up, with oxygen, to form the complex layers shown in Fig. 7.7. The superconductivity may be destroyed by partial reduction of the structure: oxygens are removed, from the O(4) sites, and the average oxidation state of copper is reduced. Writing the formula in general terms as $YBa_2Cu_3O_\delta$, δ appears to have a maximum value of 7, but it can be lowered smoothly to 6 or below by heating the sample in a reducing atmosphere (Fig. 7.8a); with decreasing δ, the value of T_c drops (Fig. 7.8b). The structural consequences of the partial reduction are that the chains of square planar copper running parallel to b are destroyed, leaving linear copper bridging units, similar to those seen in Fig. 7.7(b).

Currently, great efforts are being made to obtain materials with higher T_c, with the goal of achieving superconductivity at room temperature. There is some evidence from the so-called a.c. Josephson effect of superconductivity at

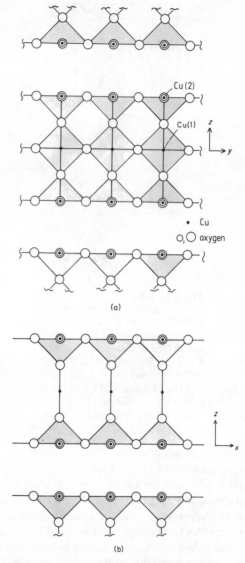

Fig. 7.7 Projections of the $YBa_2Cu_3O_7$ structure (a) down x, (b) down y

temperatures as high as 240 K, in samples which have undergone a prolonged low-temperature anneal, but these results need to be verified.

Applications

The possible applications for the ceramic superconductors are associated with their unusual electrical and magnetic properties. For example, associated with

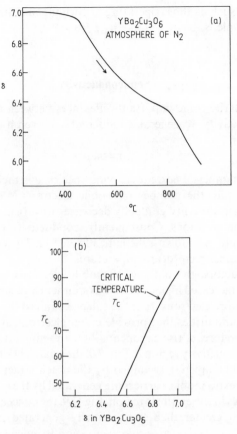

Fig. 7.8 (a) Variation in oxygen content, δ, as a function of temperature in an atmosphere of pure N_2; (b) variation of T_C with δ

the zero electrical resistance are the possibilities for power transmission over large distances without significant I^2R heating losses. The use of superconducting cables would increase greatly the efficiency of the electricity supply industry. At the other extreme, the miniaturization of circuit boards and electronic components would be aided greatly by the use of superconducting wires and contacts; the heat generated from conventional metallic materials limits the degree of miniaturization that is possible.

Associated with the perfect diamagnetism of superconductors are potential applications for shielding magnetic fields. SQUIDS (superconducting quantum interference devices) have potential for detecting minute magnetic fields, such as those associated with the human brain, with applications in medicine. Levitation is readily induced in a superconducting material, associated with the diamagnetism and this may have a variety of applications, including transport systems with

'Maglev' trains such as those operating at Birmingham airport in the UK. The levitation principle is illustrated in Fig. 7.5(b).

Semiconductivity

Let us first clarify some important differences between metallic conductivity and semiconductivity. In general, conductivity is given by

$$\sigma = ne\mu \qquad (7.1)$$

where n is the number of current-carrying species, e is their charge and μ their mobility. In metals, the number of mobile electrons is large and essentially constant, but their mobility gradually decreases with rising temperature due to electron–phonon collisions. Consequently, conductivity gradually drops with rising temperature, as shown schematically in Fig. 7.9 which is a plot of log conductivity against reciprocal temperature.

In semiconductors, the number of mobile electrons is usually small. This number may be increased in one of two ways, either by raising the temperature so as to promote more electrons from the valence band to the conduction band, or by doping with impurities that provide either electrons or holes. In the first of these, n and therefore, σ, rise exponentially with temperature, as shown for the intrinsic semiconductivity region in Fig. 7.9; the relatively small changes in μ with temperature are completely swamped by the much larger changes in n. In the second of these, extra mobile carriers are generated by the addition of dopants; at low temperatures in the extrinsic region, Fig. 7.9, the concentration of these extra carriers is much greater than the thermally generated intrinsic carrier concentration. Consequently, carrier concentration is independent of temperature

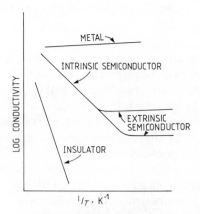

Fig. 7.9 Conductivity characteristics of metals, semiconductors and insulators

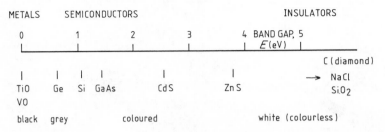

Fig. 7.10 Relation between electronic properties and magnitude of the band gap

and σ shows a slight decrease with temperature due to the mobility effect mentioned above.

Insulators differ from semiconductors only in the magnitude of their conductivity, which is usually several orders of magnitude lower. Thus, the conductivity of insulators is also sensitive to both temperature (shown in Fig. 7.9) and dopants (not shown).

The application of band theory to describing the electronic properties of metals, semiconductors and insulators is outlined in Chapter 2. The key parameter governing their electrical properties is the band gap, E, as indicated in Fig. 7.10 and Table 2.17 for a selection of materials. In order to promote electrons across the band gap, absorption of energy is required. For small band gaps, $<1\,\text{eV}$, some promotion due to thermal excitation occurs, especially with increasing temperature; materials with $E \lesssim 0.01\,\text{eV}$ are essentially metallic. For larger band gaps, radiation of appropriate wavelength can cause promotion and the property of *photoconductivity* results. Thus CdS, $E = 2.45\,\text{eV}$, absorbs visible light and is being used in the development of photocells, as a means of converting sunlight into other forms of energy.

Doped silicon

Many of the technological applications of semiconductors are associated with doped or extrinsic materials. Silicon may be converted into an extrinsic semiconductor by doping with an element from either Group III or Group V of the periodic table. First, let us see the effect of doping with a small amount, say 0.02 atom%, of a trivalent element, e.g. gallium. The gallium atoms replace silicon atoms in the tetrahedral sites of the diamond structure to form a substitutional solid solution. In pure silicon, using covalent bond theory, all the Si—Si bonds may be regarded as normal, electron pair covalent single bonds since Si has four valence electrons and is bonded to four other Si atoms (Fig. 7.11). Gallium has only three valence electrons, however, and in gallium-doped silicon, one of the Ga—Si bonds must be deficient by one electron. Using band theory, it is found that the energy level associated with each single-electron Ga—Si bond does not form part of the valence band of silicon. Instead, it forms a discrete level or atomic orbital just above the top of the valence band. This level is known as an *acceptor*

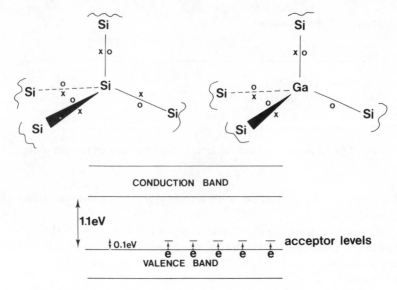

Fig. 7.11   p-type semiconductivity in gallium-doped silicon

level because it is capable of accepting an electron. The gap between the acceptor levels and the top of the valence band is small, ∼ 0.1 eV. Consequently, electrons from the valence band may have sufficient thermal energy to be promoted readily into the acceptor levels. The acceptor levels are discrete if the concentration of gallium atoms is small and it is not possible for electrons in the acceptor levels to contribute directly to conduction. The positive holes that are left behind in the valence band are able to move, however, and gallium-doped silicon is a positive hole or *p-type semiconductor*.

At normal temperatures, the number of positive holes created by the presence of gallium dopant atoms far exceeds the number created by the thermal promotion of electrons into the conduction band, i.e. the extrinsic, positive hole concentration far exceeds the intrinsic concentration of positive holes. Hence, the conductivity is controlled by the concentration of gallium atoms. With increasing temperature, the concentration of intrinsic carriers increases rapidly. At sufficiently high temperatures, the intrinsic carrier concentration may exceed the extrinsic value, in which case, a changeover to intrinsic behaviour would be observed (Fig. 7.9).

Let us consider now the effect of doping silicon with a pentavalent element such as arsenic. The arsenic atoms again substitute for silicon in the diamond-like structure, but now for each arsenic atom there is one electron more than needed to form four Si—As covalent bonds (Fig. 7.12). On the band description, this extra electron occupies a discrete level that is found to lie about 0.1 eV below the bottom of the conduction band. Again, the electrons in these levels cannot move directly as there are insufficient of them to form a continuous band. Instead these levels act as *donor levels* because the electrons in them have sufficient thermal

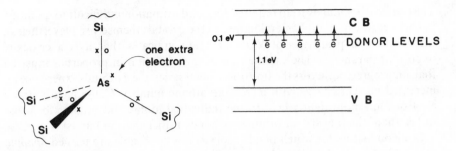

Fig. 7.12 n-type semiconductivity in arsenic-doped silicon

energy to get up into the conduction band where they are free to move. Such a material is known as an *n-type semiconductor*.

It is worthwhile at this stage to summarize the differences between extrinsic and intrinsic semiconductors. These are principally:

(a) Extrinsic semiconductors have much higher conductivities than similar intrinsic ones at normal temperatures. For example, at 25 °C, pure silicon has an intrinsic conductivity of only $\sim 10^{-2}$ ohm^{-1} cm^{-1}. By appropriate doping, however, its conductivity is increased by several orders of magnitude.

(b) The conductivity of extrinsic semiconductors may be accurately controlled by controlling the concentration of dopant. Materials with desired values of conductivity may therefore be designed and manufactured. With intrinsic semiconductors, the conductivity is strongly dependent on temperature and the presence of stray impurities.

The extrinsic region can be extended to cover a wide range of temperatures by choosing (a) a material with a large intrinsic band gap, and (b) dopants whose associated energy level is close to the relevant valence or conduction band.

Other semiconductors

Band theory is able to account satisfactorily for the properties of semi-conductors such as Si and Ge. In certain transition metal compounds, however, it appears not to be applicable, mainly because the outer valence orbitals on adjacent atoms do not overlap sufficiently strongly to form bands of delocalized energy levels. For instance, in MnO which has the rock salt structure, the t_{2g} orbitals on manganese (a d^5 ion) contain three d electrons only since it is in the high spin configuration (the other two electrons are in the e_g orbitals). If overlap of the t_{2g} orbitals on adjacent atoms occurred, then a half-filled band would be expected, with consequent metallic conductivity. This does not occur and MnO is instead a semiconductor.

A similar situation exists with FeO and CoO, which are also semiconductors. In the monoxides of the early transition elements, TiO and VO, however, significant overlap of the t_{2g} orbitals must occur since these materials are metallic.

The degree of orbital overlap required for band formation is difficult to quantify, in part because of the ephemeral nature of the orbitals themselves. Nevertheless, the contrast in properties of VO and MnO indicates that it is a critically important parameter. The explanation of this variation in properties must be that, on progressing across the transition metal series, the d orbitals experience an increased nuclear charge with increasing atomic number. Consequently, the d electrons are more tightly held to the individual atoms and become localized rather than itinerant (i.e. in orbitals or bonds rather than in bands).

Semiconductors for which band theory is not applicable are termed *hopping semiconductors*. In these, the electrons are localized on individual atoms but are able to hop to adjacent atoms provided they gain sufficient energy. There is an activation energy to be overcome, therefore, in much the same way that the band gap of intrinsic semiconductors may be regarded as the activation barrier to conduction.

The conductivity of semiconducting transition metal compounds is often increased when the transition element is present in more than one oxidation state. Such materials are known as *controlled valency* or *mixed valency semiconductors*; a good example is provided by the stoichiometric and non-stoichiometric oxides of nickel. Stoichiometric nickel(II) oxide, NiO, is a pale green solid of very low conductivity. Its colour is due to internal d–d transitions within the octahedrally coordinated Ni^{2+} ions and hence, the d electrons must be localized on the individual cations. On heating NiO in air at, e.g. 800 °C, it is oxidized to an oxygen-rich material of empirical stoichiometry, $Ni_{1-x}O$, where $x \simeq 0.1$. This material is black and a moderately good semiconductor. It has the same basic rock salt structure as NiO, but distributed over the cation sties is a mixture of Ni^{2+} and Ni^{3+} ions and cation vacancies, corresponding to the formula

$$Ni^{2+}_{1-3x} Ni^{3+}_{2x} V_x O$$

where V is a cation vacancy. The significant degree of electronic conductivity arises because electrons can transfer from Ni^{2+} to Ni^{3+} ions. Although the nickel ions themselves do not move, the net result of this electron transfer is the same as if the Ni^{3+} ions were moving, but in the opposite direction to the electron transfer. The Ni^{3+} ions are effectively positive holes and hence, black nickel oxide is a p-type semiconductor. In contrast to gallium-doped silicon, which is also p-type, the Ni^{2+}/Ni^{3+} exchange is thermally activated and therefore, highly temperature dependent.

The disadvantage of using oxidized NiO as a semiconductor is that its conductivity is difficult to control; it depends on both temperature and the partial pressure of oxygen. To overcome this difficulty, Verwey (1948) introduced the idea of controlled valency semiconductors in which the concentration of, say, Ni^{3+} ions is dependent not on temperature but on the addition of a controlled amount of dopant. For instance, lithium oxide may be reacted with nickel oxide and oxygen to form solid solutions of formula:

$$Li_x Ni^{2+}_{1-2x} Ni^{3+}_x O$$

In these, the concentration of Ni^{3+} ions, and hence the conductivity, depend on the concentration of Li^+ ions. The magnitude of the conductivity varies enormously with x, from $\sim 10^{-10}\,ohm^{-1}\,cm^{-1}$ for $x = 0$ to $\sim 10^{-1}\,ohm^{-1}\,cm^{-1}$ for $x = 0.1$ at $25\,°C$.

Applications

The main use of semiconductors is in solid state devices such as transistors, silicon chips, photocells, etc. A simple example and the main one to be discussed here is the *pn junction*. This is the solid state equivalent of the diode rectifier valve. Suppose that a single crystal of, for example, silicon is doped in such a way that one half is n-type and the other half p-type. The band structure at the junction is approximately as shown in Fig. 7.13. The Fermi levels are at different heights in the two halves. Consequently, electrons are able to flow spontaneously from the n-type to the p-type regions across the junction. The Fermi energy of electrons is similar to their electrochemical potential: as long as a difference in potential exists, electrons flow from a region of high potential to one of low potential.

In the absence of an externally applied potential difference, a few electrons move from right to left across the junction but the band structure then rapidly adjusts itself so that E_F becomes the same on either side. An alternative explanation is that a space charge layer rapidly develops at the junction as electrons flow across it. The space charge then acts as a barrier to further electron flow.

This situation changes, however, if an external potential difference is applied such that the p-type end is positive and the n-type end is negative. A continuous current is able to flow through the circuit; electrons enter the crystal from the

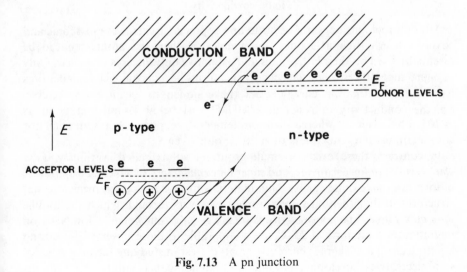

Fig. 7.13 A pn junction

right-hand electrode. They flow through the conduction band of the n-type region, drop into the valence band of the p-type region at the pn junction and then flow through the valence band, via the positive holes, to leave at the left-hand electrode. A continuous current cannot flow in the opposite direction since, under normal circumstances, with a relatively low applied voltage, electrons cannot surmount the barrier necessary in order to pass from left to right across the junction. The pn junction is therefore a rectifier in that current may pass in one direction only. It may be used to convert a.c. into d.c. electricity. Silicon-based pn junctions have largely replaced diode valves.

A more complex arrangement is the pnp or npn junction. This acts as a current or voltage amplifier. It forms the basis of the transistor which has almost completely taken the place of triode valves.

Controlled valency semiconductors find application as *thermistors*, thermally sensitive resistors. In these, use is made of the large temperature dependence of the conductivity, associated with the fact that these materials are hopping semiconductors. For example, the material $Li_{0.05}Ni_{0.95}O$ shows Arrhenius-type conductivity behaviour over a wide temperature range up to $\sim 200\,°C$. The activation energy is $\sim 0.15\,eV$. Since the conductivity behaviour is reproducible, lithium nickel oxide can be used in devices to control and measure temperature. In order to achieve reproducibility, materials that are insensitive to impurities must be used, e.g. Fe_3O_4, Mn_2O_3, Co_2O_3 doped NiO and certain spinels.

Some semiconductors are *photoconductive*, i.e. their conductivity increases greatly on irradiation with light. Amorphous selenium is an excellent photo-conductor and forms an essential component of the photocopying process (Chapter 8). Conventional band theory cannot be used to explain the properties of amorphous materials such as selenium since they lack long-range periodicity.

Ionic conductivity

Migration of ions does not occur to any appreciable extent in most ionic and covalent solids such as oxides and halides. Rather, the atoms tend to be essentially fixed on their lattice sites and can move only via crystal defects. Only at high temperatures, where defect concentrations become quite large and atoms have a lot of thermal energy, does the conductivity become appreciable, e.g. the conductivity of NaCl at $\sim 800\,°C$, just below its melting point, is $\sim 10^{-3}\,ohm^{-1}\,cm^{-1}$, whereas at room temperature, pure NaCl is an insulator with a conductivity much less than $10^{-12}\,ohm^{-1}\,cm^{-1}$.

In contrast, there exists a small group of solids called, variously, *solid electrolytes, fast ion conductors* and *superionic conductors*, in which one of the sets of ions can move quite easily. Such materials often have rather special crystal structures in that there are open tunnels or layers through which the mobile ions may move. The conductivity values, e.g. $10^{-3}\,ohm^{-1}\,cm^{-1}$ for Na^+ ion migration in β-alumina at $25\,°C$, are comparable to those observed for strong liquid electrolytes. There is currently great interest in studying the properties of solid electrolytes, developing new ones and extending their range of applications in solid state electrochemical devices. Let us consider first the behaviour of typical ionic solids and see how their electrical properties are controlled by the presence of crystal defects.

Alkali halides:vacancy conduction

In crystals of the alkali halides, e.g. NaCl, cations are usually more mobile than anions. A section through the NaCl structure is shown in Fig. 7.14 in which a Na^+ ion is moving into an adjacent vacant cation site and thereby leaving its own site vacant. The Na^+ ion that has moved can travel no further because (a) there are no other vacant sites in the vicinity for it to move into and (b) interstitial migration of Na^+ in NaCl appears not to occur to any appreciable extent. The cation vacancy may continue to move, however, because it is always surrounded by twelve Na^+ ions, one of which can jump in such a way as to occupy and effectively change places with the vacancy. It is convenient, therefore, to regard cation vacancies as the main current carriers in NaCl. Anion vacancies are present, too, in NaCl but they appear to be rather less mobile than the cation vacancies.

The magnitude of the ionic conductivity of NaCl depends on the number of cation vacancies present, which in turn depends very much on the purity and thermal history of the crystals. Vacancies are normally created by one of two methods. One method is simply to heat the crystal. The number of vacancies present in thermodynamic equilibrium increases exponentially with temperature, equation (5.12), and is the number that is *intrinsic* to the pure crystals. The other method involves addition of aliovalent impurities. As a result, vacancies may be created so as to preserve charge balance. For example, addition of a small amount of $MnCl_2$ yields, at equilibrium, a solid solution of formula

$$Na_{1-2x}Mn_xV_xCl$$

where, for each Mn^{2+} ion, there is an associated cation vacancy, V. Such

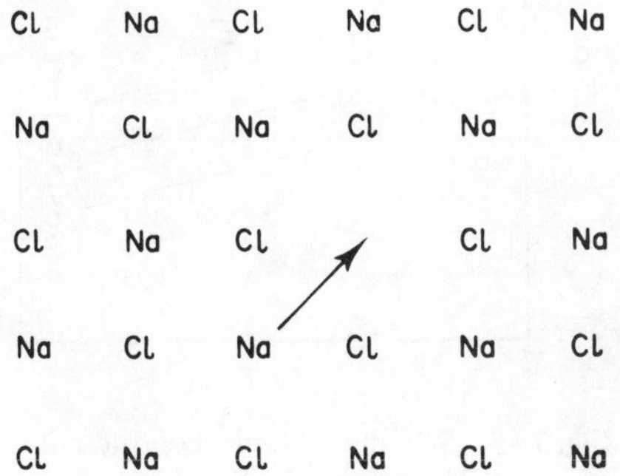

Fig. 7.14 Migration of cation vacancies, i.e. Na^+ ions, in NaCl

302

vacancies are *extrinsic* because they would not be present in pure NaCl. At low temperatures (e.g. 25 °C), the number of thermally generated intrinsic vacancies is very small and, unless the crystal is very pure, is much less than the concentration of extrinsic vacancies.

With increasing temperature, a changeover from extrinsic to intrinsic behaviour occurs when the thermally generated intrinsic vacancy concentration exceeds the impurity controlled extrinsic vacancy concentration.

The temperature dependence of ionic conductivity is usually given by the Arrhenius equation

$$\sigma = A \exp\left(\frac{-E}{RT}\right) \tag{7.2}$$

where E is activation energy, R the gas constant and T the absolute temperature. The pre-exponential factor, A, contains several constants, including the vibrational frequency of the potentially mobile ions. Graphs of $\log_e \sigma$ against T^{-1} should give straight lines of slope $-E/R$. In some treatments, a reciprocal temperature term is included in the prefactor, A, in which case it is usual practice to plot $\log \sigma T$ against T^{-1}. This may make a small difference to the value of the slope, $-E/R$. A schematic Arrhenius plot for NaCl is shown in Fig. 7.15. In the low temperature extrinsic region, the number of vacancies is dominated by the impurity level and is constant for a given impurity concentration. A set of parallel lines is shown, each of which corresponds to the conductivity of a crystal with a different amount of dopant. In the extrinsic region, the temperature dependence of σ depends only on the cation mobility, μ (equation (7.1)), whose temperature

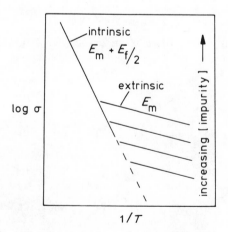

Fig. 7.15 Schematic ionic conductivity of doped NaCl crystals. Parallel lines in the extrinsic region correspond to different dopant concentrations

dependence is also given by an Arrhenius expression:

$$\mu = \mu_0 \exp\left(\frac{-E_m}{RT}\right) \tag{7.3}$$

where E_m is the activation energy for migration of cation vacancies.

In order to understand the origin of this activation energy for migration, consider the possible paths that may be taken by a Na^+ ion in jumping from its lattice site into an adjacent vacancy. A small part of the NaCl structure is shown in Fig. 7.16. It is a simple cube with Na^+, Cl^- ions at alternate corners and corresponds to one-eighth of the unit cell of NaCl. One corner Na^+ site is shown empty and a Na^+ ion from one of the other three corners moves to occupy it. The direct jump (dotted line) across the cube face is not possible because Cl^- ions 1 and 2 are very close, if not in actual contact, and an Na^+ ion would not be able to squeeze between them. Instead, Na^+ must take an indirect route (curved arrow) that passes through the middle of the cube. At the cube centre is an interstitial site that is equidistant from the eight corners. Four of the corners are occupied by Cl^- ions which are arranged tetrahedrally about the cube centre. Before the moving Na^+ ion arrives at this central interstitial site, it has to pass through a triangular window formed by the Cl^- ions 1, 2 and 3. Let us now calculate the size of this window in order to appreciate how difficult it is for Na^+ to squeeze through.

For NaCl, the unit cell parameter, a, is equal to 5.64 Å. The Na—Cl bond length in Fig. 7.16 is $a/2 = 2.82$ Å. This is equal to $(r_{Na^+} + r_{Cl^-})$, assuming the anions and cations to be in contact. Tabulated ionic radii (which vary somewhat, depending on the table that is consulted) of Na^+ and Cl^- are ~ 0.95 and ~ 1.85 Å; the value of the Na—Cl bond length calculated from the sum of these radii is ~ 2.80 Å, close to the experimentally measured value.

In close packed structures, such as NaCl, the anions are either in contact with each other or are in close proximity. The Cl^- ions 1, 2 and 3, form part of a close packed layer and the distance Cl(1)–Cl(3) is given by $[(a/2)^2 + (a/2)^2]^{1/2} = 3.99$ Å. This is about 0.3 Å larger than given by $2r_{Cl^-}$ and so adjacent Cl^- ions in NaCl are not quite in contact.

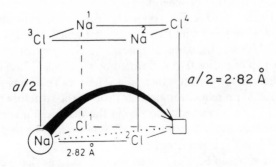

Fig. 7.16 Pathway for Na^+ migration in NaCl

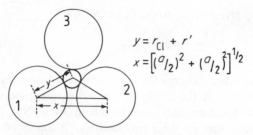

$$y = r_{Cl} + r'$$
$$x = \left[\left(\tfrac{a}{2} \right)^2 + \left(\tfrac{a}{2} \right)^2 \right]^{1/2}$$

Fig. 7.17 Triangular interstice of radius r' through which a moving Na^+ ion must pass in NaCl. Circles 1–3 are Cl^- ions

The radius r' of the 'triangular' window formed by Cl^- ions, 1, 2 and 3 may be calculated from Fig. 7.17 as follows:

$$\cos 30° = \frac{x/2}{y} = \frac{1.995}{(r_{Cl^-} + r')}$$

Therefore,

$$(r_{Cl^-} + r') = \frac{1.995}{\cos 30°} = 2.30 \, \text{Å}$$

If

$$r_{Cl^-} = 1.85 \, \text{Å}, \quad \text{then} \quad r' = 0.45 \, \text{Å}$$

The radius r'' of the interstitial site at the cube centre in Fig. 7.16 may be calculated similarly. The body diagonal of the cube, effectively given by $(2r_{Cl^-} + 2r'')$, is equal to

$$2(r_{Cl^-} + r'') = [(a/2)^2 + (a/2)^2 + (a/2)^2]^{1/2}$$
$$= 4.88 \, \text{Å}$$

Therefore,

$$r'' = 0.59 \, \text{Å}$$

We can see, therefore, that the jumping of a Na^+ ion in NaCl is a difficult and complicated process. First, the Na^+ ion has to squeeze through a narrow triangular gap of radius 0.45 Å, whereupon it finds itself in a small tetrahedral interstitial site of radius 0.59 Å. The residence time here is quite short since this site has a hostile environment with two Na^+ ions, 1 and 2, at a distance of 2.44 Å as well as four Cl^- ions at the same distance. The Na^+ ion leaves by squeezing through another gap of radius 0.45 Å (formed by Cl^- ions 1, 2 and 4) to occupy the vacant octahedral site on the other side. These calculations are inevitably somewhat idealized since relaxation or distortion of the structure must occur in the vicinity of the defects, thereby modifying the distances involved. They do nevertheless show that the migration of Na^+ ions is difficult and is associated with a considerable activation energy barrier.

In the extrinsic regions of Fig. 7.15, then, the conductivity depends on both the vacancy concentration and the mobility. It is given by combining equations (7.1) and (7.3) to give

$$\sigma = ne\mu_0 \exp\left(\frac{-E_m}{RT}\right) \tag{7.4}$$

In the intrinsic conductivity region at higher temperatures, the concentration of thermally induced vacancies is greater than the vacancy concentration associated with the dopant. Hence, the number of vacancies, n, is temperature dependent and is also given by an Arrhenius equation:

$$n = N \exp\left(\frac{-E_f}{2RT}\right) \tag{7.5}$$

This equation is the same as equation (5.12), in which $E_f/2$ is the activation energy for formation of one mole of cation vacancies, i.e. half the energy required to form one mole of Schottky defects. The vacancy mobility is again expressed by equation (7.3) and so the overall conductivity in the intrinsic region is given by

$$\sigma = Ne\mu_0 \exp\left(\frac{-E_m}{RT}\right)\exp\left(\frac{-E_f}{2RT}\right)$$

i.e.

$$\sigma = A \exp\left(-\frac{E_m + E_f/2}{RT}\right) \tag{7.6}$$

In Fig. 7.15 are shown schematic Arrhenius plots for NaCl crystals with varying degrees of purity. The set of parallel lines in the extrinsic region corresponds to different impurity levels of, for example, Mn^{2+}, whereas the single line in the intrinsic region shows that here, the conductivity is unaffected by the impurity content. This latter is reasonable if the actual impurity concentration is very small (< 1 per cent Mn^{2+}) since the presence of Mn^{2+} ions at this level does not affect significantly the activation energy for migration of the cation vacancies. The gradient of the intrinsic line is greater than that of the extrinsic lines and if both intrinsic and extrinsic gradients can be measured then E_m and E_f may be determined separately.

Good quality experimental data for NaCl single crystals (Fig. 7.18) indicate that the simple behaviour shown in Fig. 7.15 is somewhat idealized and that several complicating factors are present. Stages I and II in Fig. 7.18 correspond to the regions of intrinsic and extrinsic conductivity, respectively, of Fig. 7.15. The dashed lines in Fig. 7.18 represent extrapolations of regions I and II and show where, experimentally, deviations occur. Stage I' occurs close to the melting point (802 °C) and has been variously attributed to two possible causes. One explanation is that the upward curvature in conductivity in I' arises because anion vacancies are now becoming increasingly mobile and make a significant contribution to σ. The other is that long-range Debye–Hückel interactions between cation and anion vacancies become appreciable as the vacancy concentration increases at high temperatures. An attractive force results (similar to the Debye–Hückel interaction between ions in solutions) which has the effect

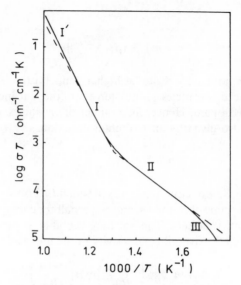

Fig. 7.18 Ionic conductivity of 'pure' NaCl
as a function of temperature

of partially compensating for the energy of formation of the vacancies. Thus, vacancy formation is easier, the vacancy concentration increases and σ increases. It is not known which explanation is correct for NaCl, although the Debye–Hückel effect is probably the cause of a similar deviation in σ at high temperatures in AgCl single crystals (see the discussion of Fig. 5.4).

Below $\sim 390\,°C$ (in this particular NaCl crystal), σ deviates downwards from the ideal extrinsic line, stage III. This is attributed to the formation of defect complexes, e.g. cation vacancy/anion vacancy pairs or cation vacancy/aliovalent cation impurity pairs. These complexes are formed by short-range attractions between defects on neighbouring or next nearest neighbour sites and are quite distinct from the Debye–Hückel interactions mentioned above which are long-range interactions and are associated with preserving electroneutrality. In stage III, cation vacancies that are present in defect complexes must first acquire the additional energy needed to dissociate themselves from the defect complexes before they can move. As a result the activation energy of stage III is greater than E_m (stage II).

Although the conductivity of NaCl crystals has been measured many times and in many different laboratories, there is not even good agreement on the value of E_m, the activation energy for Na^+ migration. Values range from 0.65 to 0.85 eV (~ 60 to $80\,kJ\,mol^{-1}$). Data for the various activation energies associated with ionic conduction in NaCl are given in Table 7.2. The probable reason for the variation in E_m values is that other defects, especially dislocations, are inevitably present in crystals and have a considerable influence on cation migration. The regions of the crystal structure that are close to a dislocation are in a stressed and distorted condition and migration of cations along 'dislocation pipes' may take

Table 7.2 *Conductivity in NaCl crystals*

Process	Activation energy (eV)
Migration of Na^+, E_m	0.65–0.85
Migration of Cl^-	0.90–1.10
Formation of Schottky pair	2.18–2.38
Dissociation of vacancy pair	~ 1.3
Dissociation of cation vacancy $- Mn^{2+}$ pair	0.27–0.50

place more easily than through regions of perfect crystal. If this is so, conductivity and the magnitude of E_m may depend on the number and distribution of dislocations and hence on the thermal history of the crystals. It is, however, difficult to make quantitative measurements of the effect of dislocations on σ because their number cannot be controlled and measured accurately.

The effect of impurities on σ is more amenable to study since their number may be controlled and measured and they have a dramatic effect on σ in the extrinsic region. The concentration of Schottky defects in NaCl in the intrinsic region and at thermodynamic equilibrium is given by the law of mass action (equation (5.5)):

$$K = \frac{[V_{Na}][V_{Cl}]}{[Na^+][Cl^-]}$$

where V_{Na} and V_{Cl} refer to vacant cation and anion sites, respectively. It is assumed that the equilibrium constant, K, is unaffected by the presence of small amounts of aliovalent impurity and that the denominator, the number of occupied sites, is essentially constant and equal to 1. Therefore,

$$[V_{Na}][V_{Cl}] = \text{constant} = x_0^2 \tag{7.7}$$

where, in the intrinsic region, $x_0 = [V_{Na}] = [V_{Cl}]$. If the number of cation vacancies in the extrinsic region is increased, e.g. by adding aliovalent cation impurity, then the number of anion vacancies must decrease in accord with equation (7.7). Let x_a and x_c be the concentration of anion and cation vacancies under extrinsic conditions (and so $x_a \neq x_c$) and c be the concentration of divalent impurity cations. Then

$$x_c = x_a + c \tag{7.8}$$

This condition arises because both anion vacancies and divalent cation impurities carry a net positive charge of $+ 1$, whereas the cation vacancy has an effective charge of $- 1$. Overall, charge balance must be retained. On combining equations (7.7) and (7.8) and solving the resulting quadratic, the following positive values of x_c and x_a are obtained:

$$x_c = \frac{c}{2}\left[1 + \left(1 + \frac{4x_0^2}{c^2}\right)^{1/2}\right] \tag{7.9}$$

$$x_a = \frac{c}{2}\left[\left(1 + \frac{4x_0^2}{c^2}\right)^{1/2} - 1\right] \tag{7.10}$$

If $x_0 \ll c$, then $x_c \to c$ and $x_a \to 0$; this result applies to the extrinsic region and is the same as that deduced above. If $x_0 \gg c$, then $x_c = x_0 = x_a$; this applies to the intrinsic region of conductivity.

Not all impurities lead to an increase in σ in the extrinsic region. Impurities of the same valence, e.g. K^+ or Br^- in NaCl, normally have no effect on σ unless they are present in large concentrations. Impurities that lead to a reduction in concentration of the mobile species cause a reduction in σ. Thus, if divalent anions could be dissolved in NaCl crystals, the concentration of the less mobile anion vacancies would increase at the expense of the more mobile cation vacancies. This effect is not observed to any large extent in NaCl because divalent anion impurities are not soluble, but it is an important effect in AgCl, as described below.

Silver chloride: interstitial conduction

The predominant defects in AgCl are cation Frenkel defects, i.e. interstitial Ag^+ ions associated with Ag^+ ion vacancies (Chapter 5). Experiments have shown that the interstitial Ag^+ ions are more mobile than the Ag^+ vacancies. Two possible mechanisms for migration of interstitial Ag^+ are shown schematically in Fig. 7.19(a). In the direct *interstitial* mechanism (1), the interstitial Ag^+ ion jumps to an adjacent empty interstitial site. In the indirect or *interstitialcy* mechanism (2), a knock-on process occurs. The interstitial Ag^+ ion causes one of its four Ag^+ neighbours to move off its normal site into an adjacent interstitial site and itself occupies the vacant lattice site thereby created. It is possible to distinguish between the interstitialcy and direct interstitial mechanisms if accurate data for both diffusion and conductivity are available. In diffusion measurements, the crystal is doped with radioactive Ag^{+*} ions and the migration of these radioactive or tracer Ag^+ ions is followed. In conductivity measurements, all the Ag^+ ions, not only the radioactive ones, contribute to the net conductivity. A relationship exists between the coefficient for self-diffusion, D, and the conductivity, σ. This is given by the Nernst–Einstein equation:

$$D = \frac{kT}{f n (Ze)^2}\sigma \tag{7.11}$$

where (Ze) is the charge on the mobile ions and n is their concentration; f is a correlation factor, the *Haven ratio*, whose value depends on the mechanism of ion migration. The value of the Haven ratio is different for the two mechanisms shown in Fig. 7.19(a). In mechanism 2, the net distance of charge displacement is greater than either of the jump distances of the individual Ag^+ ions. In mechanism 1, however, the jump distance of the Ag^+ ion equals the distance of the overall charge migration. Since diffusion measurements determine the movement of the ions themselves and conductivity measure the overall charge

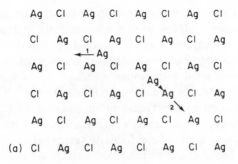

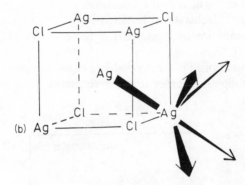

Fig. 7.19 (a) Migration of interstitial Ag^+ ions by (1) direct interstitial jump and (2) indirect interstitialcy mechanism. (b) Pathway for migration of Ag^+ in AgCl by an interstitialcy mechanism

displacement, the Haven ratio is different for the two mechanisms. It has been found experimentally that the interstitialcy mechanism (2) is the operative mechanism in AgCl.

The difference between the vacancy migration mechanism that occurs in NaCl and the interstitialcy mechanism in AgCl may be seen by comparing Figs 7.16 and 7.19(b). Both NaCl and AgCl have the same rock salt crystal structure. In the vacancy mechanism, a Na^+ ion moves from one corner of the cube to another, via an interstitial site at the cube centre which is occupied only transiently (Fig. 7.16). In the interstitialcy mechanism, a Ag^+ ion effectively moves from the interstitial site in the centre of one cube to the site in the centre of an adjacent cube by knocking on a Ag^+ ion placed at one of the corner sites (Fig. 7.19b).

The effect of aliovalent cation impurities on the extrinsic conductivity region of AgCl is different to that observed in NaCl. The presence of, say, Cd^{2+} again increases the number of cation vacancies, but because the product of the concentrations of cation vacancies and Ag^+ interstitials is constant (for dilute impurity levels; see equation (5.15)) the interstitial Ag^+ concentration must decrease with increasing Cd^{2+} concentration. Addition of Cd^{2+} therefore leads

310

to a reduction in the concentration of the more mobile species; the resulting Arrhenius conductivity plot is shown schematically in Fig. 7.20. An extrinsic region at lower temperatures is again observed but it is displaced *downwards* to lower σ values. The degree of downward displacement increases with increasing Cd^{2+} concentration until a minimum conductivity is reached at which the conductivity due to the more numerous but less mobile cation vacancies equals that due to the less numerous but more mobile interstitial Ag^+ ions. At still higher defect concentrations, cation vacancy migration predominates and σ increases. This is shown in (b) as a plot of conductivity versus defect concentration for two temperatures. In the intrinsic region, the conductivity is independent of Cd^{2+} concentration. In extrinsic region I, interstitial conduction predominates and σ decreases as $[Ag_i^+]$ decreases with increasing $[Cd^{2+}]$. In extrinsic region II, vacancy conduction predominates and σ increases as $[V_{Ag}]$ increases with increasing $[Cd^{2+}]$.

The relevant equations for the dependence of σ on impurity concentration, c, are obtained as follows. In order to preserve charge balance:

$$x_c = c + x_i \tag{7.12}$$

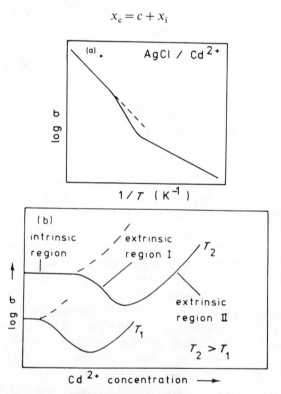

Fig. 7.20 (a) Schematic diagram showing the effect of Cd^{2+} on conductivity of AgCl crystals. (b) Effect of Cd^{2+} impurity on isothermal conductivity of AgCl. Dashed lines represent the effect of adding a divalent anionic impurity

where x_i and x_c are the concentrations of interstitial Ag^+ ions and Ag^+ ion vacancies, respectively. The conductivity in the extrinsic region is given by

$$\sigma = e(x_c\mu_c + x_i\mu_i)$$

$$= e\mu_c\frac{c}{2}\left[1 + \left(1 + \frac{4x_0^2}{c^2}\right)^{1/2}\right] + e\mu_i\frac{c}{2}\left[\left(1 + \frac{4x_0^2}{c^2}\right)^{1/2} - 1\right] \qquad (7.13)$$

Accurate measurements of σ for AgCl show that Fig. 7.20(a) is somewhat idealized and that deviations do occur. At high temperatures, long-range Debye–Hückel interactions become important and give an upward departure from the intrinsic slope, whereas at low temperatures 'complexes' form between cation vacancies and aliovalent cation impurities to give a downward deviation in the extrinsic region. Defect energies, in electronvolts, for AgCl are as follows:

Formation of Frenkel defect   1.24
Migration of cation vacancy   0.27–0.34
Migration of interstitial Ag^+ 0.05–0.16

Alkaline earth fluorides

The most important defect in this group is probably the anion Frenkel defect in which an interstitial F^- ion occupies the centre of a cube that has eight F^- ions at the corners (Fig. 1.25). Conductivity measurements have shown that the anion vacancy is more mobile than the interstitial F^- ion. This contrasts with AgCl in which the interstitial Ag^+ is more mobile than the cation vacancy. In some materials that have the fluorite structure, e.g. PbF_2, the conductivity at high temperatures becomes quite large.

Solid electrolytes (or fast ion conductors, superionic conductors)

Most crystalline materials, like NaCl or MgO, have low ionic conductivities because, although the atoms or ions undergo thermal vibrations, they cannot usually escape from their lattice sites. The small group of solid electrolytes are an exception. In solid electrolytes, one component of the structure, cationic or anionic, is not confined to specific lattice sites but is essentially free to move throughout the structure. Solid electrolytes are, therefore, intermediate in structure and property between, on the one hand, normal crystalline solids with regular three-dimensional structures and immobile atoms or ions and, on the other, liquid electrolytes which do not have regular structures but do have mobile ions. Often, solid electrolytes are stable only at high temperatures. At lower temperatures they may undergo a phase transition to give a polymorph with a low ionic conductivity and a more usual type of crystal structure (Fig. 7.21). For example, Li_2SO_4 and AgI are both poor conductors at 25 °C but at temperatures of 572 and 146 °C, respectively, their crystal structures change to give polymorphs, α-Li_2SO_4 and α-AgI, that have mobile Li^+ and Ag^+ ions (σ

312

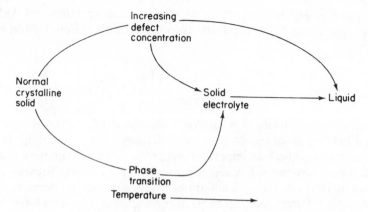

Fig. 7.21 Solid electrolytes as intermediate between normal crystalline solids and liquids

$\sim 1\,\text{ohm}^{-1}\,\text{cm}^{-1}$). On heating, the conductivity therefore increases dramatically at the phase transition.

Other solid electrolytes form as a consequence of a gradual increase in defect concentration with increasing temperature. For example, in ZrO_2, the concentration of anion vacancies above $\sim 600\,°C$ is sufficiently large that zirconia is a good, high-temperature oxide ion conductor. The distinction between normal ionic solids and solid electrolytes is often not well defined, especially for materials such as ZrO_2 which undergo a gradual change in behaviour with increasing temperature.

It is now becoming apparent from both theoretical studies and experimental results on a wide variety of materials that ionic conductivities of between 0.1 and $10\,\text{ohm}^{-1}\,\text{cm}^{-1}$ are the maximum that are likely to be obtained for any solid material. These values are obtained when a large proportion of the ions are moving at any one time. According to some authors, the names 'superionic conductor' and 'fast ion conductor' should be reserved for materials that belong to this latter category of optimized conductivity. Although these names are in common usage, they are misnomers because the mobile ions do not have any 'super' properties nor are they exceptionally mobile. Rather, the high conductivities are associated with a large concentration of mobile species and a relatively low activation energy for ion migration.

The classification of solid electrolytes as intermediate between normal ionic solids and ionic liquids (Fig. 7.21) is supported by data on the relative entropies of polymorphic transitions and of melting. For normal, monovalent ionic materials such as NaCl, disordering of both cations and anions occurs on melting; a typical entropy of fusion is $24\,\text{J mol}^{-1}\,\text{K}^{-1}$. For AgI, the $\beta \rightarrow \alpha$ transition at $146\,°C$ may be regarded as quasi-melting of the silver ions. It has an entropy of transition of $14.5\,\text{J mol}^{-1}\,\text{K}^{-1}$. At the melting point of AgI, only the iodide atoms are left to become disordered. This fits in with a much reduced value for the entropy of fusion, $11.3\,\text{J mol}^{-1}\,\text{K}^{-1}$. The combined entropies of transition and fusion in AgI

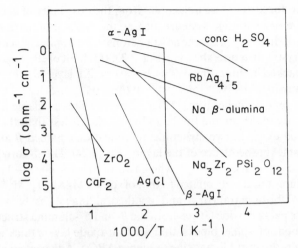

Fig. 7.22 Ionic conductivity of some solid electrolytes
with concentrated H_2SO_4 for comparison

are close to the fusion entropy of NaCl. Similar effects are observed in fluorides of some divalent metals, e.g. PbF_2. It has an entropy of fusion of only $16.4 \, J \, mol^{-1} \, K^{-1}$ whereas that for MgF_2, a typical ionic solid with low conductivity, is $\sim 35 \, J \, mol^{-1} \, K^{-1}$! This is because the fluoride ions in PbF_2 are disordered above $\sim 500 \, °C$ and the entropy of fusion corresponds to disordering of the lead ions alone.

Conductivity values for a variety of solid electrolytes are given in Fig. 7.22 in the form of Arrhenius diagrams. A selection of the more important solid electrolytes and their applications are discussed next.

β-alumina

β-alumina is the name for a family of compounds of general formula $M_2O \cdot nX_2O_3$ where n can have various values in the range 5 to 11, M is a monovalent cation, such as $alkali^+$, Cu^+, Ag^+, Ga^+, In^+, Tl^+, NH_4^+, H_3O^+, and X is a trivalent cation, Al^{3+}, Ga^{3+} or Fe^{3+}. The most important member of this family is sodium β-alumina ($M = Na^+$, $X = Al^{3+}$), which has been known for many years as a byproduct of the glass-making industry). It forms in the refractory lining of furnaces by reaction of soda from the melt with alumina in the refractory bricks. Its name is a misnomer because, although it was originally thought to be a polymorph of Al_2O_3, it is now known that additional oxides such as Na_2O are essential in order to stabilize its crystal structure.

Interest in β-alumina as a solid electrolyte began with the pioneering work of the Ford Motor Co. who, in 1966, found that the Na^+ ions are very mobile at room temperature and above. Ford also found that other cations could be ion exchanged for Na^+ and that these cations, too, are mobile. Since that time, interest in solid electrolytes has mushroomed. In our energy-conscious society much impetus for further research comes from the possibility of developing new, high density energy storage systems such as the Na/β-alumina/S cell.

The high conductivity of the monovalent ions in β-alumina is a consequence of its unusual crystal structure, shown in Fig. 7.23. It is built of close packed layers of oxide ions, stacked in three dimensions, but every fifth layer has three-quarters of its oxygens missing. The Na$^+$ ions reside in these oxygen-deficient layers and are able to move very easily because (a) there are more sites available than there are Na$^+$ ions to occupy them and (b) the radius of Na$^+$ is less than that of the O^{2-} ion. β-alumina exists in two structural modifications, named β and β'', which differ in the stacking sequence of the layers (Fig. 7.24). The β'' form occurs with more soda-rich crystals, $n \simeq 5 - 7$, whereas β occurs for $n \simeq 8 - 11$. Both the β and β'' structures are closely related to that of spinel, MgAl$_2$O$_4$; Al^{3+} ions occupy a selection of both tetrahedral and octahedral interstices between pairs of adjacent close packed oxide layers. Both the β- and β''-alumina structures may be regarded as built of 'spinel blocks' that are four oxide layers thick and in which the oxide layers are in cubic stacking sequence ABCA. Adjacent spinel blocks are separated by the oxygen-deficient layers or 'conduction planes' in which the Na$^+$ ions reside. The unit cells are hexagonal with $a = 5.60$ Å and $c = 22.5$ Å (β), 33.8 Å (β''). In the c direction, perpendicular to the oxide layers, there are two spinel blocks in the unit cell of β and three blocks in the unit cell of β''. The structure of

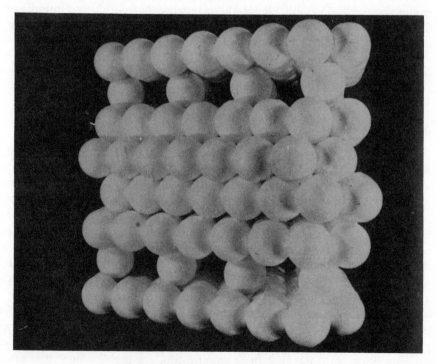

Fig. 7.23 Oxide layers in β-alumina

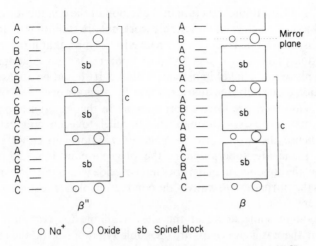

Fig. 7.24 Oxide packing arrangements in β- and β''-alumina

the 'spinel blocks' must be considered as defective in comparison with the ideal spinel structure. Thus, spinel contains both Mg^{2+} and Al^{3+} ions, in the ratio of 1:2, but the spinel blocks of β, β'' alumina contain only Al^{3+}, apart from small amounts of dopant ions such as Li^+, Mg^{2+} that are often added. In order to maintain charge balance, Al^{3+} vacancies must therefore also be present in the spinel blocks. The overall stacking sequence of oxide ions including both spinel blocks and conduction planes is cubic, ABC, in β'' but is a more complex, ten layer sequence in β, Fig. 7.24.

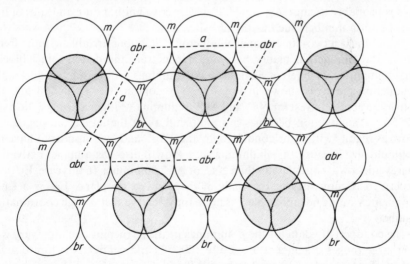

Fig. 7.25 Conduction plane in β-alumina; the base of the hexagonal unit cell is shown dashed

The details of the atomic structure in the region of the conduction plane have been the subject of much crystallographic work but are still not well understood. In Fig. 7.25 is shown one layer of close packed O^{2-} ions, that forms one wall of the conduction plane, and the 'column' or 'spacer' O^{2-} ions, (stippled) of the conduction plane in the layer immediately above. It should be apparent that in the conduction plane only one quarter of the O^{2-} sites are occupied, i.e. for every stippled O^{2-} ion there are three empty sites, m. In the β modification, a mirror plane of symmetry passes through and parallel to each conduction plane (Fig. 7.24); hence the layers of close packed O^{2-} ions on either side of the conduction plane are superposed in the projection in Fig. 7.25. In the β'' modification, the conduction plane does not coincide with a mirror plane and the oxide layers that form the two walls of the conduction plane are staggered relative to each other.

Let us look in some detail at the sites available for Na^+ ions in the β modification; there is a choice of three possible types of site: (a) the mid oxygen positions, m, (b) the Beevers–Ross sites, br, which were favoured in the original structure determination of Beevers and Ross, and (c) the anti-Beevers–Ross sites, abr. From the crystallographic results, it appears that Na^+ ions spend most of their time in br and m sites, but in order to undergo any long range migration they must pass through abr sites. Both br and m sites are large sites, e.g. Na^+ in a br site is coordinated to three oxygens in the oxide plane below, three in the plane above and three within the conduction plane. Na—O bond distances are large, $\sim 2.8\,\text{Å}$, compared with more typical values of $\sim 2.4\,\text{Å}$. The abr site is much smaller than br and m sites because two oxygens are quite close, one directly above and one directly below, giving two short Na—O distances of $2.3\,\text{Å}$. Most other monovalent cations also prefer to occupy the br and m sites in β-alumina with the exception of Ag^+ and Tl^+, which show a considerable preference for the abr sites. This is probably because Ag^+ and Tl^+ prefer covalent bonding and sites of low coordination number, such as the abr site.

β-alumina is a two-dimensional conductor. Alkali ions are able to more freely within the conduction planes but cannot penetrate the dense, spinel blocks. The conductivity of different single crystal β-aluminas in directions parallel to the conduction planes is shown in Fig. 7.26. The conductivity is highest and the activation energy lowest for Na^+ and Ag^+ β-alumina. With increasing cation size (K^+, Tl^+), conduction becomes more difficult since the larger cations cannot move as readily within the conduction planes. Ag^+ and Na^+ appear to have the optimum size because Li^+ β-alumina (not shown) also has a higher activation energy and lower conductivity than Na^+ or Ag^+ β-alumina. In this case, Li^+ ions appear to occupy sites in the walls of the conduction plane; Li^+ is a small polarizing cation and, unlike Na^+, is not happy in large sites of high coordination number.

The conductivity data for the β-aluminas fit the Arrhenius equation very well over large ranges of temperature (up to $1000\,^\circ\text{C}$) and conductivity (up to seven orders of magnitude). Contrast this with the NaCl data (Fig. 7.18) which show changes of slope several times over a range of $\sim 400\,^\circ\text{C}$. This simple behaviour

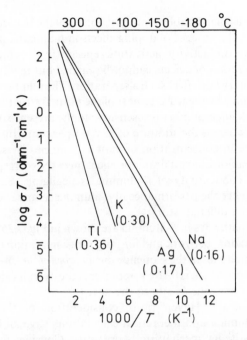

Fig. 7.26 Conductivity of some single crystal β-aluminas. Activation energies, in electron volts, are in parentheses. (From Whittingham and Huggins, *Solid State Chemistry*, 1972, p. 139)

seen in β-alumina is characteristic of solid electrolytes in general and also of many complex oxides, silicates, etc., which are not particularly good ionic conductors. It is a remarkable fact that many materials with simple crystal structures, such as NaCl, exhibit complex conductivity behaviour whereas materials with complex structures and stoichiometries, such as β-alumina, exhibit simple conductivity behaviour. This simple behaviour is observed independently of the magnitude of σ because β-alumina is equally well behaved at $300\,^\circ$C where $\sigma \sim 10^{-1}\,\text{ohm}^{-1}\,\text{cm}^{-1}$ or at $-180\,^\circ$C where $\sigma \sim 10^{-8}\,\text{ohm}^{-1}\,\text{cm}^{-1}$.

Another important feature of the conductivity of solid electrolytes is that, unlike in, for example NaCl, the σ values are usually reproducible between different samples and laboratories and appear to be largely insensitive to the presence of small amounts of impurities. The conductivity of β-alumina has been measured in many laboratories and in the large majority the data agree very well with an activation energy of 0.16 ± 0.01 eV.

It is not feasible to apply defect equilibria considerations and the law of mass action to, for example, β-alumina for several reasons. First, the equations that govern the concentration of Frenkel and Schottky defects apply only to very small defect concentrations (< 0.1 per cent of the lattice sites defective) whereas the evidence is that a considerable number, if not all, of the Na^+ ions in β-alumina are mobile. Second, the absence of doping effects in β-alumina indicates that the separation of the conductivity activation energy into components due to formation and migration of defects cannot be carried out readily and, indeed, it may be a mistake to try and effect such a separation. The number of mobile ions is so large that they must be regarded as part of the normal structure rather than as defects, and, as such, their number is insensitive to the presence of small amounts of impurity. Thus, whereas the addition of, say, 0.1 per cent $MnCl_2$ to NaCl may lead to an increase in concentration of cation vacancies by several orders of magnitude and, therefore, have a dramatic effect on σ, the corresponding effect of impurities on the conductivity of β-alumina is negligible. Small amounts of impurity may influence the performance of β-alumina ceramics used in the Na/S battery, but that is a different story.

The conduction pathways in β-alumina are shown in Fig. 7.25. The *br*, *abr* and *m* sites form a hexagonal betwork and for long range migration to occur an Na^+ ion must pass through sites in the sequence–*br*–*m*–*abr*–*m*–*br*–*m*–. The activation energy for conduction, which is 0.16 eV, represents the overall value for migration of a Na^+ ion from, say, one *br* site to the next.

Measurements of the Haven ratio, consideration of the pathways for conduction in β-alumina and theoretical calculations have all indicated that a knock-on or interstitialcy mechanism is operative. Consider first the idealized situation with β-alumina crystals of composition $NaAl_{11}O_{17}$. In this, the Na^+ ions occupy the *br* sites; *abr* and *m* sites are unoccupied. Next allow one Na^+ ion to leave its *br* site. It must pass through an adjacent *abr* site and, on the other side, it encounters an Na^+ ion on its *br* site. There is not room for the moving Na^+ to squeeze past the Na^+ on the *br* site. Neither is there room for it to stop short in the nearest *m* site because it would be too close to the Na^+ ion on the *br* site. The moving Na^+ ion has two choices, therefore. It can return to its own *br* site, in which case no net conduction occurs. Alternatively, it can knock the 'blocking' Na^+ ion out of its *br* site. If the latter occurs, the ejected Na^+ ion may simply move into one of the two adjacent *m* sites, in which case, the incoming Na^+ ion remains in the third *m* site; the net result is that a split interstitial is created. Alternatively, the ejected Na^+ ion may escape completely from its *br* and adjacent *m* sites, in which case a chain reaction begins. While the exact mechanistic details are a matter of speculation it is clear that Na^+ ions cannot move independently of each other in β-alumina but that, instead, a cooperative process must occur.

An interesting effect that has been observed in β-aluminas containing a mixture of two alkali cations is the so-called *mixed alkali effect*. In this, the mobility of both alkali ions, e.g. Na^+ and K^+, is less than that in either pure end member. Consequently, the conductivity passes through a minimum and the activation energy for conduction passes through a maximum value for intermediate compositions. The mixed alkali effect is a well known but poorly understood phenomenon in glasses and had not been encountered previously in crystalline materials. The occurrence of this rather unusual effect and the absence of any satisfactory explanation for it shows how difficult it is to understand the nature of the forces that control ion mobility, even in a crystalline material whose structure is apparently well understood.

Much of the work on β-alumina has been directed towards maximizing the conductivity of polycrystalline β-alumina ceramics. Since the conductivity of β'' is several times larger than that of β, compositions are desired that both stabilize the β'' phase (it is not very stable in the $Na_2O–Al_2O_3$ system without additives) and maximize its content relative to that of the β phase. It is found that this is best achieved by doping with small amounts of Li_2O and MgO. These appear to enter into solid solution with β''-alumina and stabilize its crystal structure.

AgI and Ag^+ ion solid electrolytes

It has been known for many years that AgI undergoes a phase transition at 146 °C and that the high-temperature form, α-AgI, has an exceptionally high conductivity, $\sim 1\ \mathrm{ohm^{-1}\,cm^{-1}}$, which is about four orders of magnitude larger than the room-temperature value (Fig. 7.27). The activation energy for conduction in α-AgI is only 0.05 eV and the structure of α-AgI is so suited for easy motion of Ag^+ that the ionic conductivity actually *decreases* slightly on melting at 555 °C.

β-AgI, stable below 146 °C, has the würtzite structure with hexagonal close packed I^- ions and Ag^+ in tetrahedral sites. Another low-temperature polymorph is γ-AgI; its structure is that of sphalerite. The highly conducting, high-temperature α polymorph is body centred cubic. In it, iodide ions lie at the corner and body centre positions and Ag^+ ions appear to be distributed statistically over

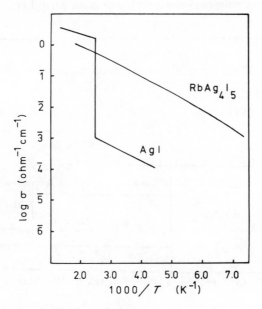

Fig. 7.27 Conductivity of Ag^+ in AgI and $RbAg_4I_5$

a total of thirty-six sites of tetrahedral and trigonal coordination. The tetrahedral sites are linked together by sharing faces and the trigonal sites lie at the centres of the faces of the AgI_4 tetrahedra. The iodide ions are essentially fixed and Ag^+ ions can readily move from one site to the next in a liquid-like manner. The disordered Ag^+ arrangements and the easy motion of Ag^+ between sites must be related to the nature of the bonding between silver and iodine. Silver is a polarizing cation since its outer $4d$ electrons are relatively ineffective in shielding nuclear charge. Iodide is a large and polarizable anion and so covalent bonds readily form between Ag^+ and I^- that are characterized by structures with low coordination numbers. During conduction, silver can readily move from one tetrahedral site to the next via an intermediate, three-coordinate site; covalent bonding at the intermediate site helps to stabilize it and reduce the activation energy for conduction. It is interesting that AgCl and AgBr have reasonable conductivities at high temperatures—but nothing like the values exhibited by AgI. It is expected that the bonding in these is less covalent than in AgI. However, they have a different crystal structure to that of α-AgI (they have the rock salt structure) and this is also undoubtedly an important factor.

In an attempt to stabilize the highly conducting α-AgI phase at lower temperatures, various anionic and cationic substitutions have been tried. The most successful so far has been the partial replacement of silver by rubidium in $RbAg_4I_5$. This material has the highest ionic conductivity at room temperature of any known crystalline substance, $0.25\,\text{ohm}^{-1}\,\text{cm}^{-1}$. Its activation energy for conduction is $0.07\,\text{eV}$ (Fig. 7.27). The amount of electronic conductivity in $RbAg_4I_5$ is negligibly small, $\sim 10^{-9}\,\text{ohm}^{-1}\,\text{cm}^{-1}$ at $25\,^\circ\text{C}$.

The phase diagram for the system RbI–AgI is shown in Fig. 7.28. Two binary

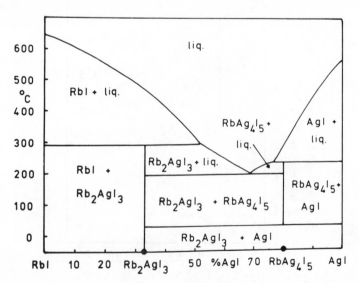

Fig. 7.28 Phase diagram for AgI–RbI. (From Takahashi, *J. Electrochem. Soc.*, **3**, 84, 1973)

compounds are present, Rb_2AgI_3 and $RbAg_4I_5$. The latter is thermodynamically stable over the temperature range 27 to 230 °C. It melts incongruently to AgI and liquid at ~ 230 °C and has a lower limit of stability at 27 °C. Decomposition of $RbAg_4I_5$ to AgI and Rb_2AgI_3 below 27 °C takes place only slowly although it is hastened by the presence of moisture or iodine vapour. With care, $RbAg_4I_5$ may be readily cooled to below room temperature without decomposing and its conductivity can then be measured over a wide temperature range.

The crystal structure of $RbAg_4I_5$ is rather different to that of α-AgI, but it also contains a random arrangement of silver atoms distributed over a network of face-sharing tetrahedral sites. Again there are many more available sites than there are silver atoms to fill them. Rubidium atoms are immobilized in sites that have a distorted octahedral environment of I^- ions. Conductivity data for $RbAg_4I_5$, Fig. 7.27, fall on a smooth curve that shows a small departure from linear Arrhenius behaviour.

A disordered, α-AgI-like structure can be stabilized at low temperatures by a variety of cations, notably large alkalis, NH_4^+, substituted NH_4^+ and certain organic cations. Some examples, all of which have conductivities at 25 °C in the range 0.02 to 0.20 ohm^{-1} cm^{-1} are $[(CH_3)_4N]_2Ag_{13}I_{15}$, $PyAg_5I_6$ where Py is the pyridinium ion $(C_5H_5NH)^+$, and $(NH_4)Ag_4I_5$.

A range of anions may be substituted partially for iodine and some of the phases formed have high conductivities, e.g. Ag_3SI, $Ag_7I_4PO_4$ and $Ag_6I_4WO_4$. The latter mixed iodide/oxysalt phases have considerable thermal stability and are unaffected by iodine vapour or moisture. They are suitable materials for use as the solid electrolyte in electrochemical cells. An interesting recent development is the preparation of glassy solid electrolytes in which molten mixtures of AgI with salts such as Ag_2SeO_4, Ag_3AsO_4 and $Ag_2Cr_2O_7$ can be preserved to room temperature, as glasses, by rapid quenching. The Ag^+ ion conductivities are again very high, e.g. 0.01 ohm^{-1} cm^{-1}, for the composition $Ag_7I_4AsO_4$. Obviously, it is not possible to know the structure of the iodide and oxysalt framework in non-crystalline materials such as these, but it seems reasonable to assume that silver can again move through a network of face-sharing tetrahedra.

It is also possible to replace partially both Ag^+ and I^- at the same time and prepare materials that have a high conductivity at room temperature; e.g.

(a) Replacement of Ag^+ by Hg^{2+} and I^- by Se^{2-}: the composition $Ag_{1.8}Hg_{0.45}Se_{0.70}I_{1.30}$ has a conductivity of 0.1 ohm^{-1} cm^{-1} at 25 °C.
(b) Replacement of Ag^+ by Rb^+ and I^- by CN^- in $RbAg_4I_4CN$.

All of the silver iodide derivative phases described above have high silver ion mobilities but very small electronic conductivities. They may therefore be used as solid electrolytes without any danger of short-circuits through the electrolyte due to electronic conductivity. On the other hand, a very desirable if not essential attribute of solid *electrode* materials is that they should have high ionic *and* electronic conductivities. Silver chalcogenides, such as α-Ag_2S which is stable at high temperatures, exhibit mixed ionic/electronic conductivity and, as in the case of α-AgI, it is possible to stabilize the high-temperature, disordered phase at lower temperatures by making various substitutions. For example, in the system $Ag_2Se–Ag_3PO_4$, α-Ag_2Se is able to dissolve 5 to 10 per cent Ag_3PO_4 in solid solution formation and the α phase is then stable down to room temperature, where it has an ionic conductivity of 0.13 ohm^{-1} cm^{-1} and an electronic conductivity of between 10^4 and 10^5 ohm^{-1} cm^{-1}.

Anion conductors

Several oxides and halides that have the fluorite (CaF$_2$) structure may be classified as solid electrolytes at high temperatures because they have high anion conductivity. One of the best examples is PbF$_2$. At room temperature, PbF$_2$ has a very low ionic conductivity and as such is a typical ionic solid. With rising temperature its conductivity increases smoothly and rapidly until at $\sim 500\,°C$ a limiting value of ~ 5 ohm^{-1} cm^{-1} is reached (Fig. 7.29). Above this temperature σ increases only slowly and there is little, if any, change in σ on melting at 822 °C. It is interesting that some materials such as PbF$_2$ arrive in the highly conducting condition gradually on increasing the temperature, whereas other materials such as AgI do so by undergoing an abrupt change in crystal structure (Fig. 7.27). Other materials that behave like PbF$_2$ are SrCl$_2$, which has a very high σ between $\sim 700\,°C$ and the melting point 873 °C, and CaF$_2$, which arrives in the highly conducting condition just as the melting point, 1418 °C, is approached.

The fluorite structure may be described in various ways, one of which is as primitive cubes of F$^-$ ions with Ca^{2+} ions at the body centres of alternate cubes (Fig. 1.29). The sites available for interstitial F$^-$ are then at the centres of the set of empty alternate cubes; these sites are coordinated by six calciums in an

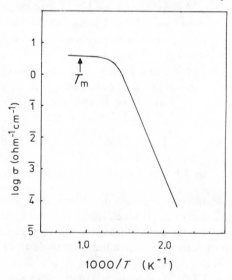

Fig. 7.29 Conductivity of PbF$_2$. (From Derrington and O'Keeffe, *Nature Phys. Sci.*, **246**, 44, 1973)

octahedral arrangement and eight fluorines at the cube corners. In creating an interstitial F^- ion, one of the corner F^- ions must move off its corner site and into the body of the cube. Defect complexes probably form but the details of the sites occupied are not known, especially in the more disordered structures at high temperatures.

The high-temperature, cubic polymorph of zirconia has the fluorite crystal structure and may be stabilized to room temperature by formation of solid solutions with CaO, Y_2O_3, etc. Such 'stabilized zirconias' are good O^{2-} ion conductors at high temperatures, mainly because the mechanism of solid solution formation involves the creation of vacant O^{2-} sites in order to preserve electroneutrality, e.g. lime-stabilized zirconia has the formula $(Ca_x Zr_{1-x})O_{2-x}$: $0.1 \lesssim x \lesssim 0.2$. With each Ca^{2+} ion that is introduced one O^{2-} ion vacancy is created. Typical conductivities in stabilized zirconias (e.g. 85 mol% ZrO_2, 15% CaO) are 5×10^{-2} ohm^{-1} cm^{-1} at 1000 °C with an activation energy for conduction of ~ 1.3 eV. At lower temperatures, of course, stabilized zirconias have conductivities that are many orders of magnitude less than those of the Na^+ and Ag^+ ion solid electrolytes. Their usefulness stems from the fact that they are refractory materials, can be used up to very high temperatures (e.g. 1500 °C) and have good oxide ion conductivity, which is an unusual property. Thoria (ThO_2) and hafnia (HfO_2) may be doped and are good O^{2-} ion conductors at high temperatures, similar to ZrO_2.

Requirements for high ionic conductivity: other ionic conductors

In order for ionic conduction to occur in crystals to an appreciable extent, certain structural conditions must be satisfied. These are:

(a) A large number of the ions of one species should be mobile (i.e. a large value of n in the equation $\sigma = ne\mu$).

(b) There should be a large number of empty sites available for the mobile ions to jump into. This is essentially a corollary of (a) because ions can be mobile only if there are empty sites available for them to occupy.

(c) The empty and occupied sites should have similar potential energies with a low activation energy barrier for jumping between neighbouring sites. It is no use having a large number of available interstitial sites if the moving ion cannot get into them.

(d) The structure should have a framework, preferably three dimensional, permeated by open channels through which mobile ions may migrate.

(e) The anion framework should be highly polarizable.

β-alumina meets the first four of these conditions, as does stabilized zirconia. The good Ag^+ ion conductors meet all five conditions. Materials which are not good solid electrolytes may satisfy some of the conditions but not all. For example, may silicates have framework structures but the cations are trapped in relatively deep potential wells. Poorly conducting β- and γ-AgI meet condition (e) but, more important, do not meet condition (c).

The zeolites with their large cavities would appear to be obvious candidates for solid electrolyte behaviour but they appear to have only moderate conductivities. The cations are usually present as hydrated complexes, which allows for ready ion exchange but not necessarily for high cation mobility. It appears that in, for example, dehydrated zeolites the channels are too large and the cations tend to stick at sites in the channel walls. A similar effect occurs in Li^+ β-alumina. The Li^+ conductivity is considerably less than in Na^+ β-alumina and the Li^+ ions occupy sites in the walls of the conduction plane; at high pressures the conductivity of Li^+ β-alumina increases because the size of the channels in the conduction plane is reduced.

There is still a great need for new materials that have high conductivity of Li^+ ions. This is because cells that contain lithium anodes have a higher e.m.f. than, for example, corresponding cells containing sodium anodes. Li_2SO_4 undergoes a phase transition at $572\,°C$ and has $\sigma \simeq 1\ ohm^{-1}\,cm^{-1}$ above this temperature; below this temperature σ is much smaller and Li_2SO_4 is of no interest as a low-temperature solid electrolyte. Various substituted lithium sulphates have been studied but so far it has not been possible to reduce the temperature of the phase transition below about $40\,°C$.

The structures of Li_4SiO_4 and Li_4GeO_4 have moderately good Li conductivity, $\sim 10^{-4}\ ohm^{-1}\,cm^{-1}$ at 300 to $400\,°C$, and are versatile host structures for doping. They are built of isolated SiO_4 and GeO_4 tetrahedra and the Li^+ ions are distributed through a network of face-sharing polyhedral sites. Substitutions

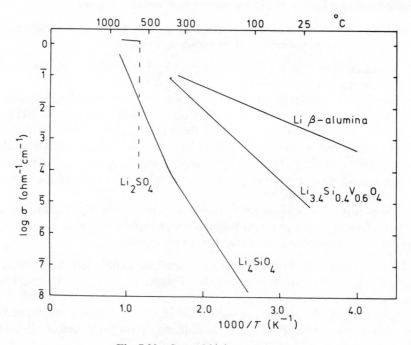

Fig. 7.30 Some Li^+ ion conductors

such as

$$\text{Li}^+ + \text{Si}^{4+} \rightleftharpoons \text{V}^{5+} \qquad \text{in Li}_4\text{SiO}_4$$

$$\text{Si}^{4+} \rightleftharpoons \text{Al}^{3+} + \text{Li}^+ \qquad \text{in Li}_4\text{SiO}_4$$

and

$$2\text{Li}^+ \rightleftharpoons \text{Zn}^{2+} \qquad \text{in Li}_4\text{GeO}_4$$

have all led to an improvement in conductivity of several orders of magnitude (Fig. 7.30), especially at lower temperatures (~ 25 to $300\,°\text{C}$). The best conductor at medium temperatures that has been found so far is $\text{Li}_{14}\text{ZnGe}_4\text{O}_{16}$ for which $\sigma = 10^{-1}\,\text{ohm}^{-1}\,\text{cm}^{-1}$ at $300\,°\text{C}$. It has been named LISICON. The highest room temperature conductivity occurs in $\text{Li}_{3.5}\text{V}_{0.5}\text{Ge}_{0.5}\text{O}_4$ for which $\sigma = 5 \times 10^{-5}\,\text{ohm}^{-1}\,\text{cm}^{-1}$.

An unusual effect has been found in mixtures of LiI and Al_2O_3. Although there is no evidence that LiI and Al_2O_3 react chemically, the conductivity of equimolar mixtures of LiI and Al_2O_3 is several orders of magnitude higher than that in pure LiI (pure Al_2O_3 is an insulator). The conductivity at $25\,°\text{C}$ is $\sim 10^{-5}\,\text{ohm}^{-1}\,\text{cm}^{-1}$. The origin of this effect is not understood but is probably due to surface conductivity at the interface between LiI and Al_2O_3 grains. It may also involve the presence of moisture: LiI is an extremely hygroscopic material.

Various materials have been examined for protonic conductivity. Two of the best that have been found are H_3O^+ β-alumina and hydrogen uranyl phosphate, $\text{HUO}_2\text{PO}_4 \cdot 4\text{H}_2\text{O}$, which has $\sigma \sim 4 \times 10^{-2}\,\text{ohm}^{-1}\,\text{cm}^{-1}$ at $25\,°\text{C}$.

Applications

Cells that contain solid electrolytes instead of liquid electrolytes may be constructed. A wide range of scientific and technological applications of such cells has been suggested, many of which are not possible with liquid electrolyte-containing cells. The schematic cell shown in Fig. 7.31 contains a solid electrolyte membrane separating two electrode compartments. The latter may contain solids, liquids or gases, which may be similar or dissimilar, e.g. oxygen gas at two different pressures or the two components of a cell, e.g. sodium and sulphur.

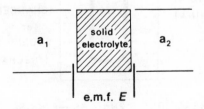

Fig. 7.31 Electrochemical cell containing a solid electrolyte

The e.m.f. of a cell reaction is given by the *Nernst equation*:

$$E = E_0 + \frac{RT}{nF} \log_e \frac{[\text{Ox}]}{[\text{Red}]} \tag{7.14}$$

Two such equations are usually needed, one for the reaction occurring at each electrode. Suppose that at the anode the following oxidation reaction occurs:

$$M \rightarrow M^+ + e$$

Then

$$E_1 = E_{0_{M/M^+}} + \frac{RT}{F} \log_e \frac{[M^+]}{[M]} \tag{7.15}$$

in which $E_{0_{M/M^+}}$ is the standard redox potential for that reaction. $[M^+]$ and $[M]$ are the concentrations of the two species and F is the Faraday unit, 96,500 Coulombs. The M^+ ions that are generated at the anode diffuse through the solid electrolyte and react at the cathode with X^-, produced according to

$$X + e \rightarrow X^-$$

and for which

$$E_2 = E_{0_{x-x}} + \frac{RT}{F} \log_e \frac{[X]}{[X^-]}$$

Summation of E_1 and E_2 gives the overall e.m.f. for the reaction: $M + X \rightarrow MX$, and this is related to the free energy of formation of MX by

$$\Delta G = -nEF \tag{7.16}$$

Electrochemical cells embodying solid electrolytes may be used to obtain thermodynamic data about materials, in particular free energies of formation of

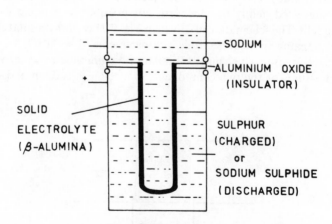

CELL REACTION: 2Na + 5S $\rightleftharpoons$ Na$_2$S$_5$ 2·08 V

Fig. 7.32   The sodium–sulphur cell

compounds, e.g. the cell $Ag(s)/AgI(s)/Ag_2S(s),S(l),C(s)$ has been used to measure the free energy of formation of Ag_2S. The cell reaction is $2Ag + S \rightarrow Ag_2S$ and since the reactants, silver and sulphur, are in their standard states

$$\Delta G^0(Ag_2S) = -2EF \qquad (7.17)$$

If the temperature dependence of the e.m.f. can be determined, one can go further and calculate entropies and enthalpies of reaction.

Much of the impetus for research on solid electrolytes has come from their possible use in new types of battery. The *sodium–sulphur cell* utilizes Na^+ β-alumina solid electrolyte and is probably the most important one (Fig. 7.32). It is a high-density secondary battery, i.e. it has a high energy/power to mass ratio and is undergoing extensive development and testing in several countries for use in, for example, electric cars and for power station load levelling. Basically, it consists of a molten sodium anode and a molten sulphur cathode separated by β-alumina solid electrolyte. Usually, the β-alumina is fabricated in the form of a tube closed at one end with the sodium inside and the sulphur outside (or vice versa). Since molten sulphur is a covalently bonded solid it is a non-conductor of electricity and the cathode material that is used consists of graphite felt impregnated with sulphur. The outer casing of the cell is made of stainless steel and serves as the current collector. The cell discharge reaction is:

$$2Na + xS \rightarrow Na_2S_x$$

where x depends on the level of charge in the cell. In the early stages of discharge, x is usually given as 5, which corresponds to the formula of the most sulphur-rich sodium sulphide, Na_2S_5, but x decreases as discharge continues. The phase diagram for the sodium–sulphur system is shown in Fig. 7.33. The sodium–sulphur cell operates at 300 to 350 °C, which is the lowest temperature at which the discharge products are molten for a large range of compositions. From the phase diagram it can be seen that when discharge reaches the stage at which $x \leqslant 3$ (i.e. 60% S, 40% Na), the liquidus curve rises rapidly and crystalline Na_2S_2 begins to form; further discharge would lead to a gradual freezing of the cathode material. The open-circuit voltage of the cell depends on both the level of charge and temperature. The maximum open-circuit voltage for a fully charged cell is 2.08 V at 300 °C. This decreases on discharge to ~ 1.8 V for $x = 3$. Thermodynamic calculations give a theoretical energy storage in the cell of 750 Wh kg^{-1}. Experimental values of the current density are usually 100 to 200 Wh kg^{-1} and are unlikely to be improved upon because most batteries give only a small percentage of their theoretical maximum output.

Other types of cells that are finding applications are miniature primary cells which operate at room temperature and which have a long life rather than a high power output. They are used in electronic watches, heart pacemakers and in military applications. Various cells have been used satisfactorily, e.g. $Ag/RbAg_4I_5/I_2$ (0.65 V) and $Li/LiI/I_2$ (2.8 V). In both of these, iodine alone cannot be used as the cathode because it does not have sufficient electronic conductivity to sustain a discharge current; instead complexed iodides are used,

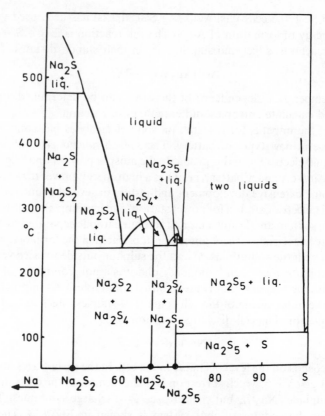

Fig. 7.33 Phase diagram for the sodium–sulphur system

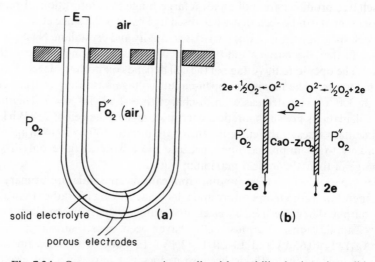

Fig. 7.34 Oxygen concentration cell with stabilized zirconia solid electrolyte

e.g. $(CH_3)_4NI_5$ containing polyiodide anions and iodine poly-2-vinyl pyridine charge transfer complex, respectively. The Li/I_2 cell is widely used for pacemaker applications. When it is operated at $37\,°C$ and gives current densities of 1 to $10\,\mu A\,cm^{-2}$, it is estimated that energy densities as high as $0.8\,Wh\,cm^{-3}$ are possible and that the batteries should operate for at least ten years.

Electrochemical cells containing solid electrolytes may be used for the measurement of partial pressures of gases or the concentrations of gases dissolved in liquids. An oxygen concentration cell that utilizes a stabilized zirconia solid electrolyte in the form of an open-ended tube is shown schematically in Fig. 7.34. Inside the tube is a reference gas such as air. The tube is coated with porous metal electrodes to allow absorption and liberation of oxygen gas. If the partial pressure of oxygen that is to be measured, P'_{O_2}, is less than the reference pressure, P''_{O_2}, the electrode reactions shown in (b) take place and oxide ions migrate through the solid electrolyte from right to left. The Nernst equations for the reactions at each electrode may be combined into a single equation that relates the difference in oxygen partial pressure to the cell voltage, i.e.

$$E = \frac{RT}{4F}\log_e\left(\frac{P''_{O_2}}{P'_{O_2}}\right) \tag{7.18}$$

The cell operates at temperatures between ~ 500 and $1000\,°C$ (in order that transport of O^{2-} ions through the electrolyte occurs sufficiently rapidly) and can be used to measure oxygen partial pressures as low as 10^{-16} atm. At lower pressures, the zirconia becomes an electronic conductor and the cell short-circuits. Stabilized thoria, ThO_2, solid electrolyte is ionically conducting over a wider range of oxygen partial pressures and may be used for $P_{O_2} < 10^{-16}$ atm. Oxygen concentration cells such as the zirconia probe have various uses, e.g. the analysis of exhaust gases and pollution, measurement of the consumption of oxygen gas during respiration and study of equilibria such as CO/CO_2, H_2/H_2O and metal/metal oxide. They may also be used to probe the oxygen activity in molten metals, e.g. steel, by dipping the probe into the melt. Oxygen concentrations are read from a calibrated meter and the response of the probe is usually very rapid.

Stabilized zirconia is used in either tube or disc form as the solid electrolyte or separator in some types of high-temperature fuel cells currently being developed. The two electrode compartments contain (a) air or O_2 and (b) a fuel gas, e.g. H_2, CO. The zirconia is again coated with porous metal electrodes and the cell reaction is, for example, $H_2 + \frac{1}{2}O_2 \rightarrow H_2O$ or $CO + \frac{1}{2}O_2 \rightarrow CO_2$. The advantages of this type of fuel cell are that electrode polarization problems are minimal and high current densities, e.g. $0.5\,A\,cm^{-2}$ and $0.5\,W\,cm^{-2}$, can be achieved. However, constructional difficulties sometimes arise and the cell can be heated up only slowly to the working temperature.

The fuel cell can also be used as an electrolyser and as a means of storing energy by reversing the direction of current flow. Thus, steam can be decomposed and the products, H_2 and O_2, stored. It could also be used to regenerate oxygen from CO_2 in, for example, spaceships and to deoxidize liquid metals.

Intercalation electrodes

Many of the difficulties associated with use of electrochemical cells and batteries are associated with the interfacial region between the electrode and electrolyte. With metal electrodes and ionically conducting electrolytes, interfacial polarization occurs associated with charge build-up at the interface; usually, a continuous current cannot flow across such an interface. These problems may be avoided by the use of *reversible electrodes*, i.e. electrodes that allow conduction by both electrons and the mobile ions of the electrolyte. Thus, with a reversible sodium electrode, Na^+ ions may move across the interface between electrode and electrolyte with little or no polarization problems. Molten sodium is a suitable reversible electrode for β-alumina because the reaction

$$Na^+ + e \rightleftharpoons Na$$

occurs at the sodium/β-alumina interface.

There is much interest in *solid solution electrodes*, first suggested by Steele, which are solid reversible electrodes in which the electrode composition may vary by addition or removal of ions. Tungsten bronzes, e.g. Na_xWO_3, have good Na^+ mobility because the parent WO_3 structure has a framework built of corner-sharing WO_6 octahedra similar to that in perovskite, (Fig. 1.36); a three-dimensional network of interconnected channels permeates the structure along which Na^+ (or other alkali cations) may migrate. The oxidation state of tungsten may vary between $+V$ and $+VI$ and the chemical reactions involved in the operation of the tungsten bronze solid solution electrode may be written as

$$xNa^+ + WO_3 + xe^- \rightleftharpoons Na_xW_x^VW_{1-x}^{VI}O_3$$

Layered, transition metal dichalcogenide structures, such as TiS_2 (cadmium chloride structure; Fig. 1.35), are promising solid solution electrode materials because 'intercalation' of alkali metal ions may occur. Alkali ions enter the structure by pushing the TiS_2 layers apart to give solid solutions such as Li_xTiS_2 in which $0 \lesssim x \lesssim 1$. The reaction is reversible and the TiS_2 is recovered as Li^+ ions diffuse out.

Dielectric materials

Dielectric materials are electrical insulators. They are used principally in capacitors and electrical insulators. They should possess the following properties in order to be of use in practical applications. They should possess high *dielectric strength*, i.e. they should be able to withstand high voltages without undergoing degradation and becoming electrically conducting. They should have low *dielectric loss*, i.e. in an alternating electric field, the loss of electrical energy, which appears as heat, should be minimized.

Application of a potential difference across a dielectric does lead to a polarization of charge within the material although long-range motion of ions or electrons should not occur. The polarization disappears when the voltage is removed. Ferroelectric materials are a special type of dielectric in that they retain

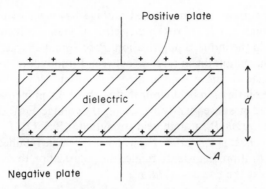

Fig. 7.35 Dielectric material between the plates
of a parallel plate capacitor

a large, residual polarization of charge after the electric field has been removed.

Dielectric properties may be defined by the behaviour of the material in a parallel plate capacitor. This is a pair of conducting plates, parallel to one another and separated by a distance, d, that is small compared with the linear dimensions of the plates (Fig. 7.35). With a vacuum between the plates, the capacitance C_0 is defined as

$$C_0 = \frac{e_0 A}{d} \qquad (7.19)$$

where e_0 *is the permittivity of free space*, 8.854×10^{-12} F m^{-1}, and A is the area of the plates. Since e_0 is constant, the capacitance depends only on the dimensions A and d of the capacitor. On applying a potential difference, V, between the plates, a quantity of charge, Q_0, is stored on them, given by

$$Q_0 = C_0 V \qquad (7.20)$$

If a dielectric substance is now placed between the plates and the same potential difference applied, the amount of charge stored increases to Q_1 and the capacitance therefore increases to C_1. The dielectric constant or relative permittivity, ε', of the dielectric is related to this increase in capacitance by

$$\varepsilon' = \frac{C_1}{C_0} \qquad (7.21)$$

The magnitude of ε' depends on the degree of polarization or charge displacement that can occur in the material. For air, $\varepsilon' \simeq 1$. For most ionic solids, $\varepsilon' = 5$ to 10. For ferroelectric materials such as $BaTiO_3$, $\varepsilon' = 10^3$ to 10^4.

Ferroelectricity

Ferroelectric materials are distinguished from ordinary dielectrics by (a) their extremely large permittivities and (b) the possibility of retaining some residual electrical polarization after an applied voltage has been switched off. As the potential difference applied across a dielectric substance is increased, a proportional increase in the induced polarization, P, or stored charge, Q, occurs. With ferroelectrics, this simple linear relation between P and V does not hold, as shown in Fig. 7.36. Instead, more complicated behaviour with a *hysteresis loop* is observed. The polarization behaviour that is observed on increasing the voltage is not reproduced on subsequently decreasing the voltage. Ferroelectrics exhibit a *saturation polarization*, P_s, at high field strength (for $BaTiO_3$, $P_s = 0.26\,C\,m^{-2}$ at 23 °C) and a *remanent polarization*, P_R, which is the value retained as V is reduced to zero after saturation. In order to reduce the polarization to zero, a reverse field is required; this is the *coercive field*, E_c.

Some common ferroelectric materials are listed in Table 7.3. All are characterized by structures in which one type of cation present, e.g. Ti^{4+} in $BaTiO_3$, can undergo a significant displacement, e.g. 0.1 Å, relative to its anionic neighbours. These charge displacements give rise to dipoles and the high dielectric constants that are characteristic of ferroelectrics.

The unit cell of $SrTiO_3$, which has the same perovskite structure as $BaTiO_3$, is shown in Fig. 1.36. In the primitive cubic unit cell, titanium ions occupy corner

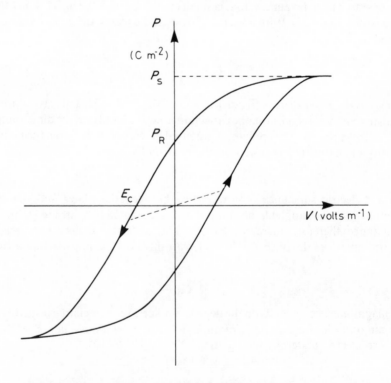

Fig. 7.36 Hysteresis loop of a ferroelectric. The dashed line passing through the origin represents the behaviour of normal dielectric materials

Table 7.3 *Some ferroelectric materials*

	$T_c(°C)$
Barium titanate, $BaTiO_3$	120
Rochelle salt, $KNaC_4H_4O_6.4H_2O$	Between -18 and $+24$
Potassium niobate, $KNbO_3$	434
Potassium dihydrogen phosphate, KDP, KH_2PO_4	-150
Lead titanate, $PbTiO_3$	490
Lithium niobate, $LiNbO_3$	1210
Bismuth titanate, $Bi_4Ti_3O_{12}$	675
Gadolinium molybdate, GMO, $Gd_2(MoO_4)_3$	159
Lead zirconate titanate, PZT, $Pb(Zr_xTi_{1-x})O_3$	Depends on x

positions, oxygen ions occupy the cube edge centres and strontium is at the cube body centre. The structure is composed of (TiO_6) octahedra which link together by sharing corners to form a three-dimensional framework; strontium ions occupy twelve coordinate cavities within this framework.

This ideal, cubic, perovskite structure, which is stable above 120 °C in $BaTiO_3$, does not possess a net dipole moment since the charges are symmetrically positioned. The material therefore behaves as a normal dielectric, albeit with a very high dielectric constant. Below 120 °C, structural distortion occurs in $BaTiO_3$. The TiO_6 octahedra are no longer regular because titanium is displaced off its central position and in the direction of one of the apical oxygens (Fig. 7.37). This gives rise to a spontaneous polarization. If a similar parallel displacement occurs in all of the TiO_6 octahedra, a net polarization of the solid results.

In ferroelectric $BaTiO_3$, the individual TiO_6 octahedra are polarized all of the time; the effect of applying an electric field is to persuade the individual dipoles to align themselves with the field. When complete alignment of all the dipoles occurs the condition of saturation polarization is reached. From the observed magnitude of P_s, it has been estimated that titanium is displaced by ~ 0.1 Å off the

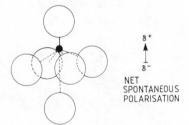

NET
SPONTANEOUS
POLARISATION

Fig. 7.37 Spontaneous polarization of a TiO_6 octahedron in $BaTiO_3$. Note that in principle, there are six equivalent orientations in which this can occur

centre of its octahedron and in the direction of one of the oxygens. This has been confirmed by X-ray crystallography. This distance of 0.1 Å or 10 pm is fairly small when compared with the average Ti—O bond distance of ~ 1.95 Å in TiO_6 octahedra. Alignment of dipoles is shown schematically in Fig. 7.38(a); each arrow represents one distorted TiO_6 octahedron and all are shown with a common direction of distortion.

In ferroelectrics such as $BaTiO_3$, *domain structures* form because adjacent TiO_6 dipoles tend to align themselves parallel to each other (Fig. 7.39). The domains are of variable size but are usually quite large, tens or hundreds of

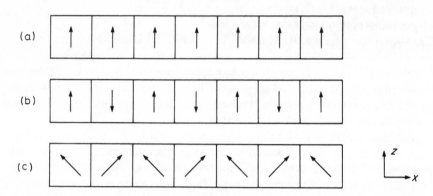

Fig. 7.38 Dipole orientation (schematic) in (a) a ferroelectric, (b) an antiferroelectric and (c) a ferrielectric

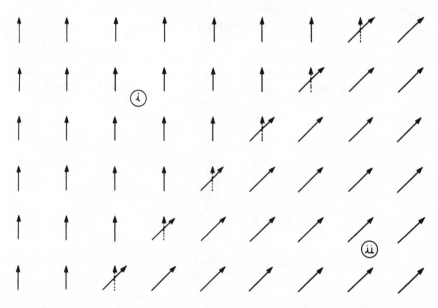

Fig. 7.39 Ferroelectric domains separated by a domain wall or boundary

ångstroms across. Within a single domain, the polarization of the dipoles has a common crystallographic direction. The net polarization of a piece of ferroelectric material is the vector resultant of the polarizations of the individual domains.

Application of an electric field across a ferroelectric leads to a change in the net polarization. This can arise from several possible processes:

(a) The direction of polarization of the domains may change. This would happen if all the TiO_6 dipoles within a domain were to change their orientation, e.g. if all the dipoles in domain (ii) in Fig. 7.39 changed their orientation so as to be parallel to the dipoles in domain (i).

(b) The magnitude of P within each domain may increase, especially if some randomness in dipole orientation is present before the field is applied.

(c) *Domain wall* migration may occur such that favourably oriented domains grow in size at the expense of unfavourably oriented ones. For example, domain (i) in Fig. 7.39 may grow by migration of the domain wall one step to the right. To effect this, the dipoles at the edge of domain (ii) change their orientation to the positions shown dashed.

The ferroelectric state is usually a low-temperature condition since the effect of increasing thermal motions at high temperatures is sufficient to break down the common displacement in adjacent octahedra and destroy the domain structure. The temperature at which breakdown occurs is the ferroelectric Curie temperature, T_c (Table 7.3). Above T_c, the material is paraelectric (i.e. non-ferroelectric). High dielectric constants still occur above T_c (Fig. 7.40), but no residual polarization is retained in the absence of an applied field. Above T_c ε' is usually given by the Curie–Weiss Law, $\varepsilon' = C/(T - \theta)$, where C is the Curie constant and θ the Curie–Weiss temperature. Usually, T_c and θ either coincide or differ by only a few degrees. The ferroelectric–paraelectric transition, at T_c, is an example of an order–disorder phase transition. However, unlike order–disorder phenomena in, say, brass, no long-range diffusion of ions occurs. Rather, the ordering that occurs below T_c involves preferential distortion or tilting of polyhedra and is therefore an example of a *displacive phase transition*. In the high temperature paraelectric phase, the distortions or tilts of the polyhedra, if they occur at all, are randomized.

A necessary condition for a crystal to exhibit spontaneous polarization and be ferroelectric is that its space group should be non-centrosymmetric. Often the symmetry of the paraelectric phase stable above T_c is centrosymmetric and the ordering transition that occurs on cooling simply involves a lowering of symmetry to that of a non-centric space group.

Several hundred ferroelectric materials are now known, including a large number of oxides that have distorted (non-cubic) perovskite structures. These contain cations that are happy in a distorted octahedral environment—Ti, Nb, Ta—and the asymmetric bonding within the MO_6 octahedron gives rise to spontaneous polarization and a dipole moment. Not all perovskites are ferroelectric, e.g. $BaTiO_3$ and $PbTiO_3$ are whereas $CaTiO_3$ is not, and this may be correlated with the ionic radii of the ions involved. It appears that the larger

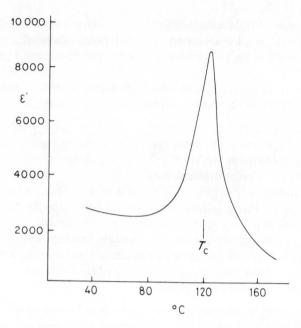

Fig. 7.40 Dielectric constant of barium titanate ceramic

Ba^{2+} ion causes an expansion of the unit cell relative to Ca^{2+}; this results in longer Ti—O bonds in $BaTiO_3$ and allows the Ti^{4+} ions more flexibility to move within the TiO_6 octahedra. Other ferroelectric oxides contain cations that are asymmetrically bonded because of the presence of a lone pair of electrons in their outer valence shell. These are the cations of the heavy p-block elements that are in oxidation states two less than the group valency, e.g. Sn^{2+}, Pb^{2+}, Bi^{3+}, etc.

Ferroelectric oxides are used in capacitors because of their high dielectric constants, especially near to T_c (Fig. 7.40). In order to maximize ε', for practical applications, it is therefore necessary to displace the Curie point so that it is close to room temperature. The Curie point of $BaTiO_3$, 120 °C (Fig. 7.40), may be lowered and broadened when either Ba^{2+} or Ti^{4+} are partially replaced by other ions. The substitution $Ba^{2+} \rightleftharpoons Sr^{2+}$ causes a unit cell contraction and reduction in T_c; replacement of 'active' Ti^{4+} by 'non-active' tetravalent ions such as Zr^{4+} and Sn^{4+} causes a rapid decrease in T_c.

A related type of spontaneous polarization occurs in *antiferroelectric* materials. In this, individual dipoles again occur but they generally arrange themselves so as to be antiparallel to adjacent dipoles (Fig. 7.38b). As a result, the net spontaneous polarization is zero. Above the antiferroelectric Curie temperature, the materials revert to normal paraelectric behaviour. Examples of antiferroelectrics, with their Curie temperatures, are: lead zirconate, $PbZrO_3$, 233 °C; sodium niobate, $NaNbO_3$, 638 °C; and ammonium dihydrogen phosphate, $NH_4H_2PO_4$, -125 °C.

The electrical characteristics of antiferroelectrics are rather different to those of ferroelectrics. The antiferroelectric state is a non-polar one and no hysteresis loop occurs, although a large increase in permittivity may occur close to T_c (for $PbZrO_3$, $\varepsilon' \simeq 100$ at $200\,°C$ and $\simeq 3000$ at $230\,°C$). Sometimes, the antiparallel arrangement of dipoles in the antiferroelectric state is only marginally more stable than the parallel arrangement in the ferroelectric state and a small change in conditions may lead to a phase transition. For example, application of an electric field to $PbZrO_3$ causes it to change from an antiferroelectric to a ferroelectric structure, Fig. 7.41(a); the magnitude of the field required depends on temperature. The polarization behaviour is then as shown in Fig. 7.41(b). At low fields, no hysteresis occurs and $PbZrO_3$ is antiferroelectric; at high positive and negative fields, hysteresis loops occur and $PbZrO_3$ is ferroelectric.

A related type of polarization phenomenon in which the structure is antiferroelectric in certain direction(s) only is shown in Fig. 7.38(c); in the x direction, the net polarization is zero and the structure is antiferroelectric, but in the z direction, a net spontaneous polarization occurs. This type of structure is known as *ferrielectric*; it occurs in, for example, $Bi_4Ti_3O_{12}$ and lithium ammonium tartate monohydrate.

The importance of hydrogen bonding in certain ferroelectric and antiferroelectric materials is shown in Fig. 7.42. Ferroelectric KH_2PO_4 and antiferroelectric $NH_4H_2PO_4$ are both built of isolated PO_4 tetrahedra that are linked together by K^+ and NH_4^+ ions and hydrogen bonds. These hydrogen bonds link up oxygens in adjacent PO_4 tetrahedra. The two structures differ mainly in the position of hydrogen in the hydrogen bonds.

Each PO_4 tetrahedron forms four hydrogen bonds with adjacent PO_4 tetrahedra. In each of these hydrogen bonds, the hydrogens are displaced so as to be nearer to one oxygen or the other, i.e. in each hydrogen bond, the hydrogen atom has a choice of two positions, neither of which is midway along the bond.

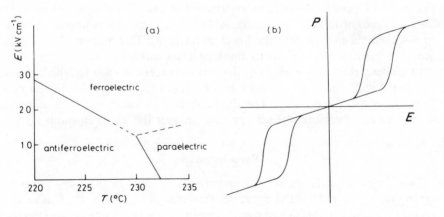

Fig. 7.41 (a) Antiferroelectric–ferroelectric transition in $PbZrO_3$ as a function of the applied field, E. (b) Polarization behaviour across this transition

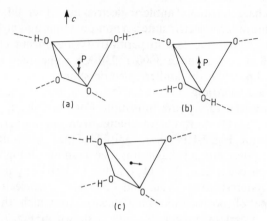

Fig. 7.42 Displacement of phosphorus within a $PO_2(OH)_2$ tetrahedron giving rise to spontaneous polarization

For each PO_4 tetrahedron, therefore, two hydrogens are close and two are somewhat further away. In the high temperature, paraelectric forms of KH_2PO_4 and $NH_4H_2PO_4$, the hydrogen positions are randomized over the two positions in each bond and a disordered structure obtains. In ferroelectric, low-temperature KH_2PO_4, the hydrogens order themselves so that both are associated with the upper edge of each PO_4 tetrahedron (a). The hydrogens are responsible indirectly for spontaneous polarization within the PO_4 tetrahedra since the phosphorus atoms are displaced downward away from the hydrogen atoms. This generates dipoles whose directions are parallel to the c crystallographic axis. In order to reverse the directions of the dipoles it is not necessary to bodily invert the tetrahedra. Instead, a simple movement of hydrogen atoms within the H bond achieves the same effect. The two hydrogens associated with the upper oxygens in (a) more away laterally to associate themselves with the lower oxygens of adjacent tetrahedra. At the same time, two hydrogens move in to associate themselves with the lower oxygens (b). This motion of hydrogen atoms *perpendicular* to c leads to dipole reversal *parallel* to c.

In antiferroelectric $NH_4H_2PO_4$, the two hydrogens of each tetrahedron are associated with one upper and one lower oxygen (c); this creates dipoles in a direction perpendicular to c. The dipole directions are reversed in adjacent tetrahedra and therefore, the net polarization over the whole crystal is zero.

Pyroelectricity

Pyroelectric crystals are related to ferroelectric ones in that they are non-centrosymmetric and exhibit a net spontaneous polarization, P_s. Unlike ferroelectrics, however, the direction of P_s cannot be reversed by an applied electric field. P_s is usually temperature dependent:

$$\Delta P_s = \pi \Delta T \qquad (7.22)$$

where π is the pyroelectric coefficient. This is mainly because the thermal expansion that occurs on heating changes the sizes (i.e. lengths) of the dipoles. A good example of a pyroelectric crystal is ZnO, which has the wurtzite structure. It contains a hexagonal close packed array of O^{2-} ions with Zn^{2+} ions located in one set of tetrahedral sites, say the T_+ sites. The ZnO_4 tetrahedra all point in the same direction and since each tetrahedron possesses a dipole moment, the crystal has a net polarization. Opposite, (001) surfaces of a ZnO crystal must contain, respectively, Zn^{2+} and O^{2-} ions as the outermost layers of ions. Usually, however, polar impurity molecules are absorbed onto the crystal in order to neutralize the surface charges. Consequently, the pyroelectric effect in a crystal is often not detectable under constant temperature conditions but becomes apparent only when the crystal is heated, thereby changing P_s.

Piezoelectricity

Under the action of an applied mechanical stress, piezoelectric crystals polarize and develop electrical charges on opposite crystal faces. As is the case for ferro- and pyroelectricity, the crystal must belong to one of the non-centrosymmetric point groups. The occurrence of piezoelectricity depends on the crystal structure of the material and the direction of applied stress, e.g. quartz develops a

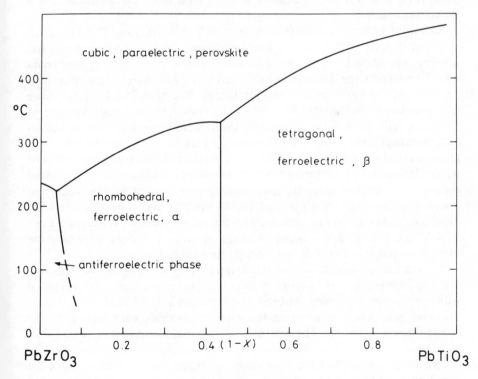

Fig. 7.43 Phase diagram for the PZT system. (Data from E. Sawaguchi, *J. Phys. Soc. Japan*, **8**, 615, 1953)

polarization when subjected to a compressive stress along [100] but not when stressed along [001]. The polarization, P, and stress, σ, are related to the piezoelectric coefficient, d, by

$$P = d\sigma$$

Many crystals that contain tetrahedral groups, e.g. ZnO, ZnS, are piezoelectric since application of a shearing stress distorts the tetrahedra. One of the most important piezoelectrics is PZT, lead zirconate titanate, which is a series of solid solutions between $PbZrO_3$ and $PbTiO_3$. These solid solutions are also antiferroelectric and ferroelectric at certain compositions, as shown by a partial phase diagram, Fig. 7.43. The best piezoelectric compositions occur at $x \sim 0.5$.

Applications of ferro-, pyro- and piezoelectrics

The main commercial application of ferroelectrics is in capacitors. Because ferroelectrics have a high permittivity or dielectric constant, ε', usually in the range 10^2 to 10^4, they can be used in the construction of large capacitors. The main commercial materials are $BaTiO_3$ and PZT (lead zirconate titanate) which are used in the form of dense, polycrystalline ceramics. By contrast, conventional dielectrics such as TiO_2 or $MgTiO_3$ have ε' in the range of 10 to 100. Hence, for a given volume, a $BaTiO_3$ capacitor has 10 to 1000 times the capacitance of a dielectric capacitor.

An important use of certain ferroelectrics such as $BaTiO_3$ and $PbTiO_3$, which does not depend directly on their ferroelectricity, is in PTC thermistors (i.e. positive temperature coefficient thermally sensitive resistors). In most non-metallic materials, the electrical resistivity decreases with increasing temperature, i.e. the resistivity has a negative temperature coefficient (NTC). However, some ferroelectrics, including $BaTiO_3$, show an anomalous and large increase in resistivity as the temperature approaches the ferroelectric–paraelectric transition temperature, T_c (Fig. 7.44). This increase in ρ matches approximately the large increase in ε' close to T_c although the reasons for the increase in ρ appear not to be well understood. PTC thermistors are used as switches; when a current is passed through any resistive material, joule heating losses, given by $I^2 R$, cause the material to heat up. With a $BaTiO_3$ thermistor, the resistivity increases dramatically as it heats and consequently the current switches off. Applications include (a) thermal and current overload protection devices, in which the thermistor acts as a reusable fuse, and (b) time delay fuses.

Pyroelectric crystals are used mainly in infrared radiation detectors. If desired, they can be made spectrally sensitive by coating the probe surface of the crystal with appropriate absorbing material. For detectors, it is desirable to maximize the ratio π/ε', which means that ferroelectric materials with high dielectric constant are not suitable. The best detector material found to date is triglycine sulphate.

Piezoelectric crystals have been used for many years as transducers for converting mechanical to electrical energy, and vice versa. Applications are

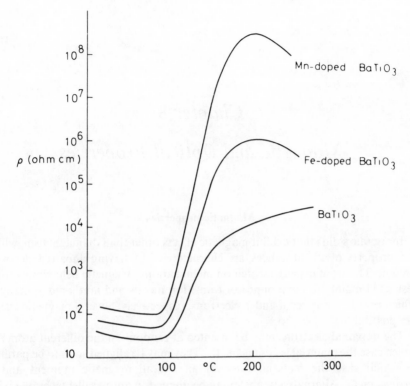

Fig. 7.44 Positive temperature coefficient resistivity in semiconducting BaTiO₃ ceramic with various dopants. (Data from H. Ueoka, *Ferroelectrics*, 7, 351, 1974)

diverse, e.g. as bimorphs in microphones, earphones, loudspeakers and stereo pick-ups; as fuses, solenoid ignition systems and cigarette lighters, sonar generators and ultrasonic cleaners. More complex devices are used in transformers, filters and oscillators. Most of these applications use PZT ceramics, quartz, Rochelle salt or $Li_2SO_4 \cdot H_2O$.

Chapter 8

Magnetic and Optical Properties

Magnetic properties

Inorganic solids that exhibit magnetic effects other than diamagnetism, which is a property of all substances, are characterized by having unpaired electrons present. These are usually located on metal cations. Magnetic behaviour is thus restricted mainly to compounds of transition metals and lanthanides, many of which possess unpaired d and f electrons, respectively. Several magnetic effects are possible.

The unpaired electrons may be oriented at random on the different atoms, in which case the material is *paramagnetic*. They may be aligned so as to be parallel, in which case the material posseses an overall magnetic moment and is *ferromagnetic*. Alternatively, they may be aligned in antiparallel fashion, giving zero overall magnetic moment and *antiferromagnetic* behaviour. If the alignment of the spins is antiparallel but with unequal numbers in the two orientations, a net magnetic moment results and the behaviour is *ferrimagnetic*.

Clearly, there are strong analogies between these magnetic properties and corresponding electrical properties such as ferroelectricity. One difference, of course, is the absence of the magnetic monopole. This would be the magnetic equivalent of an electrical charge, as possessed by an ion or an electron.

Magnetic oxides, especially ferrites such as $MgFe_2O_4$, are modern-day materials with uses in transformer cores, magnetic recording and information storage devices, etc. The theory of magnetic behaviour is, unfortunately, rather complicated. There is a plethora of terms, symbols and units that is bewildering to the uninitiated. To make matters worse, there appear to be two different and rather contradictory ways of evaluating the magnetic moments of ions containing unpaired electrons. Here we shall use a bare minimum of theory in order to be able to appreciate the different kinds of magnetic behaviour and how they are related to crystal structure.

Behaviour of substances in a magnetic field

When a substance is placed in a magnetic field, H, the density of lines of force in the sample, known as the *magnetic induction*, B, is given by H plus a contribution

342

$4\pi I$ due to the sample itself:

$$B = H + 4\pi I \tag{8.1}$$

where I is the *magnetic moment* of the sample per unit volume. The *permeability*, P, and *susceptibility*, κ, are defined as

$$P = \frac{B}{H} = 1 + 4\pi\kappa \tag{8.2}$$

$$\kappa = \frac{I}{H} \tag{8.3}$$

The *molar susceptibility*, χ, is given by

$$\chi = \frac{\kappa F}{d} \tag{8.4}$$

where F is the formula weight and d the density of the sample.

Table 8.1 *Magnetic susceptibilities*

Behaviour	Typical χ value	Change of χ with increasing temperature	Field dependence?
Diamagnetism	-1×10^{-6}	None	No
Paramagnetism	0 to 10^{-2}	Decreases	No
Ferromagnetism	10^{-2} to 10^{6}	Decreases	Yes
Antiferromagnetism	0 to 10^{-2}	Increases	(Yes)

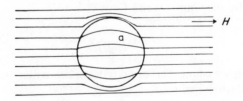

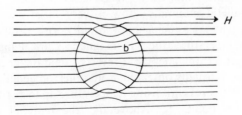

Fig. 8.1 Behaviour of (a) diamagnetic and (b) paramagnetic substances in a magnetic field

The different kinds of magnetic behaviour may be distinguished by the values of P, κ, χ and by their temperature and field dependences (Table 8.1).

Diamagnetic substances are those for which $P < 1$ and κ, χ are small and slightly negative. For *paramagnetic* substances, the reverse is the case; $P > 1$ and κ, χ are positive. When placed in a magnetic field, the number of lines of force passing through a substance is greater if it is paramagnetic and slightly less if it is diamagnetic, than would pass through a vacuum (Fig. 8.1). Consequently, paramagnetic substances are attracted by a magnetic field whereas diamagnetic substances experience a slight repulsion. Superconductors show perfect diamagnetism, χ has the value -1 and a magnetic field is expelled completely by them.

In *ferromagnetic* substances, $P \gg 1$ and large values of κ, χ are observed. Such materials are strongly attracted to a magnetic field. In *antiferromagnetic* substances, $P > 1$ and κ, χ are positive; values of κ, χ are comparable to or somewhat less than those for paramagnetic substances.

Effects of temperature: Curie and Curie–Weiss laws

The susceptibilities of the different kinds of magnetic material are distinguished by their different temperature dependences as well as by their absolute magnitudes. Many paramagnetic substances obey the simple Curie law, especially at high temperatures. This states that the magnetic susceptibility is inversely proportional to temperature:

$$\chi = \frac{C}{T} \tag{8.5}$$

where C is the Curie constant. Often, however, a better fit to the experimental data is provided by the Curie–Weiss Law:

$$\chi = \frac{C}{T + \theta} \tag{8.6}$$

where θ is the Weiss constant. These two types of behaviour are shown in Fig. 8.2 in which χ^{-1} is plotted against T.

For ferro- and antiferromagnetic substances, the temperature dependence of χ does not fit the simple Curie/Curie–Weiss laws, as shown schematically in Fig. 8.3. Ferromagnetic materials show a very large susceptibility at low temperatures that decreases increasingly rapidly with rising temperature (b). Above a certain temperature (the ferromagnetic Curie temperature, T_C), the material is no longer ferromagnetic but reverts to paramagnetic, where Curie–Weiss law behaviour is usually observed. For antiferromagnetic materials (c), the value of χ actually increases with rising temperature up to a critical temperature, known as the Néel point, T_N. Above T_N, the material again reverts to paramagnetic behaviour.

The magnitude of χ in the different materials and its variation with temperature may be explained as follows.

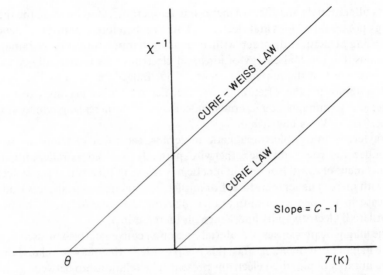

Fig. 8.2 Plot of reciprocal susceptibility against temperature showing Curie and Curie–Weiss law behaviour

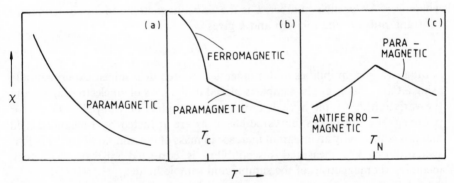

Fig. 8.3 Temperature dependence of the magnetic susceptibility for (a) paramagnetic, (b) ferromagnetic and (c) antiferromagnetic materials

The paramagnetic χ values correspond to the situation where unpaired electrons are present in the material and show some tendency to align themselves in a magnetic field. In ferromagnetic materials, the electron spins are aligned parallel due to cooperative interactions between spins on neighbouring ions in the crystal structure. The large χ values represent this parallel alignment of a large number of spins. In general, not all spins are parallel in a given material unless (a) very high magnetic fields and (b) low temperatures are used. In antiferromagnetic materials, the electron spins are aligned antiparallel and have a cancelling effect on χ. Hence small χ values are expected. Residual χ values may be associated with disorder in the antiparallel spin arrangement.

For all materials, the effect of increasing temperature is to increase the thermal energy possessed by ions and electrons. There is, therefore, a natural tendency for increasing structural disorder with increasing temperature. For paramagnetic materials, the thermal energy of ions and electrons acts to partially cancel the ordering effect of the applied magnetic field. Indeed, as soon as the magnetic field is removed, the orientation of the electron spins becomes disordered. Hence, for paramagnetic materials, χ decreases with increasing temperature, in Curie/Curie–Weiss law fashion.

For ferro- and antiferromagnetic materials, the effect of temperature is to introduce disorder into the otherwise perfectly ordered parallel/antiparallel arrangement of spins. For ferromagnetic materials, this leads to a rapid decrease in χ with increasing temperature. For antiferromagnetic materials, this leads to a decrease in the degree of antiparallel ordering, an increase in the number of 'disordered' electron spins and hence an increase in χ.

The magnetic properties of materials are often conveniently expressed in terms of the magnetic moment, μ, since this is a parameter that may be related directly to the number of unpaired electrons present. The relationship between χ and μ is

$$\chi = \frac{N\beta^2\mu^2}{3kT} \tag{8.7}$$

where N is Avogrado's number, β is the Bohr magneton and k is Boltzmann's constant. Substituting for N, β and k gives

$$\mu = 2.83\sqrt{\chi T} \tag{8.8}$$

Magnetic susceptibilities and moments are often determined experimentally using a Gouy balance. The sample is placed in the jaws of an electromagnet and the variation in sample mass is monitored as a function of applied field. For paramagnetic substances, unpaired electrons are attracted by a magnetic field and this is shown by an apparent increase in mass of the sample when the field is switched on. The measured susceptibility is corrected for various factors, including diamagnetism of the sample and sample holder.

Calculation of magnetic moments

The approach generally used by spectroscopists and chemists for the calculation of magnetic moments of ions that contain unpaired electrons is as follows.

The magnetic properties of unpaired electrons are regarded as arising from two causes, electron spin and electron orbital motion. Of most importance is the spin component. An electron may usefully be visualized as a bundle of negative charge spinning on its axis. The magnitude of the resulting spin moment, μ_s, is 1.73 Bohr magnetons (BM), where the Bohr magneton is defined as

$$1 \text{ BM} = \frac{eh}{4\pi mc} \tag{8.9}$$

where e = electron charge
  h = Planck's constant
  m = electron mass
  c = velocity of light

The formula used for calculating μ_s for a single electron is

$$\mu_s = g\sqrt{s(s+1)} \tag{8.10}$$

where s is the spin quantum number, $\frac{1}{2}$, and g is the gyromagnetic ratio, ~ 2.00. Substituting for s and g gives $\mu_s = 1.73$ BM for a single electron.

For atoms or ions that contain > 1 unpaired electron, the overall spin moment is given by

$$\mu_s = g\sqrt{S(S+1)} \tag{8.11}$$

where S is the sum of the spin quantum numbers of the individual unpaired electrons. Thus, for high spin Fe^{3+}, containing five unpaired $3d$ electrons, $S = \frac{5}{2}$ and $\mu_s = 5.92$ BM. Calculated values of μ_s for different numbers of unpaired electrons are given in Table 8.2.

The motion of an electron around the nucleus may, in some materials, give rise to an *orbital moment* which contributes to the overall magnetic moment. In cases where the orbital moment makes its full contribution,

$$\mu_{S+L} = \sqrt{4S(S+1) + L(L+1)} \tag{8.12}$$

where L is the orbital angular momentum quantum number for the ion. Equations (8.10) to (8.12) are applicable to free atoms or ions. In practice, in solid materials, equation (8.12) does not hold because the orbital angular momentum is either wholly or partially *quenched*. This happens when the electric fields of the surrounding atoms or ions may restrict the orbital motion of the electrons.

Table 8.2 *Experimental and theoretical magnetic moments for some transition metal ions.* (Data taken from Cotton and Wilkinson, 1966)

Ion	Number of unpaired electrons	$\mu_{s(\text{calc})}$	$\mu_{S+L(\text{calc})}$	$\mu_{(\text{observed})}$
V^{4+}	1	1.73	3.00	~ 1.8
V^{3+}	2	2.83	4.47	~ 2.8
Cr^{3+}	3	3.87	5.20	~ 3.8
Mn^{2+}	5 (high spin)	5.92	5.92	~ 5.9
Fe^{3+}	5 (high spin)	5.92	5.92	~ 5.9
Fe^{2+}	4 (high spin)	4.90	5.48	$5.1–5.5$
Co^{3+}	4 (high spin)	4.90	5.48	~ 5.4
Co^{2+}	3 (high spin)	3.87	5.20	$4.1–5.2$
Ni^{2+}	2	2.83	4.47	$2.8–4.0$
Cu^{2+}	1	1.73	3.00	$1.7–2.2$

Experimentally observed magnetic moments for a variety of ions are given in Table 8.2, together with the values calculated using equations (8.11) and (8.12). In most cases, the observed moments are similar to or somewhat larger than the calculated, spin-only values.

The methods outlined above for the calculation of magnetic moments have their origin in quantum mechanics. The details of the methods are actually quite involved, but, even so, agreement between theory and experiment is often not good (Table 8.2). An alternative, much simpler method is often used, especially by people working with properties such as ferro- and antiferromagnetism and their applications. In this, the magnetic moment of a single unpaired electron is set equal to one Bohr magneton. For an ion that possesses n unpaired electrons, the magnetic moment is given by n BM. Thus high spin Mn^{2+}, Fe^{3+} would both have a magnetic moment of 5 BM. This method may be quantified by the simple equation

$$\mu = gS \tag{8.13}$$

where $g \simeq 2.00$ and S, the spin state of the ion, equals $n/2$. The values obtained in this way underestimate the true values (compare columns 2 and 5 of Table 8.2); nevertheless, they provide a rough and useful guide to the magnitude of μ. A modification to equation (8.13) is to allow g to become an adjustable parameter. Equation (8.13) is regarded as a spin-only formula and by allowing g to exceed 2, a contribution to μ arising from orbital momentum is effectively allowed for. Thus, for Ni^{2+}, a g value in the range 2.2 to 2.3 is often used. In discussing the magnetic properties of ferrite phases, we shall use equation (8.13) to evaluate μ because of its simplicity.

Mechanisms of ferro- and antiferromagnetic ordering: superexchange

In the paramagnetic state, the individual magnetic moments of the ions containing unpaired electrons are arranged at random. Alignment occurs only on application of a magnetic field. The energy of interaction between dipoles and a magnetic field may be calculated readily, although details are not given here. It is, however, generally greater than the thermal energy, kT, possessed by the ions or dipoles.

In the ferro- and antiferromagnetic states, alignment of magnetic dipoles occurs spontaneously. There must therefore be some positive energy of interaction between neighbouring spins that allows this to occur, either in parallel or antiparallel fashion. The origin of this coupling of spins or cooperative interaction is quantum mechanical. Qualitatively, the effect is understood, although a complete rationale for the behaviour of, for example, ferromagnetic iron or cobalt is still needed.

One process by which coupling of spins occurs to give rise to antiferromagnetism in, for example, NiO, is known as *superexchange*. It is shown schematically in Fig. 8.4. The Ni^{2+} ion has eight d electrons. In an octahedral environment, two of these electrons singly occupy the e_g orbitals,

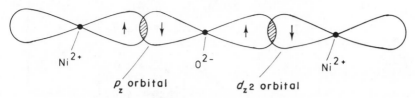

Fig. 8.4 Antiferromagnetic coupling of spins of d electrons on Ni^{2+} ions through p electrons of oxide ions

d_{dz^2} and $d_{x^2-y^2}$. These orbitals are oriented so as to be parallel to the axes of the unit cell and therefore point directly at adjacent oxide ions. The unpaired electrons in the e_g orbitals of Ni^{2+} ions are able to couple magnetically with electrons in the p orbitals of the O^{2-} ions. This coupling may well involve the formation of an excited state in which the electron transfers from the e_g orbital of the Ni^{2+} ion to the oxygen p orbital. The p orbitals of the O^{2-} ion contain two electrons each, which are also coupled antiparallel. Hence, provided Ni^{2+} and O^{2-} ions are sufficiently close that coupling of their electrons is possible, a chain coupling effect occurs which passes through the crystal structure (Fig. 8.4). The net effect of this is that neighbouring Ni^{2+} ions, separated by intervening O^{2-} ions, are coupled antiparallel.

Some more definitions

Ferromagnetic materials have a domain structure, similar to the domain structure of ferroelectric materials. Within each domain all the spins are aligned parallel but unless the material is in the saturation condition, different domains may have different spin orientations.

The response of ferromagnetic materials to an applied magnetic field is similar to that of ferroelectrics in an applied electric field. A hysteresis loop occurs in the plot of magnetization, M, or induction, B, against the applied field, H. A similar loop (Fig. 7.36) is observed for the polarization of a ferroelectric plotted against voltage. At sufficiently high fields the saturation magnetization condition is reached when the spins of all the domains are parallel. During the processes of magnetization and demagnetization in an alternating magnetic field, energy is dissipated, usually as heat. During one complete cycle this amount of energy, the *hysteresis loss*, is proportional to the area inside the hysteresis loop. For certain applications, low loss materials are required, one essential for which is that the area encompassed by the hysteresis loop should be as small as possible.

Materials that are magnetically *soft* are those of low *coercivity*, H_c. The coercivity is the magnitude of the reverse field required to achieve demagnetization. Soft materials also have low *permeability* and hence a hysteresis loop that is 'narrow at the waist' and of small area. Materials that are magnetically *hard* are those with a high coercivity and a high *remanent magnetization*, M_r. This latter is the magnetization that remains after the field is switched off. Hard materials are not easily demagnetized, therefore, and find uses as permanent magnets.

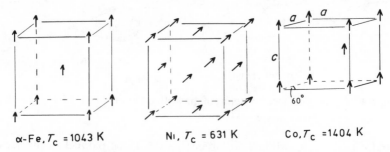

α-Fe, T_C = 1043 K   Ni, T_C = 631 K   Co, T_C = 1404 K

Fig. 8.5 Ferromagnetic ordering in body centred cubic α-Fe, face centred cubic Ni and hexagonal close packed Co

Ferromagnetic materials have a preferred or 'easy' direction of magnetization; in iron, this is parallel to the axes of the cubic unit cell (Fig. 8.5). The *magnetocrystalline anisotropy* is the energy required to rotate the magnetization out of this preferred direction.

An additional source of energy loss in an alternating magnetic field is associated with electrical currents called *eddy currents* that are induced in the material. The varying magnetic field induces a varying voltage and the eddy current losses are given by I^2R or V^2/R. Eddy currents are therefore minimized in highly resistive materials. One advantage that most of the magnetic oxides have over metals is their much higher electrical resistance.

Most magnetic materials exhibit the property of *magnetostriction*, i.e. they change their shape on magnetization. For instance, nickel and cobalt both contract in the direction of magnetization but expand in the perpendicular directions. With iron the reverse happens at low fields. At high fields, iron behaves like nickel and cobalt. The dimensional changes involved are small. The coefficient of magnetostriction, λ_s, defined as $\lambda_s = \Delta l / l_0$, increases with H up to a maximum value in the range 1 to 60×10^{-5} for saturation magnetization. The effect is therefore comparable to changing the temperature of the material by a few degrees.

Selected examples of magnetic materials, their structures and properties

Metals and alloys

Five transition elements—Cr, Mn, Fe, Co, Ni—and most of the lanthanide elements exhibit either ferro- or antiferromagnetism. A large number of alloys and intermetallic compounds also show some kind of magnetic ordering.

Iron, cobalt and nickel are ferromagnetic, as shown in Fig. 8.5. In body centred cubic α-Fe, the spins point in a [100] direction, parallel to a cubic cell edge, whereas in face centred cubic nickel, they point in a [111] direction, parallel to a cube body diagonal. Cobalt has a hexagonal close packed structure and the spins are oriented parallel to the c axis of the unit cell. These examples demonstrate

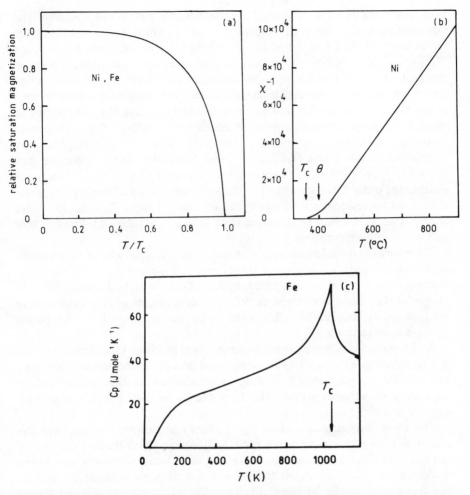

Fig. 8.6 Some properties of ferromagnetic materials: (a) saturation magnetization relative to value at absolute zero, as a function of reduced temperature. (b) Curie–Weiss Law plot for nickel of inverse susceptibility against temperature, showing a deviation close to T_C, (c) heat capacity of iron as a function of temperature. Data taken from Wert and Thomson, *Physics of Solids*, original references given therein

clearly that ferromagnetism is not associated with a particular type of crystal structure!

Chromium and manganese are both antiferromagnetic at low temperatures, $T_N = 95$ K (Mn), 313 K (Cr). Manganese has a complex crystal structure, but chromium is body centred cubic, similar to α-Fe. In chromium the spins are arranged in antiparallel fashion, parallel to one of the cube unit cell axes.

Some of the characteristics of ferromagnetic materials are given in Fig. 8.6. In (a) is shown the temperature dependence of the magnetic susceptibility or magnetic moment. The vertical axis represents the saturation magnetization of iron, relative to its maximum possible value, which occurs at absolute zero. The horizontal axis has a 'reduced temperature' scale in which the ratio of the actual temperature to the Curie temperature is plotted. Therefore, at the Curie point, $T/T_C = 1$. Use of reduced axes such as these facilitates comparison between materials with different Curie points and different magnetic moments. When plotted in this way, iron and nickel are found to behave very similarly: with increasing temperature the saturation magnetization stays almost constant at small T/T_C and then begins to decrease increasingly rapidly as T_C is approached.

Above the Curie point, iron, cobalt and nickel are paramagnetic. At temperatures well above T_C, Curie–Weiss law behaviour is observed but deviations occur close to T_C(b). These deviations are attributed to the existence of short range order between the spins. The long range order of the ferromagnetic state is lost but residual, short range order remains just above T_C; hence the Weiss temperature, θ, is somewhat removed from T_C. Data are shown for nickel; iron and cobalt behave similarly.

The transition from ferromagnetic to paramagnetic behaviour at T_C has many of the characteristics of a second-order or lambda phase transition and is a classic example of an order–disorder phase transition. Order is perfectly attainable only at absolute zero; at all real temperatures, disorder is present and increases rapidly with increasing temperature. This is shown by the heat capacity which passes through a maximum at T_C(c).

At this stage, let us give an airing to an unsolved problem. One of the mysteries of ferromagnetism concerns its dependence on position in the periodic table and in particular the question of how many unpaired electrons there are available to contribute to ferromagnetism. The facts are as follows. Provide your own explanation!

The three ferromagnetic elements in the first transition series have the electronic configurations shown in Table 8.3. Column 2 gives the configuration of the free ion in its ground state; the $4s$ level is fully occupied in each case. In the ferromagnetic state (column 4), the $4s$ band is not full but some of its electrons are transferred to the $3d$ band. Evidence for this comes from band theory calculations and the value of the saturation magnetization which is proportional to the number of unpaired spins (column 3). Thus, iron has a net moment of 2.2 BM per atom and hence has, on average, 2.2 unpaired d electrons per iron atom; i.e. of the $7.4d$ electrons, 4.8 are of one sign and 2.6 of the other. So far, we

Table 8.3 *Electronic constitution of iron, cobalt and nickel*

Metal	Free ion configuration	Ferromagnetic state	
		Number of unpaired spins	Configuration
Fe	$d^6 s^2$	2.2	$d^{7.4} s^{0.6}$
Co	$d^7 s^2$	1.7	
Ni	$d^8 s^2$	0.6	

are on familiar ground because chemists are used to the idea that electrons are transferrable between $4s$ and $3d$ levels, depending on circumstances. The sticking point concerns the maximum number of unpaired electrons that can contribute to ferromagnetism in elements or alloys of the $3d$ series. The answer, apparently, is 2.4 per atom but no satisfactory explanation of the significance of this number 2.4 is available at present. The effective number of unpaired electrons varies with total electron content across the $3d$ series. The maximum value of 2.4 is found for an alloy of composition $Fe_{0.8}Co_{0.2}$. With increasing total electron content, the number of unpaired spins decreases gradually, passing through cobalt and nickel before decreasing to zero in the alloy $Ni_{0.4}Cu_{0.6}$. Thus pure copper is paramagnetic. To the other side of the $Fe_{0.8}Co_{0.2}$ composition, the number of unpaired spins also drops systematically, passing through iron, manganese and chromium. Both manganese and chromium are antiferromagnetic at low temperatures.

One factor that appears to be important is the degree of overlap of the d orbitals and the breadth of the d band. This is directly related to the interatomic separation in the metals, which increases with atomic number across the $3d$ transition series. At short distances, overlap is strong and quantum mechanical exchange forces dictate an antiparallel coupling of spins, as found in antiferromagnetic Cr, Mn. With increasing interatomic distance, overlap is still strong but the coupling now gives a parallel arrangement, as in ferromagnetic iron, cobalt and nickel. At larger separation, coupling is weaker and paramagnetic behaviour is observed as in copper.

The lanthanide elements have magnetically ordered structures that are associated with unpaired $4f$ electrons. Exceptions are those elements for which the $4f$ shell is empty: La, $4f^{\circ}$; or full: Yb, $4f^{14}$; Lu, $4f^{14}$. Most of the lanthanides are antiferromagnetic below room temperature. Some, especially the later lanthanides, form both ferro- and antiferromagnetic structures at different temperatures. For these, the sequence with decreasing temperature is always:

$$\text{Paramagnetic} \rightarrow \text{antiferromagnetic} \rightarrow \text{ferromagnetic}$$

Table 8.4 *Néel (antiferromagnetic) and Curie (ferromagnetic) temperatures (K) in lanthanides.* (Data taken from Taylor, 1970)

Element	Néel temperature T_N	Curie temperature T_C
Ce	12.5	
Pr	25	
Nd	19	
Sm	14.8	
Eu	90	
Gd	—	293
Tb	229	222
Dy	179	85
Ho	131	20
Er	84	20
Tm	56	25

Néel and Curie temperatures are given in Table 8.4. The exception appears to be gadolinium which does not form an antiferromagnetic structure.

Several antiferromagnetic lanthanides exhibit the additional property known as *metamagnetism*. In this, they can be induced to switch over to a ferromagnetic state by application of a suitably high magnetic field. For instance, dysprosium is ferromagnetic below 85 K but at higher temperatures changes to antiferromagnetic. By application of a magnetic field, the ferromagnetic state can be preserved above 85 K and up to the Néel temperature, 179 K.

Transition metal oxides

The oxides of the first transition series show very large, systematic changes in properties with atomic number and number of d electrons. The variation in electrical conductivity of the divalent oxides, MO, has been discussed in Chapter 7, with, for instance, the extremes of metallic conductivity in TiO and insulating behaviour in NiO. The oxides MO also show a large range of magnetic behaviour which parallels roughly the electrical behaviour. The oxides of the early elements, TiO, VO and CrO are diamagnetic. In these oxides, the d electrons are not localized on the individual M^{2+} ions but are delocalized over the entire structure in a partly filled t_{2g} band. There appear to be no magnetic interactions between these delocalized electrons; consequently, the materials are diamagnetic as well as being conductors of electricity. The oxides of the later elements, MnO, FeO, CoO and NiO are paramagnetic at high temperatures and exhibit ordered magnetic structures at low temperatures. In these oxides, the d electrons are localized on the individual M^{2+} ions. This localization of the unpaired electrons is responsible for both the observed magnetic properties and the virtual absence of electrical conductivity.

The oxides MnO, FeO, CoO and NiO are all antiferromagnetic at low temperatures and change to paramagnetic above the Néel temperature, T_N. Values of T_N, in degrees celsius, are MnO, -153; FeO, -75; CoO, -2; NiO, $+250$. All have similar structures in both the antiferromagnetic and paramagnetic forms. Using nickel oxide as an example, at high temperatures it has the rock salt crystal structure. The structure may be viewed and described in various ways. For our purposes, if the structure is viewed along any of the four equivalent [111] directions, parallel to the body diagonals of the face centred cubic unit cell, alternate layers of Ni^{2+} and O^{2-} ions are found.

Below 250 °C, the structure of NiO undergoes a rhombohedral distortion: a slight contraction of the structure occurs along one of the threefold axes parallel to a [111] direction. A similar contraction occurs in MnO but in FeO the structure is, instead, slightly elongated. The symmetry of the structures is lowered since all of the fourfold axes and three of the threefold axes are lost, leaving a single threefold axis. In practice, the degree of distortion of the structures from cubic symmetry is small and the effects are barely noticeable in, for example, the X-ray powder pattern of NiO.

The cause of the rhombohedral distortion in NiO is antiferromagnetic ordering of the Ni^{2+} ions. Within a given layer of Ni^{2+} ions, the spins of all the Ni^{2+} ions are aligned parallel but in adjacent layers they are all antiparallel (Fig. 3.19). Magnetic superstructures such as these may be studied very elegantly using neutron diffraction. Two types of scattering contribute to the observed diffraction pattern: scattering by atomic nuclei and scattering by unpaired electrons. The former type gives a diffraction pattern similar to that observed using X-rays, although the intensities may be somewhat different. In antiferromagnetic structures, the second type of scattering gives rise to extra lines in the neutron powder diffraction pattern. This is because (a) cooperative interactions between unpaired electrons may give rise to a superstructure and (b) neutrons are scattered strongly by unpaired electrons whereas X-rays are not.

Neutron powder diffraction patterns for MnO below and above the Néel temperature, together with a schematic X-ray powder pattern at room temperature, are shown in Fig. 8.7. Comparison of the two patterns above T_N (b and c) shows that lines appear in the same positions but are of very different intensities. In the rock salt structure, the condition for reflection is that h, k, l should be either all odd or all even. Hence, the first four lines to be expected in the powder pattern

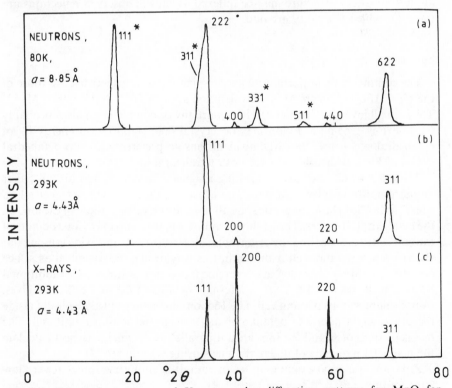

Fig. 8.7 Schematic neutron and X-ray powder diffraction patterns for MnO for $\lambda = 1.542\,\text{Å}$. Peaks are assigned Miller indices for the cubic unit cells given. Neutron data are adapted from Shull, Strauser and Wollan, *Phys. Rev.* **83**, 333 (1951). X-ray data are from Powder Diffraction File, Card No. 7-230

are 111, 200, 220 and 311. All four lines appear in both patterns but 200 and 220 are weak in the neutron pattern (b). The small intensity of 200 and 220 in (b) is largely because the neutron scattering powers of Mn^{2+} and O^{2-} are opposite in sign although slightly different in magnitude. Partial cancellation therefore occurs for the 200 and 220 reflections since Mn^{2+} and O^{2-}. ions on the same planes scatter out of phase with each other. This is precisely the opposite of the case with scattering of X-rays, for which the scattering factors of all elements have the same sign and therefore for the 200 and 220 reflections, Mn^{2+} and O^{2-} scatter in phase with each other.

Comparison of (a) and (b) shows that below T_N extra lines (asterisked) appear in the neutron diffraction pattern. These extra lines are associated with the antiferromagnetic superstructure. Although, as mentioned above, the true symmetry of the antiferromagnetic structure is rhombohedral, to a first approximation it can be treated as cubic with cell dimensions that are twice the value for the high temperature paramagnetic structure, i.e. $a = 8.85$ Å at 80 K ($< T_N$) whereas at 293 K ($> T_N$), $a = 4.43$ Å. The volumes of the unit cells are therefore, in the ratio of 8:1. The extra lines in the powder pattern of the antiferromagnetic structure may be indexed as shown; observed reflections are those for which h, k and l are odd.

Spinels

The commercially important magnetic spinels, known as ferrites, are those of the type MFe_2O_4, where M is a divalent ion such as Fe^{2+}, Ni^{2+}, Cu^{2+}, Mg^{2+}. They are all inverse, (Chapter 1), either partially or completely. This is probably because Fe^{3+}, being a d^5 ion, has no crystal field stabilization energy in an octahedral site; hence, the large divalent ions go preferentially into octahedral sites and Fe^{3+} is distributed over both tetrahedral and octahedral sites.

These ferrite phases have interesting magnetic structures and are all either antiferromagnetic or ferrimagnetic. This is because the ions on the tetrahedral, 8a sites (Table 1.19b) have magnetic spins that are antiparallel to those of the ions on the octahedral, 16d sites. This is shown in Fig. 8.8; the contents of four octants of the unit cell are shown and the origin coincides with an 8a ion. Oxide ions are not shown. When the unit cell is drawn in this way, it may be described as a face centred cubic array of 8a ions, at corner and face centre positions, with additional 8a ions in the centre of one set of alternate octants of the unit cell. This gives, overall, eight 8a ions per unit cell. The 16d ions are arranged tetrahedrally inside the other set of alternate octants, giving sixteen 16d ions per unit cell. The magnetic spins of 8a and 16d ions are antiparallel, as shown. Let us now calculate the magnetic moments of different spinels using equation (8.13).

$ZnFe_2O_4$ is a good example to begin with. It is an inverse spinel at very low temperatures, i.e.

$$[Fe^{3+}]^{tet}[Zn^{2+}, Fe^{3+}]^{oct}O_4$$

Since equal numbers of Fe^{3+} ions are on tetrahedral 8a and octahedral 16d sites, with opposed spins, the net moment of the Fe^{3+} ions is zero. Since Zn^{2+} has no

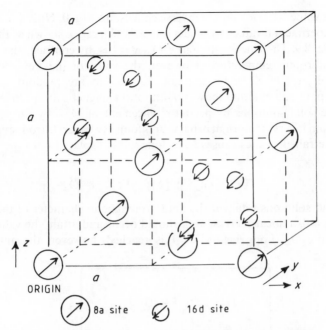

ORIGIN

⊘ 8a site   ⊘ 16d site

Fig. 8.8 Magnetic structure of antiferromagnetic and ferromagnetic spinels. Cations only are shown

magnetic moment, $ZnFe_2O_4$ has zero overall moment and is antiferromagnetic. Experimental measurements confirm this ($T_N = 9.5$ K).

A similar result would be expected for $MgFe_2O_4$ but it has, in fact, a residual, overall moment and is ferrimagnetic. Two explanations are possible. Either the spinel is not completely inverse and there are more Fe^{3+} ions on the octahedral sites than on the tetrahedral ones. This would give only partial cancellation of the spins. Or, the effective magnetic moment per Fe^{3+} ion is different for the two types of site. Experimental measurements indicate that the first explanation is correct. At high temperatures, $MgFe_2O_4$ gradually transforms to the normal spinel structure. The degree of inversion in a sample at room temperature depends very much on its thermal history, especially the rate at which it was cooled from high temperature. Thus, rapidly quenched samples have a lower degree of inversion and hence a higher magnetic moment than slowly cooled ones.

Manganese ferrite, $MnFe_2O_4$, is about 80 per cent normal, 20 per cent inverse, but since both cations, Mn^{2+} and Fe^{3+}, are d^5, the overall magnetic moment is insensitive to the degree of inversion and to heating/thermal history effects. $MnFe_2O_4$ is expected to be ferrimagnetic with an overall moment of ~ 5 BM, and this is found to be the case.

An interesting example of the importance of cation site occupancies and solid solution effects is provided by the mixed ferrites, $M_{1-x}Zn_xFe_2O_4$: M = Mg, Ni, Co, Fe, Mn. These ferrites are largely inverse for $x = 0$, i.e.

$$[Fe^{3+}]^{tet}[M^{2+}, Fe^{3+}]^{oct}O_4$$

Expected values of μ for purely inverse spinels are Mg: 0, Ni:2, Co:3, Fe:4 and Mn:5. Experimental values are in fact somewhat larger, as shown by the left-hand axis of Fig. 8.9. The zinc ferrite, $ZnFe_2O_4$, $x = 1$ is, by contrast, almost entirely normal at room temperature. However, the spins of the Fe^{3+} ions on the octahedral sites of $ZnFe_2O_4$ are not aligned but are random. $ZnFe_2O_4$ is therefore paramagnetic and shows no saturation magnetization. On formation of the ferrite solid solutions by partial replacement of M^{2+} by Zn^{2+}, a gradual change from inverse to normal behaviour is found to occur. Introduction of Zn^{2+} into the tetrahedral sites causes Fe^{3+} ions to be displaced onto the octahedral sites, i.e.

$$[Fe_{1-x}^{3+}Zn_x^{2+}]^{tet}[M_{1-x}^{2+}Fe_{1+x}^{3+}]^{oct}O_4$$

If the solid solutions retained the antiferromagnetic character of the MFe_2O_4 ferrites, $x = 0$, a linear increase in μ should occur and attain the value of 10 for $x = 1$, $ZnFe_2O_4$. Long before $x = 1$ is reached, however, the antiferromag-

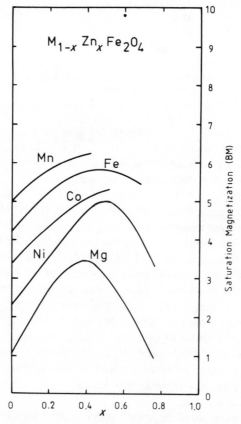

Fig. 8.9 Variation in saturation magnetization with composition for ferrite solid solutions (From Gorter, *Nature*, **165**, 798, 1950)

netic coupling between 16d and 8a sites is destroyed and the saturation magnetization values drop (Fig. 8.9). For small values of x, the experimental values of the saturation magnetization increase, consistent with retention of the antiferromagnetic/ferrimagnetic ordering, but pass through a maximum for $x = 0.4$ to 0.5.

In addition to the magnetic moment of a ferrite, other parameters are important in controlling the magnetic properties. These include the saturation magnetization, M_{sat}, the magnetostrictive constant, λ_s, the permeability, P, and the magnetocrystalline anisotropy constant, K_1. Suffice it to say here that the values of these parameters vary widely between the different ferrite phases. A particular ferrite may be chosen depending on the values of these properties and the application in mind. Further variation in the magnetic properties may be achieved by making mixed ferrites which are solid solutions of two or more pure ferrites. For instance, replacement of Mn^{2+} by Fe^{2+} in $MnFe_2O_4$, giving a solid solution $Mn^{2+}_{1-x}Fe^{2+}_xFe^{3+}_2O_4$, reduces the magnetic anisotropy parameter to zero. This parameter measures the ease with which the orientation of the magnetic moment may be altered in an applied magnetic field. Reduction of the magnetic anisotropy yields an increased permeability which is usually desirable in commercial ferrites. An undesirable side-effect, however, is an increase in electrical conductivity with increasing Fe^{2+} content.

Garnets

The garnets are a large family of complex oxides, some of which are important ferrimagnetic materials. They have the general formula $A_3B_2X_3O_{12}$. A is a large ion with a radius of ~ 1 Å and has a coordination number of eight in a distorted cubic environment. B and X are smaller ions which occupy octahedral and tetrahedral sites, respectively. The garnets with interesting magnetic properties have A = Y or rare earth, Sm, Gd, Tb, Dy, Ho, Er, Tm, Yb, Lu; B, X = Fe^{3+}. One of the most important is yttrium iron garnet (YIG), $Y_3Fe_5O_{12}$. Many other A, B, X combinations are possible, e.g.:

	A	B	X	O
Grossular:	Ca_3	Al_2	Si_3	O_{12}
Uvarovite:	Ca_3	Cr_2	Si_3	O_{12}
Pyrope:	Mg_3	Al_2	Si_3	O_{12}
Andradite:	Ca_3	Fe_2	Si_3	O_{12}
	Ca_3	$CaZr$	Ge_3	O_{12}
	Ca_3	Te_2	Zn_3	O_{12}
	Na_2Ca	Ti_2	Ge_3	O_{12}
	$NaCa_2$	Zn_2	V_3	O_{12}

Crystallographic data for YIG are given in Table 8.5. The body centred cubic cell is large, $a = 12.376$ Å, and contains eight formula units. No attempt is made to give a drawing of the structure here, but the structure may be regarded as a framework built of corner-sharing BO_6 octahedra and XO_4 tetrahedra. The

Table 8.5 *The garnet structure: crystallographic data*
Space group: Ia3d, No 230, body centred cubic
Atomic coordinates for $Y_3Fe_5O_{12}$(YIG)

Atom	Position	Fractional coordinates
Y	24c	$\frac{1}{8}, 0, \frac{1}{4}$
Fe(1)	24d	$\frac{3}{8}, 0, \frac{1}{4}$
Fe(2)	16a	$0, 0, 0$
O	96h	$u, v, w{:}u = -0.0275,$ $v = 0.0572,\ w = 0.1495$

Coordination numbers:

Y	24c	CN = 8,	distorted cube
Fe(1)	24d,	CN = 4,	tetrahedron
Fe(2)	16a,	CN = 6,	octahedron
O	96h	CN = 4,	distorted tetrahedron, $2Y^{3+}$, $1Fe^{3+}(1)$, $1Fe^{3+}(2)$

larger A ions occupy eight coordinate cavities within this framework. In YIG and the rare earth garnets the B and X ions are the same, Fe^{3+}.

YIG and the rare earth garnets are all ferrimagnetic with a Curie temperature in the range 548 to 578 K. In order to evaluate the overall magnetic moment of these garnets, the three types of ion, on sites 24c, 24d and 16a have to be considered. It appears that the spins of the 24d ions are coupled so as to be antiparallel to those of both the 24c and 16a ions. If we consider the two types of Fe^{3+} ion first, their spins partially cancel, giving a net moment of one Fe^{3+} ion per formula unit $M_3Fe_5O_{12}$, i.e. 5 BM. Since Y^{3+} is a d° ion, it has no magnetic moment. Consequently, the expected moment of YIG is 5 BM, which is in excellent agreement with the measured value.

For the rare earth garnets, the overall moment is expected to be given by

$$(3\mu_M - 5)\quad BM$$

where μ_M is the moment of the 24c ion. For Gd^{3+}, which is an f^7 ion, $\mu_{Gd} = 7\,BM$ and the net moment for GdIG is calculated to be 16 BM, which also agrees with experiment. For Lu^{3+}, which is f^{14}, $\mu_{Lu} = 0$ and the net moment is therefore 5 BM, in agreement with experiment. For the other ions, Tb to Yb, it appears that the orbital moment is not completely quenched and the μ_M values are larger than given by the spin-only formula with $g = 2.00$. Comparison between experimental and calculated values is shown in Fig. 8.10. For the latter, two curves are given, corresponding to the spin-only formula and the (spin + orbital moment) formula. Experimental values generally fall between the two theoretical curves and show that the orbital moment is only partially quenched.

The magnetic moments of the rare earth garnets show an interesting and unusual temperature dependence. The spontaneous moments at absolute zero, shown in Fig. 8.11, decrease with rising temperature and fall to zero at the *compensation temperature*. They then rise again, but in the opposite direction, and fall to zero a second time at the Curie temperature, as shown for dysprosium iron

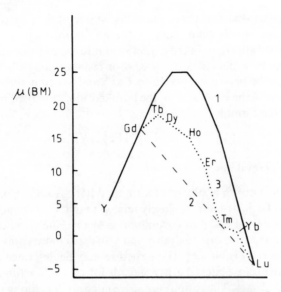

Fig. 8.10 Variation of magnetic moment at 0 K of garnets. Curve 1; calculated, spin + orbital formula. Curve 2: calculated, spin only formula. Curve 3: experimental. (Data from Standley, Clarendon Press, *Oxide magnetic materials*, 1972)

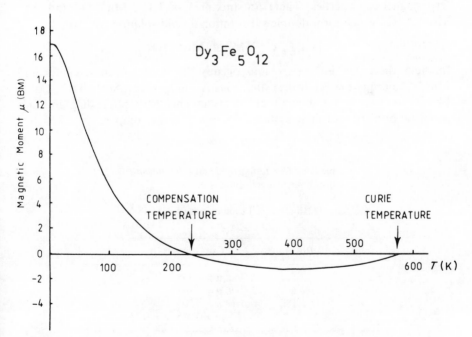

Fig. 8.11 Spontaneous magnetization in dysprosium iron garnet as a function of temperature

garnet. This effect occurs because the spins on one of the rare earth sublattices randomize more rapidly than those on the Fe^{3+} sublattice.

Many ionic substitutions in the garnet structure are possible and the magnetic properties may be systematically varied. For instance, the large trivalent ion in the 24c site may be partially replaced by Ca^{2+} and to compensate for the charge mismatch, some of the Fe^{3+} ions on the tetrahedral sites may be replaced by V^{5+} according to the formula:

$$[Y_{3-2x}^{3+}Ca_{2x}^{2+}]Fe_2^{3+}[Fe_{3-x}^{3+}V_x^{5+}]O_{12}$$

Ilmenites and perovskites

Ilmenites are a group of phases of formula ABO_3: A = Fe, Co, Ni, Cd, Mg; B = Ti, Rh, Mn. Their structure is closely related to that of chromium sesquioxide, Cr_2O_3; haematite, α-Fe_2O_3; or corundum, α-Al_2O_3. The symmetry is rhombohedral but the structure can be drawn and visualized more easily using a larger, hexagonal unit cell (Table 8.6). The structure may be described as an approximately hexagonal close packed array of oxide ions with the cations in two-thirds of the octahedral sites. The cations are segregated so that, along the c axis, there are alternate layers of A and B cations. Another way of looking at the ilmenite structure is as a derivative of NiAs with one-third of the octahedral sites vacant.

The perovskite structure of $SrTiO_3$ has been described in Chapter 1. Some oxides containing Fe^{3+} and $Mn^{3+,4+}$ have a perovskite structure and interesting ferromagnetic properties. These are mixtures of La^{3+} $Mn^{3+}O_3$ and A^{2+} $Mn^{3+,4+}O_3$ which form double substitutional solid solutions of formula:

$$[La_{1-x}^{3+}A_x^{2+}][Mn_{1-x}^{3+}Mn_x^{4+}]O_3$$

In these, the larger La^{3+}, A^{2+} ions occupy the twelve-coordinate sites and $Mn^{3+,4+}$ occupy the octahedral sites. Ions A can be Ca^{2+}, Sr^{2+}, Ba^{2+}, Cd^{2+}, Pb^{2+}. Closer study has shown that the crystal chemistry, phase diagrams and magnetic properties of these systems are, in fact, quite complex.

Table 8.6 *Crystallographic data for ilmenite*
Rhombohedral cell dimensions, $a = 5.538$ Å
$\alpha = 54.41°$
Hexagonal cell dimensions, $a = 5.048$ Å
$c = 14.026$ Å

Atom	Position*	Coordinates
Fe	6c	0, 0, u: $u = 0.358$
Ti	6c	0, 0, u: $u = 0.142$
O	18f	$uv\,w$: $u = 0.305$, $v = 0.015$, $w = 0.25$

*This is a label used in *International Tables for Crystallography*. The number refers to the number of equivalent positions in the unit cell.

Magnetoplumbites

Magnetoplumbite is the mineral $PbFe_{12}O_{19}$; its barium analogue $BaFe_{12}O_{19}$, known as BaM, is an important component of permanent magnets. The structure of magnetoplumbite is closely related to that of β-alumina, '$NaAl_{11}O_{17}$' (Chapter 7). The latter is basically a close packed structure with five oxide layers in the subrepeat unit. Each layer contains four oxide ions per unit cell and the distinctive feature of the structure is that every fifth layer has three-quarters of the oxide ions missing. Hence there are $(4 \times 4) + 1 = 17$ oxide ions to each five layer subrepeat unit. In β-alumina, the empty spaces in the fifth layer are partially occupied by Na^+ ions. Magnetoplumbite has a similar five-layer repeat unit. Four of the layers contain close packed oxide ions, as in β-alumina. The fifth layer contains three-quarters of its quota of oxide ions with a large divalent ion— Pb^{2+}, Ba^{2+} —in the other oxide ion site. Hence, the repeat unit contains $(4 \times 4) + (1 \times 3) = 19$ oxide ions with one Ba^{2+}, Pb^{2+} ion completing the fifth layer.

The magnetic structure of BaM is complex because there are Fe^{3+} ions in five different sets of crystallographic sites. However, the net effect is that in the formula unit $BaFe_{12}O_{19}$, eight of the Fe^{3+} ions have their spins oriented in one direction and the remaining four are antiparallel, giving a resultant of four Fe^{3+} ions with a moment of 20 BM.

Applications: structure–property relations

A large number of parameters affect the magnetic properties of materials. By careful control of composition and fabrication procedures, it is now possible to carry out 'crystal engineering' and deliberately prepare materials with a desired set of properties. In this section, we shall briefly summarize some applications of magnetic materials and the factors that influence the selection of a material for a particular application.

Transformer cores

A major application of ferro- and ferrimagnetic materials is in transformer and motor cores. Materials are required that are magnetically soft with large power-handling capacity and low losses. Magnetically soft materials have a high permeability—they are magnetized easily at low applied fields—and a low coercive field. They also tend to have low hysteresis losses. All of these properties are favoured in materials that have a small magnetostriction coefficient, λ_s, and a low magnetocrystalline anisotropy coefficient, K_1. Soft magnetic materials are, mechanically, those in which the domain walls are able to migrate easily. These various parameters can all be optimized by attention to the details of composition and fabrication.

As well as hysteresis losses, eddy current losses are a serious problem at high frequencies and especially in materials of low resistivity. This is because eddy currents are proportional to $(\text{frequency})^2$. Eddy currents in metals such as iron can be reduced by alloying the iron with, for example, nickel or silicon. This is because alloys generally have a much higher resistivity than the pure component metals. One great advantage of the ferrimagnetic oxides such as Mn, Zn ferrite, Ni, Zn ferrite and YIG is that, provided they are prepared correctly, they have very high resistivity and negligibly small eddy currents. This is particularly so for the garnets such as YIG; these contain trivalent cations only and so there is no easy mechanism for electronic conduction to occur. The conductivity of YIG at room temperature is only 10^{-12} ohm^{-1} cm^{-1}. With ferrites, it is necessary to ensure that all the iron present is in the $+3$ oxidation state; otherwise $Fe^{2+} \rightleftharpoons Fe^{3+} + e^-$ redox transfer may give rise to a high conductivity. For instance, magnetite, Fe_3O_4 or $Fe^{2+}Fe_2^{3+}O_4$, has a conductivity of 10^2 to 10^3 ohm^{-1} cm^{-1} at room temperature which is ~ 15 orders of magnitude higher than that of YIG. This high conductivity is associated with the mixture of valence states of iron.

Information storage

A critical requirement for magnetically based information storage components is that they should be soft with low eddy current losses and a certain type of hysteresis loop, either square or rectangular (Fig. 8.12). With this characteristic, a reverse field may be applied to a magnetized sample and it should undergo no change until the coercive field H_c, is exceeded, at which point a sudden switch in magnetization occurs. The two orientations of magnetization, $+$ and $-$, can be used to represent 0 and 1 in the binary digital system. Certain of the magnetic ferrites have the required characteristics for this application and, with switching times of 10^{-6} seconds or less, they are essential components to modern computer technology.

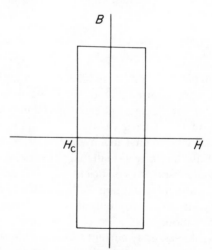

Fig. 8.12 Rectangular hysteresis loop required for information storage devices

Magnetic bubble memory devices

An interesting recent development in information storage uses thin films of garnet, a few micrometers thick, which are deposited epitaxially onto a non-magnetic substrate. The films are deposited at high temperature and, by careful choice of the composition of the garnet and, in particular, its lattice parameters, a slight difference in thermal contraction occurs on cooling the sample to room temperature. The stresses that are generated are sufficient to induce a preferred direction of magnetization in the garnet film which is perpendicular to the plane of the film. The resulting domain structure of the film has spins pointing up or down and these appear as bubbles when viewed in a polarizing microscope. These magnetic bubble materials can be used as memory components for binary digital computers.

Permanent magnets

For use in permanent magnets, the following properties are desired: high saturation magnetization, high energy product BH, high coercive field, high remanent magnetization, high Curie temperature, high magnetocrystalline anisotropy. Materials that are competing for applications as permanent magnets are likely to be based on either the metals Fe, Co, Ni or 'hard' oxides such as BaM, barium magnetoplumbite. Let us see how the performance of these two types of material can be optimized.

The hardness of magnets can be increased if ways can be found of either pinning or reducing the ease of motion of the domain walls. This may be achieved in steels by adding suitable dopants such as chromium or tungsten that cause either precipitation of, say, a carbide phase or a martensitic transformation on cooling. A novel feature of the Alnico group of magnets is that the ferromagnetic Co, Ni-based material is present as a large number of small crystalline regions embedded in an aluminium-based matrix. These small regions are all magnetized in the same direction and it is very difficult to demagnetize them or change their magnetic orientation.

The oxide magnets such as BaM are relatively light and cheap. Although their intrinsic magnetic properties are generally inferior to those of Alnico magnets, they can be improved if prepared with a magnetically aligned texture. To achieve this, the powdered starting materials are subjected to a magnetic field while they are being prepared into a shape and subsequently sintered at high temperature. The effect of the applied field is to cause magnetic alignment of the grains and this increases the remanent magnetization of the material.

Optical properties: luminescence, lasers

Luminescence and phosphors

Luminescence is the name generally given to the emission of light by a material as a consequence of it absorbing energy. Various types of excitation source may

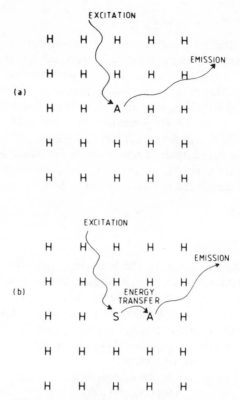

Fig. 8.13 Schematic representation of the luminescence process involving (a) an activator, A, in a host lattice, H, and (b) both a sensitizer, S, and an activator, A. (Adapted from DeLuca, *J. Chem. Educ.*, **57**, 541, 1980)

be used and are indicated as a prefix. *Photoluminescence* uses photons or light, often UV, for excitation. *Electroluminescence* uses an electrical energy input. *Cathodoluminescence* uses cathode rays or electrons to provide energy. Two types of photoluminescence may be distinguished. For a short time lapse, $\lesssim 10^{-8}$ sec, between excitation and emission, the process is known as *fluorescence*. Fluorescence effectively ceases as soon as the excitation source is removed. For much longer decay times, the process is known as *phosphorescence*. This may continue long after the source of excitation is removed.

Photoluminescent materials generally require a *host* crystal structure, ZnS, $CaWO_4$, Zn_2SiO_4, etc., which is doped with a small amount of an *activator*, a cation such as Mn^{2+}, Sn^{2+}, Pb^{2+}, Eu^{2+}. Sometimes, a second type of dopant is added to act as a *sensitizer*. The mode of operation of inorganic luminescent materials, known generally as *phosphors*, is shown schematically in Fig. 8.13. Note that the energy of the emitted light is generally less than that of the exciting radiation and is, therefore, of longer wavelength. This effective increase in

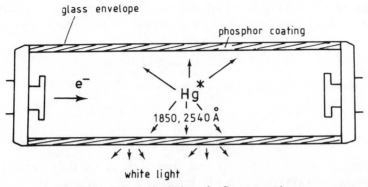

Fig. 8.14 Schematic design of a fluorescent lamp

wavelength is known as the *Stokes shift*. In fluorescent lamps, which provide the most important application of phosphors, the exciting radiation is UV light from a mercury discharge. Phosphor materials are required that absorb this UV radiation and emit 'white' light. The construction of a fluorescent lamp is shown schematically in Fig. 8.14. It consists of a glass tube lined on the inside with a coating of phosphor material and filled with a mixture of mercury vapour and argon. On passage of an electric current through the lamp, the atoms of mercury are bombarded by electrons and are excited into upper electronic energy states. They can then return to the ground state, accompanied by the emission of UV light of two characteristic wavelengths, 2540 and 1850 Å. This light irradiates the phosphor coating on the inner surface of the glass envelope which subsequently emits white light.

Examples of emission spectra of ZnS phosphors activated with a variety of

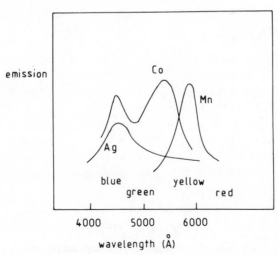

Fig. 8.15 Luminescence spectra of activated ZnS phosphors after irradiation with UV light

cations are shown in Fig. 8.15. Each activator gives a characteristic spectrum and colour to ZnS. Various types of electronic transition occur in activator ions, e.g.:

Ion	Ground state	Excited state
Ag^+	$4d^{10}$	$4d^9 5p$
Sb^{3+}	$4d^{10} 5s^2$	$4d^{10} 5s 5p$
Eu^{2+}	$4f^7$	$4f^6 5d$

The host materials used for phosphors fall into two main categories:

(a) Ionically bonded, insulating materials, such as $Cd_2B_2O_5$, Zn_2SiO_4 and apatite, $3Ca_3(PO_4)_2 \cdot Ca(Cl, F)_2$. In these, a set of discrete energy levels is associated with the activator ion and these levels are modified by the local environment of the host crystal structure. For ionic phosphors, the *configurational coordinate model* provides a useful way of representing qualitatively the luminescence processes.
(b) Covalently bonded, semiconducting sulphides, such as ZnS. In these, the energy band structure of the host is modified by the addition of localized energy levels associated with the activator ions.

Configurational coordinate model

The potential energy of the ground and excited electronic states of the luminescent centre is plotted against a general coordinate which is often the internuclear distance. This is shown schematically for the ground state in Fig. 8.16. The solid curve shows qualitatively how the potential energy varies as a function of interatomic distance. It passes through a minimum at the equilibrium

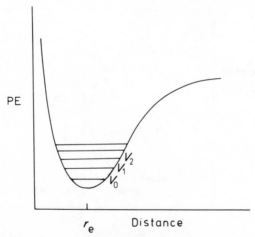

Fig. 8.16 Ground state potential energy diagram for a luminescent centre in an ionic host crystal

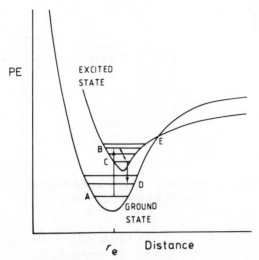

Fig. 8.17 Ground and excited state potential
energy diagrams for a luminescent centre

bond length, r_e. Within this electronic ground state, different quantized vibrational states of the ion are possible, as shown by the horizontal lines V_0, V_1, etc.

Each electronic state for the luminescent centre has a potential energy curve roughly similar to that in Fig. 8.16. Typical curves for a ground state and an excited state are shown in Fig. 8.17; using this diagram, many of the features of luminescence can be explained.

First, the process of excitation involves raising the active centre from its ground state, level A, into a higher vibrational level of the excited state, B. Second, some energy is then dissipated as the ion quickly relaxes to a lower level, C, in the excited state. This energy is lost to the host lattice and appears as heat. Third, the active centre returns to its ground state, level D or A, and, in so doing, emits light. Since the energy of excitation A → B is greater than that of emission C → D, the emitted radiation is of longer wavelength than the exciting radiation. This therefore accounts for the Stokes shift.

An effect known as *thermal quenching* in which the luminescence efficiency decreases markedly above a certain temperature can also be explained with the aid of Fig. 8.17. The potential energy curves for the ground and excited states cross over at point E. At this point, an ion in the excited state can transfer back to its ground state, at the same energy. It can then return to the lower vibrational levels of the ground state by means of a series of vibrational transitions. Point E represents a kind of spillover point, therefore. If an ion in the excited state can acquire sufficient vibrational energy to reach point E, it can spill over into the vibrational levels of the ground state. If this happens, all the energy is released as vibrational energy and no luminescence occurs. The energy of point E is obviously critical. In general, it is likely to be reached as a consequence of

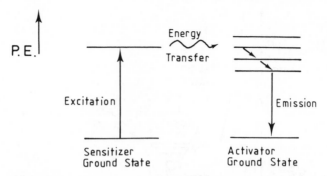

Fig. 8.18 Non-radiative energy transfer involved in operation of a sensitized phosphor

increasing the temperature since, with rising temperature, ions have increasing thermal energy and are able to move to progressively higher vibrational levels.

The type of transition described above to explain thermal quenching is an example of a *non-radiative transition*. In this the excited ion gets rid of some of its excess energy by imparting vibrational energy to the surrounding host lattice. In this way, the excited ion is able to return to a lower energy level but no electromagnetic radiation, i.e. light, is emitted.

Another type of non-radiative transition is involved in the operation of sensitized phosphors. This transition, known as *non-radiative energy transfer*, is shown schematically in Fig. 8.18. It depends on (a) there being similar energy levels in the excited states of both sensitizer and activator ions and (b) sensitizer and activator ions being relatively close together in the host crystal structure. In operation, the exciting radiation promotes sensitizer ions into an excited state. These then transfer energy to neighbouring activator ions, with little or no loss of energy during transfer, and at the same time the sensitizer ions return to their ground state. Finally, the activator ions return to their ground state by the emission of luminescent radiation.

Non-radiative energy transfer is also involved in the *poisoning* effect of certain impurities. In this, energy is transferred from either a sensitizer or an activator to a poison site at which the energy is lost to the host structure in the form of vibrational energy. Ions that have non-radiative transitions to the ground state and which must be avoided in the preparation of phosphors include Fe^{2+}, Co^{2+} and Ni^{2+}.

Some phosphor materials

An enormous number of host/activator combinations have been studied for luminescence, with a fair degree of success. However, future improvements and the development of new materials will probably depend on improved understanding of the relation between crystal structure and the energy levels of dopant ions.

A phosphor that is used extensively in fluorescent lamps is an apatite, doubly doped with Mn^{2+} and Sb^{3+}. Fluorapatite is $Ca_5(PO_4)_3F$. When doped with Sb^{3+} it fluoresces blue, and with Mn^{2+} it fluoresces orange-yellow; the two together give a broad emission spectrum that approximates to white light. A modification of the wavelength distribution in the emission spectrum is possible by partly replacing the F^- ions in fluorapatite by Cl^- ions. The effect of this is to modify the energy levels of the activator ions and hence their emission wavelengths. By careful control of composition in this way, the colour of the fluorescent light may be optimized. A selection of other lamp phosphor materials is given in Table 8.7.

Trivalent europium is an important activator ion, especially for use in red phosphors for colour television screens. In $YVO_4 : Eu^{3+}$, the vanadate group absorbs energy in the cathode ray tube but the emitter is Eu^{3+}. The mechanism of charge transfer between the vanadate group and Eu^{3+} appears to involve a non-radiative, superexchange process via the intervening oxide ions. The effect is related to the superexchange mechanism postulated to explain the antifer-romagnetic ordering of Ni^{2+} ions in NiO. In order to get efficient energy transfer by superexchange, it appears to be important that the metal–oxygen–metal bond should be approximately linear, so as to maximize the degree of orbital overlap. In YVO_4: Eu, the vanadium–oxygen–europium angle is 170° and the energy exchange takes place rapidly.

Transitions from and to several different f energy levels are possible, in principle, with Eu^{3+}. The ones that are actually observed, and hence the colour, depend on the host crystal structure and, in particular, on the symmetry of the site that it provides for Eu^{3+}. If the Eu^{3+} is located at a centre of symmetry, as it is when doped into $NaLuO_2$ and Ba_2GdNbO_6, the favoured transition is of the type $^5D_0 \rightarrow {}^7F_1$. The resulting colour is orange. When located on a non-centrosymmetric site, as in $NaGdO_2 : Eu^{3+}$, the preferred transition is $^5D_0 \rightarrow {}^7F_2$ and the colour of emission is red. The crystal structure of the host material therefore has a major effect on the resulting colour.

Colour television screens require three primary cathodoluminescent colours.

Table 8.7 *Some lamp phosphor materials.* (Data taken from Burrus, 1972)

Phosphor	Activator	Colour
Zn_2SiO_4, willemite	Mn^{2+}	Green
Y_2O_3	Eu^{3+}	Red
$CaMg(SiO_3)_2$, diopside	Ti	Blue
$CaSiO_3$, wollastonite	Pb, Mn	Yellow-orange
$(Sr, Zn)_3(PO_4)_2$	Sn	Orange
$Ca_5(PO_4)_3(F, Cl)$ fluorapatite	Sb, Mn	'White'

These are:

(a) red, mentioned above, for which $YVO_4:Eu^{3+}$ is often used,
(b) blue—$ZnS:Ag^+$;
(c) green—$ZnS:Cu^+$.

For black and white screens, mixtures of the blue-emitting $ZnS:Ag^+$ and yellow-emitting $(Zn, Cd)S:Ag^+$ are used.

Anti-Stokes phosphors

A relatively new class of phosphors that has aroused considerable interest is the anti-Stokes phosphors. These exhibit the remarkable property of emitting light or photons of higher energy (shorter wavelength) than the incident exciting radiation. Using these, it is possible, for instance, to convert infrared radiation into higher-energy, visible light. There must be a catch in this somewhere, of course! The law of conservation of energy cannot be violated. Instead, the process of excitation takes place in two or more stages, as shown schematically in Fig. 8.19.

The best studied anti-Stokes phosphors to data are host structures such as YF_3, $NaLa(WO_4)_2$ and α-$NaYF_4$, which have been doubly doped with Yb^{3+} as a sensitizer and Er^{3+} as an activator. These materials can convert infrared radiation into green luminescence. During irradiation, Yb^{3+} ions transfer two photons of infrared radiation to nearby Er^{3+} ions which are then raised into a doubly excited state and decay by the emission of visible light.

Lasers

The solid state laser is basically a luminescent solid in which certain special requirements have been met. The name 'laser' stands for light amplification by stimulated emission of radiation. The excitation process involves pumping the

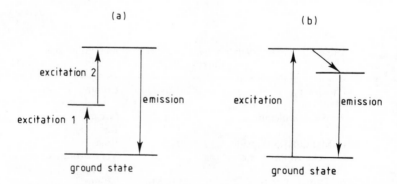

Fig. 8.19 Schematic representation of (a) anti-Stokes and (b) normal luminescence phenomena

active centres into an excited state that has a reasonably long lifetime; a situation can then be reached in which a 'population inversion' occurs and there are more active centres in the excited state than in the ground state. During luminescence, the light emitted from one centre stimulates others to decay in phase with the radiation emitted from the first centre. In this way, an intense beam or pulse of coherent radiation is built up.

The first laser system, the ruby laser, was reported in 1960 by Maiman; from this has grown a large area of modern science and technology with applications in photography, surgery, communications and precise measurements, to name but a few. Many types of laser system have been discovered. The commercially available ones fall into three main categories: gas lasers, dye lasers, with their wavelength tunability and solid state lasers. We are here concerned only with solid state lasers and the chemistry involved in their operation. It is not intended to give a general review of lasers.

The ruby laser

This was the first laser system to be discovered and more than twenty years later it is still an important one. The essential component to the ruby laser is a single crystal of Al_2O_3 doped with a small amount, 0.05 wt%, of Cr^{3+}. The Cr^{3+} ions substitute for Al^{3+} ions in the distorted octahedral sites of the corundum crystal structure (this structure is similar to that of ilmenite, $FeTiO_3$). On addition

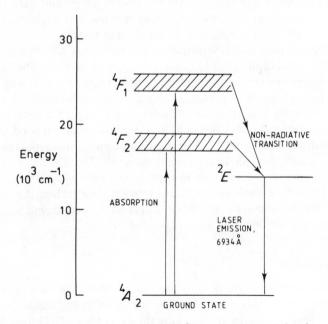

Fig. 8.20 Energy levels of the Cr^{3+} ion in ruby crystal and laser emission

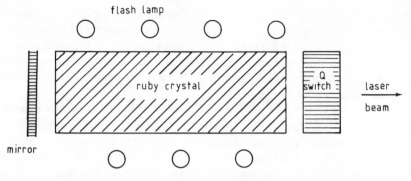

Fig. 8.21　Design of a ruby laser

of Cr_2O_3 to Al_2O_3, the colour changes from white, in Al_2O_3, to red, at low Cr^{3+} levels, to green, for larger Cr^{3+} contents; these solid solutions are discussed in general terms in Chapter 5.

The energy levels of the Cr^{3+} ion in ruby are shown schematically in Fig. 8.20. On shining intense visible light from, for example, a xenon flash lamp, onto a ruby crystal, d electrons on Cr^{3+} ions may be promoted from the $^4A_2$ ground state into $^4F_2$ and $^4F_1$ upper states. These then decay rapidly, by a non-radiative process, into the ^2E level. The lifetime of the ^2E excited state is fairly long, $\sim 5 \times 10^{-3}$ sec, which means that a considerable population inversion has time to build up. Laser action then occurs by transition from the ^2E level to the ground state. During this transition many ions are stimulated to decay, in phase with each other, giving an intense, coherent pulse of red light, of wavelength 6934 Å.

The design of a ruby laser is shown schematically in Fig. 8.21. It contains a ruby crystal rod, several centimetres long and 1 to 2 cm in diameter. The flash lamp is shown as being wrapped around the ruby rod. Alternatively, it may be placed alongside the rod; the two are then arranged inside a reflection cavity such that the rod is effectively irradiated from all sides. At one end of the rod is a mirror for reflecting the light pulse back through the rod. At the other end is a device known as a Q switch which may either allow the laser beam to pass out of the system or may reflect it back through the rod for another cycle. The Q switch may simply be a rotating mirror timed to allow out the laser beam when it has reached its optimum intensity: as the light pulse passes back and forth along the rod, it builds up in intensity as more of the active centres are stimulated to emit radiation that is coherent with the initial pulse.

Neodymium lasers

The host material for the laser active Nd^{3+} ion is either a glass or yttrium aluminium garnet (YAG), $Y_3Al_5O_{12}$. The energy levels and transitions involved in the operation of neodymium lasers are shown in Fig. 8.22. During irradiation from a high energy lamp, several absorption transitions occur, although only one

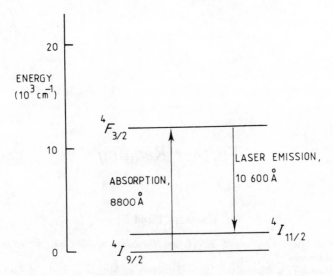

Fig. 8.22 Energy levels of the Nd^{3+} ion in neodymium lasers

is shown. These excited states all decay non-radiatively to the $^4F_{3/2}$ level from which laser action occurs to the $^4I_{11/2}$ level with a wavelength of 10,600 Å for Nd–glass and 10640 Å for Nd^{3+} – YAG. The $^4F_{3/2}$ state is long lived, $\sim 10^{-4}$ sec, but depends somewhat on the Nd^{3+} concentration. This again allows a large population inversion to build up and allows Nd^{3+} to be used in high power lasers.

Further Reading

Chapters 1 and 2

D. M. Adams (1974). *Inorganic Solids, An Introduction to Concepts in Solid-State Structural Chemistry*, Wiley.

C. S. Barrett and T. B. Massalski (1966). *Structure of Metals*, 3rd ed., McGraw-Hill.

L. Bragg, G. F. Claringbull and W. H. Taylor (1965). *Crystal Structure of Minerals*, Cornell University Press, Ithaca, N.Y.

G. M. Clark (1972). *The Structures of Non-Molecular Solids. A Coordinated Polyhedron Approach*, Applied Science Publishers.

R. C. Evans (1964). *An Introduction to Crystal Chemistry*, 2nd ed., Cambridge University Press.

H. Krebs (1968). *Inorganic Crystal Chemistry*, McGraw-Hill.

F. Liebau (1985). *Structural Chemistry of Silicates*, Springer-Verlag.

H. D. Megaw (1973). *Crystal Structures, A Working Approach*, Saunders.

I. Naray-Szabo (1969). *Inorganic Crystal Chemistry*, Akademiai Szabo.

E. Parthé (1964). *Crystal Chemistry of Tetrahedral Structures*, Gordon and Breach.

C. N. R. Rao and J. Gopalakrishnan (1986). *New Directions in Solid State Chemistry*, Cambridge University Press.

R. W. G. Wyckoff (1971). *Crystal Structures*, Vols 1 to 6, Wiley.

Chapter 3

L. V. Azaroff (1968). *Elements of X-ray Crystallography*, McGraw-Hill.

G. E. Bacon (1975). *Neutron Diffraction*, Oxford.

F. Donald Bloss (1971). *Crystallography and Crystal Chemistry*, Holt, Rinehart and Winston.

M. J. Buerger (1960). *Crystal Structure Analysis*, Wiley.

C. W. Bunn (1961). *Chemical Crystallography, An Introduction to Optical and X-ray Methods*, Clarendon Press.

B. D. Cullity (1978). *Elements of X-ray Diffraction*, Addison Wesley.

L. S. Dent Glasser (1977). *Crystallography and its Applications*, Van Nostrand Reinhold.

N. F. M. Henry, H. Lipson and W. A. Wooster (1960). *The Interpretation of X-ray Diffraction Photographs*, Macmillan.

R. Jenkins and J. L. DeVries (1970). *Worked Examples in X-ray Analysis*, Springer-Verlag.

H. P. Klug and L. E. Alexander (1974). *X-ray Diffraction Procedures for Polycrystalline and Amorphous Materials*, 2nd ed., Wiley.

M. F. C. Ladd and R. A. Palmer (1978). *Structure Determination by X-ray Crystallography*, Plenum Press.

D. Mckie and C. Mckie (1986). *Essentials of Crystallography*, Blackwell.

E. W. Nuffield (1966). *X-ray Diffraction Methods*, Wiley.
H. S. Peiser, H. P. Rooksby and A. J. C. Wilson (1960). *X-ray Diffraction by Polycrystalline Materials*, Chapman and Hall.
D. E. Sands (1969). *Introduction to Crystallography*, W. A. Benjamin.
G. H. Stout and L. H. Jensen (1968). *X-ray Structure Determination: A Practical Guide*, Macmillan.
B. K. Vainshtein (1981). *Modern Crystallography*, Springer-Verlag.
B. E. Warren (1969). *X-ray Diffraction*, Addison Wesley.
E. J. W. Whittaker (1981). *Crystallography*, Pergamon.
A. J. C. Wilson (1970). *Elements of X-ray Crystallography*, Addison Wesley.
M. M. Woolfson (1970). *An Introduction to X-ray Crystallography*, Cambridge University Press.
J. Wormald (1973). *Diffraction Methods*, Clarendon Press.

Chapter 4

G. M. Bancroft (1973). *Mössbauer Spectroscopy*, McGraw-Hill.
A. Bianconi, L. Incoccia and S. Stipcich (Eds) (1983). *EXAFS and near Edge Structure*, Springer-Verlag.
L. S. Birks (1969). *X-ray Spectrochemical Analysis*, 2nd ed., Interscience.
D. Briggs (1977). *Handbook of X-ray and Ultraviolet Photoelectron Spectroscopy*, Heyden.
A. K. Cheetham and Peter Day (1987). *Solid State Chemistry, Techniques*, Oxford.
T. Daniels (1973). *Thermal Analysis*, Kogan Page.
LeRoy Eyring (1980). The application of high-resolution electron microscopy to problems in solid state chemistry, *J. Chem. Ed.*, **57**, 565.
N. N. Greenwood (1967). The Mössbauer spectra of chemical compounds, *Chem. Brit.*, **3**, 56.
N. H. Hartshorne and A. Stuart (1971). *Practical Optical Crystallography*, Arnold.
P. W. Hawkes (1972). *Electron Optics and Electron Microscopy*, Taylor and Francis, London.
J. M. Honig and C. N. R. Rao (Eds) (1981). *Preparation and Characterisation of Materials*, Academic Press.
R. Jenkins (1974). *An Introduction to X-ray Spectrometry*, Heyden.
R. C. Mackenzie (1970). *Differential Thermal Analysis*, Vols I and II, Academic Press.
W. Wayne Meinke (1973). Characterisation of solids—chemical composition, in *Treatise on Solid State Chemistry* (Ed. N. B. Hannay), Vol. I, Plenum Press.
R. E. Newnham and Rustum Roy (1975). Structural characterisation of solids, in *Treatise on Solid State Chemistry* (Ed. N. B. Hannay), Vol. 2, p. 437, Plenum Press.
M. I. Pope and M. D. Judd (1977). *Differential Thermal Analysis*, Heyden.
M. W. Roberts (1981). Photoelectron spectroscopy and surface chemistry, *Chem. Brit.*, **1981**, 510.
K. Siegbahn *et al.* (1967). *ESCA: Atomic, Molecular and Solid State Structure Studied by Means of Electron Spectroscopy*, Almquist and Wicksells, Uppsala.
W. W. Wendlandt (1974). *Thermal Methods of Analysis*, Wiley.
J. Zussman (Ed.) (1977). *Physical Methods in Determinative Mineralogy*, Academic Press.

Chapter 5

L. W. Barr and A. B. Lidiard (1970). Defects in ionic crystals, in *Physical Chemistry* (Ed. W. Jost), Vol. 10, Academic Press, N.Y.
R. J. Brook (1974). Defect structure of ceramic materials, chap. 3 in *Electrical Conductivity in Ceramics and Glass* (Ed. N. M. Tallan), Part A, Marcel Dekker, N.Y.
LeRoy Eyring and M. O'Keeffe (Ed.) (1970). *The Chemistry of Extended Defects in Non-Metallic Compounds*, North Holland.

M. E. Fine (1973). Introduction to chemical and structural defects in crystalline solids, in *Treatise on Solid State Chemistry* (Ed. N. B. Hannay), Vol. I, Plenum Press.

N. N. Greenwood (1968). *Ionic Crystals, Lattice Defects and Nonstoichiometry*, Butterworths, London.

D. Hull (1965). *Introduction to Dislocations*, Pergamon.

A. Kelly (1966). *Strong Solids*, Clarendon Press, Oxford.

P. Kofstad (1972). *Non-Stoichiometry, Electrical Conductivity and Diffusion in Binary Metal Oxides*, Wiley, N.Y.

F. A. Kröger (1974). *The Chemistry of Imperfect Crystals*, North Holland.

George G. Libowitz (1973). Defect equilibria in solids, in *Treatise on Solid State Chemistry* (Ed. N. B. Hannay), Vol I, Plenum Press.

E. Mandelcorn (Ed.) (1964). *Non-stoichiometric Compounds*, Academic Press.

W. J. Moore (1967). *Seven Solid States*, W. A. J. Benjamin Inc.

F. R. N. Nabarro (1967). *Theory of Crystal Dislocations*, Clarendon Press, Oxford.

J. Nolting (1970). Disorder in solids, *Angew Chemie, Intern. Ed.*, **9**, 989.

A. Rabenau (1970). *Problems of Non-stoichiometry*, North Holland.

W. T. Read Jr. (1953). *Dislocations in Crystals*, McGraw-Hill, N.Y.

R. E. Reed-Hill (1964). *Physical Metallurgy Principles*, Van Nostrand Reinhold.

A. L. G. Rees (1954). *The Defect Solid State*, Methuen.

R. J. D. Tilley (1986). *Defect Crystal Chemistry and its Applications*, Blackie.

Chapter 6

A. M. Alper (Ed.) (1971). *High Temperature Oxides*, Vols 1 to 4, Academic Press, New York.

A. M. Alper (1976). *Phase Diagrams*, Vols 1 to 5, Academic Press.

W. G. Ernst (1976). *Petrologic Phase Equilibria*, W. H. Freeman & Co., San Francisco.

A. Findlay (1951). *The Phase Rule and Its Applications*, 9th ed., Dover, N.Y.

P. Gordon (1968). *Principles of Phase Diagrams in Materials Systems*, McGraw-Hill, New York.

Phase Diagrams for Ceramists 1964 ed., 1969 Suppl., 1975 Suppl., 1981 Suppl., American Ceramic Society, Columbus, Ohio. The standard work of reference for phase diagrams of non-metallic inorganic materials.

A. Reisman (1970). *Phase Equilibria*, Academic Press, New York.

J. E. Ricci (1966). *The Phase Rule and Heterogeneous Equilibrium*, Dover, New York.

F. Tamas and I. Pal (1970). *Phase Equilibria Spatial Diagrams*, Butterworths.

Chapter 7

L. W. Barr and A. B. Lidiard (1970). In *Physical Chemistry, An Advanced Treatise*, Vol. 10, *Solid State*, Academic Press.

D. Bloor (1982). Plastics that conduct electricity, *New Scientist*, **1982**, 577–580.

J. C. Burfoot and G. W. Taylor (1979). *Polar Dielectrics and Their Applications*, Macmillan.

P. A. Cox (1987) *The Electronic Structure and Chemistry of Solids*, Oxford.

S. Geller (Ed.) (1977). Solid electrolytes, in *Topics in Applied Physics*, Vol. 21, Springer-Verlag.

S. Hackwood and R. G. Linford (1981). Physical Techniques for the Study of Solid Electrolytes. *Chem. Res.*, **81**, 327.

W. Hayes (1978). Superionic conductors, *Contemp. Phys.*, **19**(5), 469–486.

G. Holzäpfel and H. Rickert (1977). Solid state electrochemistry—new possibilities for research and industry, *Die Naturwiss.*, **64**(2), 53–58.

A. Hooper (1978). Fast ionic conductors, *Contemp. Phys.*, **19**, 147–168.

A. K. Jonscher (1983). *Dielectric Relaxation in Solids*, Chelsea Dielectrics Press, London.

W. Jones and J. M. Thomas (1979). Applications of electron microscopy to organic solid state chemistry, *Progr. Solid St. Chem.*, **12**, 101–124.

Mansel Davies *Molecular Properties*.

D. V. Morgan and K. Board (1983). *An Introduction to Semiconductor Microtechnology*, Oxford.

W. J. Moore (1967). *Seven Solid States*, Benjamin.

R. E. Newnham (1975). *Structure—Property Relations*, Springer–Verlag.

P. J. Nigrey, D. MacInnes, D. P. Nairns and A. G. MacDiarmid (1981). Lightweight rechargable storage batteries using polyacetylene $(CH)_x$ as the cathode active material, *J. Electrochem. Soc.*, **128**, 1651–1654.

M. O'Keeffe and B. G. Hyde (1976). The solid electrolyte transition and melting in salts, *Phil. Mag.*, **B3**, 219–224.

K. J. Pascoe (1973). *Properties of Materials for Electrical Engineers*, Wiley.

H. Rickert (1978). Solid ionic conductors—principles and applications, *Angew. Chem. Int. Ed.*, **17**, 37–46.

R. M. Rose, L. A. Shepard and J. Wulff (1966). *The Structural and Properties of Materials.* IV. *Electronic Properties*, Wiley.

R. B. Seymour (1981). *Conductive Polymers*, Plenum Press.

B. C. H. Steele (1972). Electrical conductivity in ionic solids, in *MTP International Reviews of Science. Solid State Chemistry* (Ed. L. E. J. Roberts), p. 117, Wiley.

T. Takahashi (1973). Solid silver ion conductors, *J. Appl. Electrochem.*, **3**, 79–90.

J. M. Thomas (1974). Topography and topology in solid state chemistry, *Phil Trans. Roy Soc.*, **277**, 251–286.

D. A. Wright (1966). *Semiconductors*, Methuen.

Chapter 8

H. L. Burrus (1972). *Lamp Phosphors*, Mills and Boon.

F. A. Cotton and G. Wilkinson (1966). *Advanced Inorganic Chemistry*, Wiley.

D. J. Craik (Ed.) (1975). *Magnetic Oxides*, Parts 1 and 2, Wiley.

J. A. Deluca (1980). An introduction to luminescence in inorganic solids, *J. Chem. Ed.*, **57**, 541.

A. Earnshaw (1968). *Introduction to Magnetochemistry*, Academic Press.

T. R. Evans (1982). *Applications of Lasers to Chemical Problems*, Wiley.

J. B. Goodenough (1963). *Magnetism and the Chemical Bond*, Wiley.

H. Hibst (1982). Hexagonal ferrites from melts and aqueous solutions, magnetic recording materials, *Angew. Chemie, Int. Ed. Engl.*, **21**, 270.

W. D. Kingery, H. K. Bowen and D. R. Uhlmann (1976). *Introduction to Ceramics*, Wiley.

T. H. Maiman (1960). Stimulated optical radiation in ruby, *Nature*, **187**, 493.

C. B. Moore (1974). *Chemical and Biochemical Applications of Lasers*, Academic Press.

R. E. Newnham (1975). *Structure—Property Relations*, Springer-Verlag.

M. M. Schieber (1967). *Experimental Magnetochemistry*, North Holland.

C. G. Shull, W. A. Strauser and E. O. Wollan (1951). Neutron diffraction by paramagnetic and antiferromagnetic substances, *Phys. Rev.*, **83**, 333.

K. J. Standley (1972). *Oxide Magnetic Materials*, Clarendon Press.

R. S. Tebble and D. J. Graik (1979). *Magnetic Materials*, Wiley.

J. H. Wernick (1973). Structure and composition in relation to properties, in *Treatise on Solid State Chemistry* (Ed. N. B. Hannay), Vol. 1, Plenum Press.

Appendix 1

Interplanar Spacings and Unit Cell Volumes

The value of d, the perpendicular distance between adjacent planes in the set (hkl), may be calculated using the formulae:

Cubic
$$\frac{1}{d^2} = \frac{h^2 + k^2 + l^2}{a^2}$$

Tetragonal
$$\frac{1}{d^2} = \frac{h^2 + k^2}{a^2} + \frac{l^2}{c^2}$$

Orthorhombic
$$\frac{1}{d^2} = \frac{h^2}{a^2} + \frac{k^2}{b^2} + \frac{l^2}{c^2}$$

Hexagonal
$$\frac{1}{d^2} = \frac{4}{3}\left(\frac{h^2 + hk + k^2}{a^2}\right) + \frac{l^2}{c^2}$$

Monoclinic
$$\frac{1}{d^2} = \frac{1}{\sin^2 \beta}\left(\frac{h^2}{a^2} + \frac{k^2 \sin^2 \beta}{b^2} + \frac{l^2}{c^2} - \frac{2hl \cos \beta}{ac}\right)$$

Triclinic
$$\frac{1}{d^2} = \frac{1}{V^2}[h^2 b^2 c^2 \sin^2 \alpha + k^2 a^2 c^2 \sin^2 \beta$$

$$+ l^2 a^2 b^2 \sin^2 \gamma + 2hkabc^2(\cos \alpha \cos \beta - \cos \gamma)$$

$$+ 2kla^2 bc(\cos \beta \cos \gamma - \cos \alpha)$$

$$+ 2hlab^2 c(\cos \alpha \cos \gamma - \cos \beta)]$$

where V is the cell volume. The unit cell volumes are given by:

Cubic	$V = a^3$
Tetragonal	$V = a^2 c$
Orthorhombic	$V = abc$
Hexagonal	$V = (\sqrt{3}a^2 c)/2 = 0.866a^2 c$
Monoclinic	$V = abc \sin \beta$
Triclinic	$V = abc(1 - \cos^2 \alpha - \cos^2 \beta - \cos^2 \gamma + 2\cos \alpha \cos \beta \cos \gamma)^{1/2}$

Appendix 2

Model Building

The construction of two types of model, based on spheres and polyhedra, is outlined.

Equipment needed:

Polystyrene spheres (100 to 200), any diameter but 30 mm convenient
Solvent (5 to 10 ml) in bottle with dropper or paint brush; chloroform suitable
Card paper (2 to 3 m^2), perhaps sheets of different colour
Glue stick

Sphere packing arrangements

In order to join polystyrene spheres, glue may be used but is often messy. An alternative is to use a solvent in which polystyrene dissolves. A small amount of solvent is spotted or brushed onto a sphere and a second sphere is placed in contact with the first in the region of the solvent. If too much solvent is used it may run and spoil the appearance of the sphere. The spheres usually stick immediately on contact but the bond hardens only after leaving for a few hours. It is therefore often better to construct larger three-dimensional models in stages.

It should be relatively easy to stick together spheres to form close packed layers and then build up layers in h.c.p. and c.c.p. stacking sequences; further details will not be given. Two useful additional exercises are the following.

To show the relation between a c.c.p. structure and an f.c.c. unit cell

Stick together six spheres to form that part of a c.p. layer shown in Fig. A.1(a). Place a single sphere in the layer above, as shown in (b). Repeat steps (a) and (b) so that two identical arrangements, (b), are obtained. Allow models to harden. Orient one model as in (c) and place the second model over the first as in (d). The resulting structure should be a face centred cubic unit cell with spheres in corner and face centre positions. The f.c.c. cell therefore contains c.p. layers parallel to the (111) planes of the unit cell. On standing the model on one corner so that the c.p. layers are horizontal, it should be apparent that the stacking sequence is ABC.

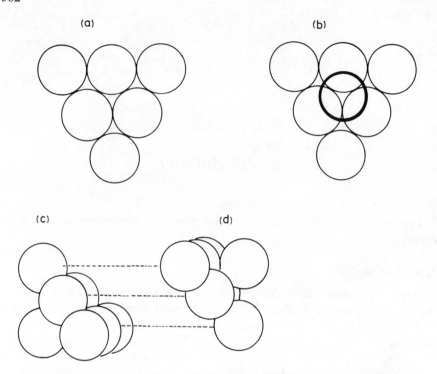

Fig. A.1 To show the relation between c.c.p. and the f.c.c. unit cell

To show the four orientations of the c.p. layers in an f.c.c. unit cell

One major difference between c.c.p. and h.c.p. is that in an h.c.p. array of spheres, only one set of c.p. layers occurs parallel to the basal plane of the hexagonal unit cell. In a c.c.p. array of spheres, c.p. layers occur in four orientations. This may be shown by constructing the pyramid shown in Fig. A.2.

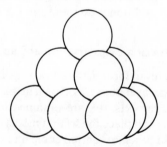

Fig. A.2 The c.p. layers in a
c.c.p. structure

First, construct a triangular base of six spheres, as in Fig. A.1; then add three spheres in the layer above and one sphere to form the apex to give the pyramid shown in Fig. A.2. Check that the packing sequence is ABC, i.e. c.c.p. The pyramid may now be reoriented so that any of its other three faces forms the base and an identical structure results. The four equivalent orientations of the pyramid correspond, therefore, to the four orientations of c.p. layers in a c.c.p. structure.

Polyhedral structures

Tetrahedra, octahedra and any other polyhedra may be made from card paper. Using templates such as in Fig. A.3, the polyhedra may be copied, cut out, folded and glued at the tabs. A convenient size for making fairly rigid polyhedra is to make the polyhedron edge $\simeq 5$ cm. The polyhedra may be linked up by glueing together corners, edges or faces. Glueing polyhedra by their corners only is a little tricky but can be done given time and patience. A selection of polyhedral linkages is given in Fig. A.4; others are shown in Figs 1.27, 1.30, 1.32, 1.34 and 1.36.

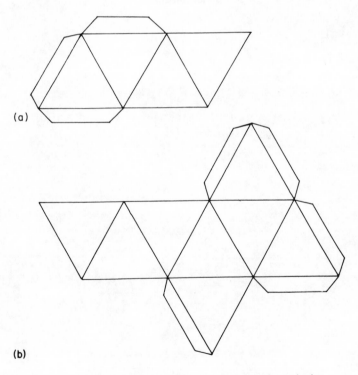

Fig. A.3 Templates for making tetrahedra and octahedra

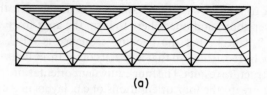

(a)

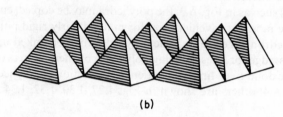

(b)

Fig. A.4 Some polyhedra linkages: (a) a row of edge-sharing octahedra; (b) a layer of corner-sharing tetrahedra

Appendix 3

Geometrical Considerations in Crystal Chemistry

Notes on the geometry of tetrahedra and octahedra

Relation of a tetrahedron to a cube

In Fig. A.5 is shown a cube with a tetrahedron inside; the centre of the tetrahedron, M, is at the cube body centre and the corners of the tetrahedron are at four alternate corners of the cube, X_1 to X_4. Using this relation it is relatively easy to make calculations on the geometry of tetrahedra.

Relation between distances M–X and X–X in a tetrahedron

Let cube edge have length l. From Pythagoras, the distance X–X is the face diagonal of the cube, i.e. $X-X = \sqrt{2}l$. Distance M–X is half the body diagonal of the cube, i.e. $M-X = \sqrt{3}l/2$. Hence, the ratio $XX/MX = \sqrt{2}l/(\sqrt{3}l/2) = \sqrt{8/3} = 1.633$. In silica structures, the basic building blocks are SiO_4 tetrahedra. In these usually $Si-O \simeq 1.62$ Å. Therefore, the tetrahedron edges, $O-O \simeq 1.633$ $(Si-O) \simeq 2.65$ Å.

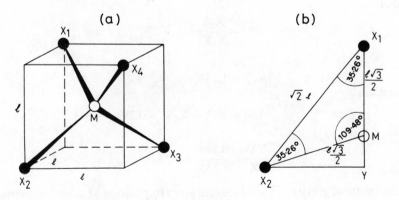

Fig. A.5 Relation of a tetrahedron to a cube

385

Angle XMX of a tetrahedron

Using the above result and the cosine equation (Fig. A.5a and b):

$$(X_1 - X_2)^2 = (M - X_1)^2 + (M - X_2)^2 - 2(M - X_1)(M - X_2)\cos\angle X_1MX_2$$

Therefore,

$$(1.633\,M - X_1)^2 = 2(M - X_1)^2 - 2(M - X_1)^2\cos\angle X_1MX_2$$

and so

$$\angle XMX = 109.48°$$

Symmetry of a tetrahedron

A tetrahedron has a threefold rotation axis along each of the $M - X$ directions, i.e. along each cube body diagonal. Both the tetrahedron and the cube therefore possess four threefold axes. The cube also possesses fourfold rotation axes passing through each pair of opposite cube faces. In the tetrahedron these axes are fourfold inversion axes, $\bar{4}$ (i.e. rotation by 90° followed by inversion through the centre of the tetrahedron).

Centre of gravity of a tetrahedron

For this, we need to know the vertical height of position M above the triangular base of the tetrahedron. Consider the triangular section X_1X_2M (Fig. A.5b). Point Y, the extension of the line $X_1 - M$, lies in the centre of the base X_2, X_3, X_4 (Fig. A.5a, not shown). Hence, distance $M - Y$ gives the height of the centre of gravity above the base. Since

$$\angle X_1MX_2 = 109.48°$$

then

$$\angle MX_1X_2 = \angle MX_2X_1 = 35.26°$$

and so

$$\cos 35.26° = \frac{X_1Y}{X_1X_2} = \frac{X_1Y}{\sqrt{2}l}$$

i.e.

$$X_1Y = 1.155l$$

Therefore,

$$YM = X_1Y - X_1M = 0.289l$$

and

$$\frac{YM}{YX_1} = \frac{0.289l}{1.155l} = 0.25$$

i.e. the centre of gravity of a tetrahedron, given by position M, is at one quarter of the vertical height above the (or any) base and in the direction of the apex.

Relation of an octahedron to a cube

In Fig. A.6 is shown an octahedron centred at the body centre of a cube with the corners of the octahedron at the face centres of the cube. If the cube has edge l, the distance MX in the octahedron is $l/2$. From Pythagoras, distance XX is $l/\sqrt{2}$.

The octahedron has fourfold rotation axes (3) parallel to each XMX straight line. These axes coincide with the fourfold axes of the cube that pass through opposite cube faces. The cube also has threefold axes (4) which pass through opposite cube corners, along the body diagonal of the cube. In the octahedron these threefold axes pass through pairs of opposite faces.

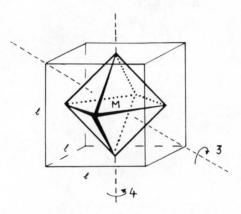

Fig. A.6 Relation of an octahedron to a cube

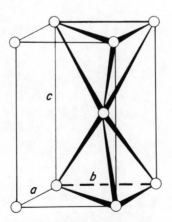

Fig. A.7 Hexagonal   unit cell; $c/a = 1.633$

Hexagonal unit cell: proof that the axial ratio, c/a, ideally equals 1.633

A hexagonal close packed array of atoms contains atoms at the corners and inside the unit cell at $\frac{1}{3}, \frac{2}{3}, \frac{1}{2}$ (Fig. 1.17). Along the a and b edges, atoms are in contact; hence a is equal to the atom diameter. For the purpose of calculating c/a, 'tetrahedral arrays' of atoms may be identified, as in Fig. A.7, in which two tetrahedra share a common corner (i.e. the atom at $\frac{1}{3}, \frac{2}{3}, \frac{1}{2}$). Hence, the c dimension equals twice the vertical height of such a tetrahedron.

Therefore:

$$\frac{c}{a} = \frac{2X_1Y}{X_1X_2} = \frac{2 \times 1.155l}{\sqrt{2}l} = 1.633$$

Appendix 4

The Elements and Some of Their Properties

Element	Symbol	Atomic weight[a]	Electronic configuration[b]	Atomic number	Main oxidation state(s)[c]	Some typical bond lengths to oxygen (Å) and coordination numbers[d]	Melting point (°C)	Crystallographic data: unit cell or structure; lattice parameters (Å); temperature[e]
Actinium	Ac	(227)	$(Rn)6d^17s^2$	89	3		1050	f.c.c., 5.311 (R.T.)
Aluminium	Al	26.98	$(Ne)3s^23p^1$	13	3	Al—O(4)1.79; Al—O(6)1.93	660	f.c.c., 4.0495, (25°C)
Americium	Am	(243)	$(Rn)5f^77s^2$	95	3,4	Am(III)—O(6)2.40; Am(IV)—O(8)2.35	850	hex, ABAC; 3.642, 11.76 (R.T.)
Antimony	Sb	121.75	$(Kr)4d^{10}5s^25p^3$	51	3,5		630	R3m; 4.5067, 57° 6.5' (25°C)
Argon	A	39.95	$(Ne)3s^23p^6$	18			−189	f.c.c., 5.42(−233°C)
Arsenic	As	74.92	$(Ar)3d^{10}4s^24p^3$	33	3,5	As(V)—O(4)1.74; As(V)—O(6)1.90	814	R3m; 4.131, 54° 10(25°C)
Astatine	At	(210)	$(Xe)4f^{14}5d^{10}6s^26p^5$	85	1			
Barium	Ba	137.34	$(Xe)6s^2$	56	2	Ba—O(6)2.76; Ba—O(12)3.00	710	b.c.c.; 5.019 (R.T.)
Berkelium	Bk	(247)	$(Rn)5f^86d^17s^2$	97	3(4)	Bk(III)—O(6)2.36; Bk(IV)—O(8)2.33		
Beryllium	Be	9.01	$(He)2s^2$	4	2	Be—O(3)1.57; Be—O(4)1.67	1280	h.c.p.; 2.2856, 3.5832 (25°C)
Bismuth	Bi	208.98	$(Xe)4f^{14}5d^{10}6s^26p^3$	83	1,3(5)	Bi(III)—O(6)2.42; Bi(III)—O(8)2.51	271	R3m; 4.7457, 57° 14.2'(31°C)
Boron	B	10.81	$(He)2s^22p^1$	5	3	B—O(3)1.42; B—O(4)1.52	2300	tet; 8.73, 5.03 (R.T.)
Bromine	Br	79.91	$(Ar)3d^{10}4s^24p^5$	35	1(3,5,7)	Br(VII)—O(4)1.66	−7	orth; 4.48, 6.67, 8.72(−150°C)
Cadmium	Cd	112.40	$(Kr)4d^{10}5s^2$	48	2	Cd—O(4)2.19; Cd—O(6)2.35	321	h.c.p.; 2.9788 5.6167 (21°C)
Caesium	Cs	132.91	$(Xe)6s^1$	55	1	Cs—O(8)3.22; Cs—O(12)3.30	29	b.c.c.; 6.14(−10°C)
Calcium	Ca	40.08	$(Ar)4s^2$	20	2	Ca—O(6)2.40; Ca—O(8)2.47	850	f.c.c.; 5.582 (18°C)
Californium	Cf	(249)	$(Rn)5f^{10}7s^2$	98	3	Cf—O(6)2.35		
Carbon	C	12.01	$(He)2s^22p^2$	6	4	C—O(3)1.32	3500	hex, graphite; 2.4612, 6.7079 (R.T.)
Cerium	Ce	140.12	$(Xe)4f^26s^2$	58	3,4	Ce(III)—O(9)2.55; Ce(IV)—O(8)2.33	804	f.c.c.; 5.1604(20°C)
Chlorine	Cl	35.46	$(Ne)3s^23p^5$	17	1(3,5,7)	Cl(V)—O(3)1.52; Cl(VII)—O(4)1.60	−101	tet; 8.56, 6.12(−185°C)
Chromium	Cr	52.00	$(Ar)3d^54s^1$	24	3,6	Cr(III)—O(6)2.02; Cr(VI)—O(4)1.70	1900	b.c.c; 2.8846(20°C)
Cobalt	Co	58.93	$(Ar)3d^74s^2$	27	2,3	Co(II)—O(6)2.05 to 2.14; Co(III)—O(6)1.93 to 2.01	1492	h.c.p; 2.507, 4.069 (R.T.)
Copper	Cu	63.54	$(Ar)3d^{10}4s^1$	29	1,2	Cu(I)—O(2)1.86; Cu(II)—O(6)1.97 to 2.66	1083	f.c.c.; 3.6147 (20°C)
Curium	Cm	(247)	$(Rn)5f^76d^17s^2$	96	3(4)	Cm(III)—O(6)2.38; Cm(IV)—O(8)2.35		
Dysprosium	Dy	162.50	$(Xe)4f^{10}6s^2$	66	3(4)	Dy(III)—O(6)2.31	1500	h.c.p.; 3.5923, 5.6545 (20°C)
Einsteinium	Es	(254)	$(Rn)5f^{11}7s^2$	99	3			
Erbium	Er	167.26	$(Xe)4f^{12}6s^2$	68	3	Er(III)—O(6)2.29	1525	h.c.p.; 3.559, 5.592 (20°C)
Europium	Eu	151.96	$(Xe)4f^76s^2$	63	2,3	Eu(III)—O(6)2.35; Eu(III)—O(7)2.43	900	b.c.c.; 4.578 (20°C)
Fermium	Fm	(253)	$(Rn)5f^{12}7s^2$	100	3			
Fluorine	F	19.00	$(He)2s^22p^5$	9	1		−220	
Francium	Fr	223	$(Rn)7s^1$	87	1			
Gadolinium	Gd	157.25	$(Xe)4f^75d^16s^2$	64	3	Gd—O(7)2.44	1320	h.c.p.; 3.6315, 5.777 (20°C)
Gallium	Ga	69.72	$(Ar)3d^{10}4s^24p^1$	31	(1)3	Ga—O(4)1.87; Ga—O(6)2.00	30	orth; 4.520, 7.661, 4.526 (20°C)
Germanium	Ge	72.59	$(Ar)3d^{10}4s^24p^2$	32	(2)4	Ge—O(4)1.79; Ge—O(6)1.94	958	diam; 5.6575 (25°C)
Gold	Au	196.97	$(Xe)4f^{14}5d^{10}6s^1$	79	(1)(3)	Au(III)—O(4)2.10	1063	f.c.c.; 4.0783 (25°C)

Element	Symbol	At. wt.	Electronic configuration	Z	Oxidation states	M—O bond lengths (Å)	m.p. (°C)	Crystal structure
Hafnium	Hf	178.49	(Xe)4f¹⁴5d²6s²	72	4	Hf—O(6)2.11; Hf—O(8)2.23	2200	h.c.p.; 3.1946, 5.0511 (24°C)
Helium	He	4.00	1s²	2	—		−270	hex; 3.75, 6.12 (−271°C)
Holmium	Ho	164.93	(Xe)4f¹¹6s²	67	3	Ho—O(6)2.30; Ho—O(8)2.42	1500	h.c.p.; 3.5761, 5.6174 (20°C)
Hydrogen	H	1.01	1s¹	1	1	H—O 1.02 to 1.22	−259	
Indium	In	114.82	(Kr)4d¹⁰5s²5p¹	49	(1)3	In—O(6)2.18; In—O(8)2.32	156	tet; 3.2512, 4.9467 (20°C)
Iodine	I	126.90	(Kr)4d¹⁰5s²5p⁵	53	1(3,5,7)	I(V)—O(3)1.83	114	orth; 4.792, 7.271, 9.773 (R.T.)
Iridium	Ir	192.22	(Xe)4f¹⁴5d⁷6s²	77	3,4(6)	Ir(III)—O(6)2.13; Ir(IV)—O(6)2.03	2443	f.c.c.; 3.8389 (R.T.)
Iron	Fe	55.85	(Ar)3d⁶4s²	26	2,3(4,6)	Fe(II)—O(6)2.18; Fe(III)—O(4)1.86	1539	b.c.c.; 2.8664 (20°C)
Krypton	Kr	83.80	(Ar)3d¹⁰4s²4p⁶	36	—		−157	f.c.c.; 5.68(−191°C)
Lanthanum	La	138.91	(Xe)5d¹6s²	57	3	La—O(6)2.46; La—O(10)2.68	920	hex; ABAC; 3.770, 12.131 (20°C)
Lawrencium	Lr	(257)	(Rn)5f¹⁴6d¹7s²	103	3			
Lead	Pb	207.19	(Xe)4f¹⁴5d¹⁰6s²6p²	82	2,4	Pb(II)—O(4)2.34; Pb(IV)—O(6)2.18	327	f.c.c.; 4.9502 (25°C)
Lithium	Li	6.94	(He)2s¹	3	1	Li—O(4)1.99; Li—O(6)2.14	180	b.c.c.; 3.5092 (20°C)
Lutetium	Lu	174.97	(Xe)4f¹⁴5d¹6s²	71	3	Lu—O(6)2.25; Lu—O(8)2.37	1700	h.c.p.; 3.5050, 5.5486 (20°C)
Magnesium	Mg	24.31	(Ne)3s²	12	2	Mg—O(4)1.89; Mg—O(6)2.12	650	h.c.p.; 3.2094, 5.2105 (25°C)
Manganese	Mn	54.94	(Ar)3d⁵4s²	25	2,3,4,7	Mn(II)—O(6)~2.10; Mn(VII)—O(4)1.66	1250	cub. 8.914 (25°C)
Mendelevium	Md	(256)	(Rn)5f¹³7s²	101	3			
Mercury	Hg	200.59	(Xe)4f¹⁴5d¹⁰6s²	80	1,2	Hg(I)—O(3)2.37; Hg(II)—O(2)2.09	−39	R̄3m; 3.005, 70°32'(−46°C)
Molybdenum	Mo	95.94	(Kr)4d⁵5s¹	42	(3,4,5)6	Mo(VI)—O(4)1.82; Mo(VI)—O(6)2.00	2620	b.c.c.; 3.1469 (20°C)
Neodymium	Nd	144.24	(Xe)4f⁴6s²	60	3(4)	Nd—O(6)2.40; Nd—O(8)2.52	1024	hex; ABAC; 3.6582, 11.802 (20°C)
Neon	Ne	20.18	(He)2s²2p⁶	10	—		−249	f.c.c.; 4.52(−268°C)
Neptunium	Np	(237)	(Rn)5f⁵7s²	93	(2,3)4(6,7)	Np(II)—O(6)2.60; Np(IV)—O(8)2.38	640	orth; 4.723, 4.887, 6.663 (20°C)
Nickel	Ni	58.71	(Ar)3d⁸4s²	28	2(3)	Ni(II)—O(6)2.10; Ni(III)—O(6)1.98	1453	f.c.c.; 3.524 (18°C)
Niobium	Nb	92.91	(Kr)4d⁴5s¹	41	(4)5	Nb(V)—O(6)2.04; Nb(V)—O(7)2.06	2420	b.c.c.; 3.006 (20°C)
Nitrogen	N	14.01	(He)2s²2p³	7	(2,3,4)5	N(V)—O(3)1.28	−210	hex; 4.03, 6.59 (−234°C)
Nobelium	No	(253)	(Rn)5f¹⁴7s²	102	3			
Osmium	Os	190.20	(Xe)4f¹⁴5d⁶6s²	76	4(6)8	Os(IV)—O(6)2.03	2700	h.c.p.; 2.7353, 4.3191 (20°C)
Oxygen	O	16.00	(He)2s²2p⁴	8	(1)2		−219	cub; 6.83 (−225°C)
Palladium	Pd	106.40	(Kr)4d¹⁰	46	2(4)	Pd(II)—O(4)2.04; Pd(IV)—O(6)2.02	1552	f.c.c.; 3.8907 (22°C)
Phosphorus	P	30.97	(Ne)3s²3p³	15	3,5	P(V)—O(4)1.57	44	orth; 3.32, 10.52, 4.39 (black) (R.T.)
Platinum	Pt	195.09	(Xe)4f¹⁴5d⁹6s¹	78	4(6)	Pt(IV)—O(6)2.03	1769	f.c.c.; 3.9239 (20°C)
Plutonium	Pu	(242)	(Rn)5f⁶7s²	94	3,4,6	Pu(III)—O(6)2.40; Pu(IV)—O(8)2.36		monocl. 6.18, 4.82, 10.97, 101.81°(21°C)
Polonium	Po	(210)	(Xe)4f¹⁴5d¹⁰6s²6p⁴	84	2,4	Po(IV)—O(8)2.50	254	cub. 3.345 (10°C)
Potassium	K	39.10	(Ar)4s¹	19	1	K—O(6)2.78; K—O(12)3.00	63	b.c.c.; 5.32 (20°C)
Praseodymium	Pr	140.91	(Xe)4f³6s²	59	3(4)	Pr(III)—O(6)2.41; Pr(IV)—O(8)2.39	935	hex; ABAC; 3.6702, 11.828 (20°C)
Promethium	Pm	(147)	(Xe)4f⁵6s²	61	3	Pm(III)—O(6)2.38		
Protactinium	Pa	(231)	(Rn)5f²6d¹7s²	91	4,5	Pa(IV)—O(8)2.41; Pa(V)—O(9)2.35	3000	tet; 3.935, 3.238 (R.T.)
Radium	Ra	(226)	(Rn)7s²	88	2	—	700	
Radon	Rn	(222)	(Xe)4f¹⁴5d¹⁰6s²6p⁶	86	—		−71	
Rhenium	Re	186.23	(Xe)4f¹⁴5d⁵6s²	75	3,4,5,7	Re(IV)—O(6)2.03; Re(VII)—O(6)1.97	3170	h.c.p.; 2.760, 4.458 (R.T.)
Rhodium	Rh	102.91	(Kr)4d⁸5s¹	45	3(4,6)	Rh(III)—O(6)2.07; Rh(IV)—O(6)2.02	1960	f.c.c.; 3.8044 (20°C)
Rubidium	Rb	85.47	(Kr)5s¹	37	1	Rb—O(6)2.89; Rb—O(12)3.13	39	b.c.c.; 5.70 (20°C)

Appendix 4 (*Contd.*)

Element	Symbol	Atomic weight[a]	Atomic number	Electronic configuration[b]	Main oxidation state(s)[c]	Some typical bond lengths to oxygen (Å) and coordination numbers[d]	Melting point (°C)	Crystallographic data: unit cell or structure; lattice parameters (Å); temperature[e]
Ruthenium	Ru	101.07	44	$(Kr)4d^7 5s^1$	4(6)8	Ru(III)—O(6)2.08; Ru(IV)—O(6)2.02	2400	h.c.p: 2.7058, 4.2816 (25°C)
Samarium	Sm	150.35	62	$(Xe)4f^6 6s^2$	(2)3	Sm(III)—O(6)2.36; Sm(III)—O(8)2.49	1052	R: 8.996, 23° 13'(20°C)
Scandium	Sc	44.96	21	$(Ar)3d^1 4s^2$	3	Sc—O(6)2.13; Sc—O(8)2.27	1400	h.c.p: 3.3080, 5.2653 (20°C)
Selenium	Se	78.96	34	$(Ar)3d^{10}4s^2 4p^4$	(2)4,6	Se(VI)—O(4)1.69	217	hex: 4.3656, 4.9590 (25°C)
Silicon	Si	28.09	14	$(Ne)3s^2 3p^2$	4	Si—O(4)1.66; Si—O(6)1.80	1410	diam; 5.4305 (R.T.)
Silver	Ag	107.87	47	$(Kr)4d^{10}5s^1$	1(2)	Ag(I)—O(2)2.07; Ag(I)—O(8)2.70	961	f.c.c; 4.0857 (20°C)
Sodium	Na	22.99	11	$(Ne)3s^1$	1	Na—O(4)2.39; Na—O(9)2.72	98	b.c.c; 4.2906 (20°C)
Strontium	Sr	87.62	38	$(Kr)5s^2$	2	Sr—O(6)2.56; Sr—O(12)2.84	770	f.c.c; 6.0849 (25°C)
Sulphur	S	32.06	16	$(Ne)3s^2 3p^4$	2,4,6	S(VI)—O(4)1.52	119	orth; 10.414, 10.845, 24.369 (R.T.)
Tantalum	Ta	180.95	73	$(Xe)4f^{14}5d^3 6s^2$	(3,4)5	Ta(III)—O(6)2.07; Ta(V)—O(6)2.04	3000	b.c.c; 3.3026 (20°C)
Technetium	Tc	98.91	43	$(Kr)4d^6 5s^1$	(3)(4,5,6)7	Tc(IV)—O(6)2.04	2700	h.c.p: 2.735, 4.388 (R.T.)
Tellurium	Te	127.60	52	$(Kr)4d^{10}5s^2 5p^4$	(2)4(6)	Te(IV)—O(3)1.92	450	hex; 4.4566, 5.9268 (25°C)
Terbium	Tb	158.93	65	$(Xe)4f^9 6s^2$	3,4	Tb(III)—O(6)2.32; Tb(IV)—O(8)2.28	1450	h.c.p: 3.599, 5.696 (20°C)
Thallium	Tl	204.37	81	$(Xe)4f^{14}5d^{10}6s^2 6p^1$	1(3)	Tl(I)—O(6)2.90; Tl(I)—O(12)3.16	304	h.c.p: 3.4566, 5.5248 (18°C)
Thorium	Th	232.04	90	$(Rn)6d^2 7s^2$	(3)4	Th—O(6)2.40; Th—O(9)2.49	1700	f.c.c; 5.0843 (R.T.)
Thulium	Tm	168.93	69	$(Xe)4f^{13}6s^2$	(2)3	Tm(III)—O(6)2.27; Tm(III)—O(8)2.39	1600	h.c.p: 3.5372, 5.5619 (20°C)
Tin	Sn	118.69	50	$(Kr)4d^{10}5s^2 5p^2$	2,4	Sn(II)—O(8)2.62; Sn(IV)—O(6)2.09	232	tet; 5.8315, 3.1814 (25°C)
Titanium	Ti	47.90	22	$(Ar)3d^2 4s^2$	(2)4	Ti(II)—O(6)2.26; Ti(IV)—O(6)2.01	1680	h.c.p: 2.9506, 4.6788 (25°C)
Tungsten	W	183.85	74	$(Xe)4f^{14}5d^4 6s^2$	(4)6	W(VI)—O(4)1.81; W(VI)—O(6)1.98	3380	b.c.c; 3.1650 (25°C)
Uranium	U	238.03	92	$(Rn)5f^3 6d^1 7s^2$	(3)4(5)6	U(IV)—O(9)2.45; U(VI)—O(4)1.88	1133	orth; 2.854, 5.869, 4.955 (27°C)
Vanadium	V	50.94	23	$(Ar)3d^3 4s^2$	2-5	V(II)—O(6)2.19; V(V)—O(4)1.76	1920	b.c.c; 3.028 (30°C)
Xenon	Xe	131.30	54	$(Kr)4d^{10}5s^2 5p^6$	2,4,6		-112	f.c.c; 6.24 (-185°C)
Ytterbium	Yb	173.04	70	$(Xe)4f^{14}6s^2$	2,3	Yb(III)—O(6)2.26; Yb(III)—O(9)2.38	824	f.c.c; 5.481 (20°C)
Yttrium	Y	88.91	39	$(Kr)4d^1 5s^2$	3	Y—O(6)2.29; Y—O(9)2.50	1500	h.c.p: 3.6451, 5.7305 (20°C)
Zinc	Zn	65.37	30	$(Ar)3d^{10}4s^2$	2	Zn—O(4)2.00; Zn—O(6)2.15	419	h.c.p: 2.6649, 4.9468 (25°C)
Zirconium	Zr	91.22	40	$(Kr)4d^2 5s^2$	4	Zr—O(6)2.12; Zr—O(8)2.24	1850	h.c.p: 3.2312, 5.1477 (25°C)

[a] Based on the relative atomic mass of $^{12}C = 12.00$. These atomic weights are for natural isotopic abundances except for those given in brackets which refer to short-lived elements without a natural abundance. For these, the mass, the mass of the isotope with longest half-life is given.

[b] These are the ground state configurations of the elements. In some cases with transition elements, lanthanides and actinides, the difference in energy between the ground state and the first excited state is small.

[c] Most elements show a variety of unstable oxidation states in addition to the stable one(s). In compiling this table, principal note is taken of the oxidation state(s) found in crystalline oxides and halides.

[d] Values are taken from Shannon and Prewitt, *Acta Cryst*, B25, 925, 1969; B26, 1046, 1970. Note that bond distances generally increase with cation coordination number and decrease with increasing cation oxidation state. Bond distances to fluorine are usually 0.05 to 0.10 Å shorter than distances to oxygen: bond distances to chlorine are usually 0.20 to 0.40 Å larger.

[e] Data taken mainly from *International Tables for X-ray Crystallography*, Vol. III, p. 278. f.c.c. = face centred cubic (cubic close packed); hex = hexagonal; h.c.p. = hexagonal close packed;
R3m, R3m = rhombohedral; b.c.c = body centred cubic

Questions

1.1 What symmetry elements do the following tetrahedral-shaped molecules possess: (a) CH_3Cl, (b) CH_2Cl_2, (c) CH_2ClBr?

1.2 What symmetry element does the following everyday objects have in common: (a) teapot, (b) a pair of trousers, (c) a tricycle?

1.3 Using the d-spacing formula for a cubic substance, equation (1.2), predict the hkl values that correspond to the three largest d-spacings. What are the d-spacing values if the cubic cell edge, a, is 5.0 Å?

1.4 What are the indices for crystallographic directions that are parallel to the unit cell b edge (a) in the positive y direction and (b) in the negative y direction?

1.5 KCl has the rock salt structure with $a = 6.2931$ Å. Calculate the density of KCl.

1.6 TlBr has a cubic unit cell with $a = 3.97$ Å and density 7.458 g/cc. How many formula units, Z, are in the unit cell? Suggest a possible structure for TlBr.

1.7 At 20 °C, Fe is body centred cubic, $Z = 2$, $a = 2.866$ Å. At 950 °C, Fe is face centred cubic, $Z = 4$, $a = 3.656$ Å. At 1425 °C, Fe is again body centred cubic, $Z = 2$, $a = 2.940$ Å. At each temperature, calculate (a) the density of iron, (b) the metallic radius of iron atoms.

1.8 Show that the density of a body centred cubic arrangement of spheres is 0.6802.

1.9 Metallic gold and platinum both have face centred cubic unit cells with dimensions 4.08 and 3.91 Å, respectively. Calculate the metallic radii of the gold and platinum atoms.

1.10 Starting with a cubic close packed array of anions, what structure types are generated by (a) filling all the tetrahedral sites with cations, (b) filling one half of the tetrahedral sites, e.g. the T_+ sites, with cations, (c) filling all the octahedral sites with cations, (d) filling alternate layers of octahedral sites with cations?

1.11 Repeat the above question but with a hexagonal close packed array of anions. Comment on the absence of any known structure in category (a).

1.12 Identify the following cubic structure types from the information on unit cells and atomic coordinates:

(i) $MX:M$ $\frac{1}{2}00, 0\frac{1}{2}0, 00\frac{1}{2}, \frac{111}{222}$; X $000, \frac{11}{22}0, \frac{1}{2}0\frac{1}{2}, 0\frac{11}{22}$

(ii) $MX:M$ $000, \frac{11}{22}0, \frac{1}{2}0\frac{1}{2}, 0\frac{11}{22}$; X $\frac{111}{444}, \frac{313}{444}, \frac{331}{444}, \frac{133}{444}$

(iii) $MX:M$ 000; X $\frac{111}{222}$

(iv) $MX_2:M$ $000, \frac{11}{22}0, \frac{1}{2}0\frac{1}{2}, 0\frac{11}{22}$; X $\frac{111}{444}, \frac{311}{444}, \frac{131}{444}, \frac{113}{444}, \frac{331}{444}, \frac{313}{444}, \frac{133}{444}, \frac{333}{444}$

1.13 Starting from the rock salt structure, what structure types are generated by the following operations:

(i) removal of all atoms or ions of one type;

(ii) removal of half the atoms or ions of one type in such a way that alternative layers only are present;

(iii) replacement of all the cations in the octahedral sites by an equal number of cations in one set of tetrahedral sites.

1.14 Explain why the NiAs structure is commonly found with metallic compounds but not with ionic ones.

1.15 Compare the packing density of the NaCl and CsCl structures for which, in both cases, anion–anion and anion–cation direct contacts occur.

1.16 What kind of complex anion do you expect in the following: (a) Ca_2SiO_4; (b) $NaAlSiO_4$(Al tetrahedral); (c) $BaTiSi_3O_9$; (d) $Ca_2MgSi_2O_7$ melilite; (e) $CaMgSi_2O_6$ diopside; (f) $Ca_2Mg_5Si_8O_{22}(OH, F)_2$ amphibole, tremolite (the OH, F ions are not bonded to Si); (g) $CaAl_2(OH)_2(Si_2Al_2)$ O_{10} mica, margarite (two Al tetrahedral, two Al octahedral); (h) $Al_2(OH)_4Si_2O_5$ kaolinite (Al octahedral, OH not bonded to Si)?

1.17 Sodium oxide, Na_2O, has the antifluorite structure, $a = 5.55$ Å. Calculate (a) the sodium–oxygen bond length, (b) the oxygen–oxygen bond length, (c) the density of Na_2O.

1.18 $KNbO_3$ has the perovskite structure, $a = 4.007$ Å. Calculate (a) the K—O bond length, (b) the Nb—O bond length, (c) the density of $KNbO_3$. What is the lattice type?

1.19 Silicon has the diamond structure, $a = 5.4307$ Å. Calculate the radius of the silicon atom in it.

1.20 PbO_2 has the fluorite structure, $a = 5.349$ Å. Assuming the oxide ion has a radius of 1.26 Å, calculate the shortest O–O distance and decide whether the oxide ions are touching. Repeat for the Pb ions. Is it reasonable to describe the structure as a close packed structure?

1.21 Silver oxide, Ag_2O, has a cubic unit cell, $Z = 2$, with atomic coordinates Ag: $\frac{111}{444}, \frac{331}{444}, \frac{313}{444}, \frac{133}{444}$; O:$000, \frac{111}{222}$. What are the atomic coordinates if the unit cell is displaced so that an Ag atom is at the origin? What is the lattice type of Ag_2O? What is the coordination number of Ag and O? Does the structure possess a centre of symmetry?

2.1 The mineral grossular, $Ca_3Al_2Si_3O_{12}$, has the garnet structure with 8-coordinate Ca, octahedral Al and tetrahedral Si. Each oxygen is co-ordinated to one Si, one Al and two Ca. Show that the structure obeys Pauling's electrostatic valency rule.

2.2 Explain, using Pauling's electrostatic valency rule, why silicate structures never contain more than two SiO_4 tetrahedra sharing a common corner.

2.3 BeF_2 has the same structure as SiO_2, MgF_2 is the same as rutile and CaF_2 has the fluorite structure. Does this seem reasonable?

2.4 In Table 1.7 are given unit cell constants for some oxides MO with the rock salt structure. Assuming that (i) $r_{O^{2-}} = 1.26$ Å; (ii) $r_{O^{2-}} = 1.40$ Å, calculate for each (a) two values of the cation radius, $r_{M^{2+}}$, (b) two values for the radius ratio, $r_{M^{2+}}/r_{O^{2-}}$. Assess the usefulness of the radius ratio rules in predicting octahedral coordination for M^{2+} for these oxides. Repeat the calculations for the two oxides with the wurtzite structure, Table 1.11, and for some oxides with the fluorite and antifluorite structures, Table 1.9

2.5 Calculate lattice energy values for the alkaline earth oxides using Kapustinskii's equation and the data given in Table 1.7. Compare your results with those given in Table 2.6. Estimate the enthalpies of formation of these oxides.

2.6 Account for the observation that whereas CuF_2 and CuI are stable compounds, CuF and CuI_2 are not stable.

2.7 The most stable oxide of lithium is Li_2O but for rubidium and caesium, the peroxides M_2O_2 and superoxides MO_2 are more stable than the simple oxides M_2O. Comment.

2.8 Using Sanderson's methods, estimate the partial ionic character of the sodium halides. Hence calculate the radii of the atoms involved and the unit cell a value (all have the rock salt structure). Compare your answers with the data given in Table 1.7.

2.9 The oxides MnO, FeO, CoO, NiO all have the cubic rock salt structure with octahedral coordination of the cations. The structure of CuO is different and contains grossly distorted CuO_6 octahedra. Explain.

2.10 While the d electrons in many transition metal compounds may not be involved directly in bond formation, they nevertheless exert a considerable influence on structure. Explain.

2.11 Show schematically the band structure of elemental magnesium and account for its metallic conductivity.

2.12 TiO and NiO both have the rock salt structure but whereas TiO is metallic, pure NiO is an electrical insulator. Explain.

2.13 Pure WO_3 (ReO_3 structure) is an electrical insulator whereas tungsten bronzes such as $Na_{0.5} WO_3$ are metallic. Sketch the probable band structure of these and indicate which of the d orbitals on W are responsible for the electrical properties.

3.1 Using the $K\alpha_1$ data of Table 3.1, verify graphically Moseley's law. What wavelength do you expect for Co $K\alpha_1$ radiation?

3.2 What is the probable lattice type of crystalline substances that give the following observed reflections?

   (a) 110, 200, 103, 202, 211

   (b) 111, 200, 113, 220, 222

   (c) 100, 110, 111, 200, 210

   (d) 001, 110, 200, 111, 201

3.3 Calculate the d and 2θ values for the 111 and 200 lines in the X-ray powder pattern, Cu $K\alpha$ radiation, of a cubic substance with $a = 5.0\,\text{Å}$.

3.4 The value of n in Bragg's Law is always set equal to 1. What happens to the higher-order diffraction peaks?

3.5 A cubic alkali halide has its first six lines with d-spacing 4.08, 3.53, 2.50, 2.13, 2.04 and 1.77 Å. Assign Miller indices to the lines and calculate the value of the unit cell dimension. The alkali halide has density $3.126\,\text{g cm}^{-3}$. Identify the alkali halide.

3.6 An imaginary orthorhombic crystal has two atoms of the same kind per unit cell located at 000 and $\frac{1}{2}\frac{1}{2}0$. Derive a simplified structure factor equation for this. Hence, show that, for a C-centred lattice, the condition for reflection is $hkl : h + k = 2n$.

3.7 Derive a simplified structure factor master equation for the perovskite structure of $SrTiO_3$. Atomic coordinates are $Sr : \frac{1}{2}\frac{1}{2}\frac{1}{2}$; $Ti : 000$; $O : \frac{1}{2}00, 0\frac{1}{2}0, 00\frac{1}{2}$.

3.8 The 111 reflection in the powder pattern of KCl has zero intensity but in the powder pattern of KF it is fairly strong. Explain.

3.9 The alloy gold–copper has a face centred cubic unit cell at high temperatures in which the Au, Cu atoms are distributed at random over the available corner and face centre sites. At lower temperatures, ordering occurs: Cu atoms are located preferentially on the corner sites and one pair of face centre sites; Au atoms are located on the other two pairs of face centre sites. What effects would you expect this ordering process to have on the X-ray powder pattern?

3.10 The X-ray powder pattern of orthorhombic Li_2PdO_2 includes the following lines: 4.68 Å(002), 3.47 Å(101), 2.084 Å(112). Calculate the values of the unit cell parameters. The density is $4.87\,\text{g cm}^{-3}$; what are the cell contents?

3.11 An ammonium halide, NH_4X, has the CsCl structure at room temperature, $a = 4.059\,\text{Å}$, and transforms to the NaCl structure at $138\,^\circ C$, $a = 6.867\,\text{Å}$.
   (a) The density of the room temperature polymorph is $2.431\,\text{g cm}^{-3}$. Identify the substance.
   (b) Calculate the d-spacings of the first four lines in the powder pattern of each polymorph.
   (c) Calculate the percentage difference in molar volume between the two polymorphs, ignoring thermal expansion effects.
   (d) Assuming an effective radius of 1.50 Å for the spherical NH_4^+ ion and that anions and cations are in contact, calculate the radius of the anion in each structure. Are the anions in contact in the two structures?

3.12 'Each crystalline solid gives a characteristic X-ray powder diffraction pattern which may be used as a fingerprint for its identification.' Discuss the reasons for the validity of this statement and indicate why two solids with similar structures, e.g. NaCl and NaF, may be distinguished by their powder patterns.

3.13 What qualitative differences would you expect between powder X-ray and neutron diffraction patterns, obtained with monochromatic radiation of similar wavelength of (a) NaCl, (b) metallic Fe, (c) antiferromagnetic NiO?

4.1 Under what conditions may (i) optical microscopy, (ii) powder X-ray diffraction, be used to determine the purity of a solid material?

4.2 Compare the methods that may be used to identify and characterize crystalline samples of (a) iodoform, CHI_3, and (b) CdI_2.

4.3 How would lightly crushed samples of the following differ when viewed in a polarizing microscope: (a) SiO_2 glass, (b) quartz crystal, (c) NaCl, (d) Cu metal?

4.4 Calculate the wavelength of the electrons accelerated through 50 kV in an electron microscope.

4.5 Explain the following abbreviations: EXAFS, EELS, XRF, ESCA, MASNMR, SEM, TEM, AEM.

4.6 What technique(s) might you use to determine the following: (a) the oxidation state and coordination number of iron in brown/green bottle glass; (b) the coordination number of Cr^{3+} in a crystal of ruby; (c) the nature of a surface layer on a piece of aluminium metal; (d) the nature of some microscopic, crystalline inclusions in a polycrystalline, silicon carbide ceramic piece; (e) the Ca:Si ratio in particles of hydrated cement; (f) the location of the hydrogen atoms in a powdered sample of palladium hydride; (g) whether or not MnO possessed a magnetically ordered superstructure.

4.7 What are the approximate energy values, in units of (a) eV, (b) kJ mole^{-1}, (c) cm^{-1}, (d) Hz, for (i) changes of electron spin orientation (ii) d–d transitions in NiO, (iii) the K absorption edge in Cu metal, (iv) lattice vibrations in KCl.

4.8 Which microscopic technique(s) would you use to study (i) the texture of a piece of metal, (ii) the homogeneity of a sample of powdered glass, (iii) defects such as dislocations, stacking faults and twinning, (iv) the possible contamination of a sample of salt by washing soda?

4.9 Explain why it is generally much easier to characterize organic materials than non-molecular inorganic materials.

4.10 Suppose that you are trying to synthesize a new perovskite and that you end up with a white solid. How would you determine if it was a new perovskite, its structure, composition and purity?

4.11 What king of DTA and TG traces would you expect to obtain on heating samples of the following until they became liquid: (i) beach sand, (ii) window glass, (iii) salt, (iv) washing soda, (v) epsom salts, (vi) metallic Ni, (vii) ferroelectric $BaTiO_3$, (viii) a clay mineral?

4.12 Which of the following would you expect to give a reversible DTA effect with or without hysteresis: (i) melting of salt, (ii) decomposition of $CaCO_3$, (iii) melting of beach sand, (iv) oxidation of metallic Mg, (v) decomposition of $Ca(OH)_2$?

4.13 Salt is often added to icy roads in winter. Could DTA be used to quantify the effects of salt on ice? What results would you expect?

5.1 Explain why crystalline solids are generally more defective as a result of increasing temperature.

5.2 Calculate the enthalpy of formation of Frenkel defects in AgCl using the data given in Fig. 5.4.

5.3 What kind of defects would you expect to predominate in crystals of the following: (a) NaCl doped with $MnCl_2$; (b) ZrO_2 doped with Y_2O_3; (c) CaF_2 doped with YF_3; (d) Si doped with As; (e) a piece of aluminium that has been hammered into a thin sheet; (f) WO_3 after heating in a reducing atmosphere?

5.4 Explain why copper is a much softer metal than tungsten.

5.5 What effect would you expect the ordering of Cu, Zn atoms in β' brass, Fig. 5.11, to have on the X-ray powder pattern, i.e. how do you expect the X-ray powder patterns of β and β' brass to differ?

5.6 The law of mass action can be used to analyse defect equilibria in systems of low defect concentrations. What difficulties are likely to be encountered if this method is applied to systems with larger defect concentrations?

5.7 What kind of dislocations are characterized by the following: (a) the Burger's vector is parallel to the direction of shear and perpendicular to the line of the dislocation; (b) the Burger's vector is perpendicular to the direction of shear and parallel to the line of the dislocation?

5.8 In which directions would you expect slip to occur most readily in (a) Zn, (b) Cu, (c) α-Fe, (d) NaCl?

5.9 Assuming that the enthalpy of creation of Schottky defects in NaCl is 2.3 eV and that the ratio of vacancies to occupied sites at 750 °C is 10^{-5}, estimate the equilibrium concentration of Schottky defects in NaCl at (a) 300 °C, (b) 25 °C.

5.10 For the solid solutions of YF_3 in CaF_2, calculate the density as a function of composition for (a) a cation vacancy model, and (b) an interstitial F^- model. Unit cell data for CaF_2 are given in Chapter 1. Assume that the unit cell volume is independent of solid solution composition.

5.11 Show by means of qualitative sketches the essential differences between the X-ray powder diffraction patterns of (a) a 1:1 mechanical mixture of powders of NaCl and AgCl, and (b) a sample of (a) that has been heated to produce a homogeneous solid solution.

5.12 A sample of aluminium hydroxide was shown by chemical analysis to contain a few per cent of Fe^{3+} ions as impurity. What effect, if any, would the Fe^{3+} ions have on the powder pattern if it was present (a) as a separate iron hydroxide phase, and (b) substituting for Al^{3+} in the crystal structure of $Al(OH)_3$.

5.13 Sketch the kind of DTA trace expected on heating (a) a sample of pure Fe, and (b) an Fe–C alloy containing 1 wt% C.

6.1 What is (i) the mole per cent and (ii) the weight per cent of (a) Al_2O_3 in mullite, $Al_6Si_2O_{13}$; (b) Na_2O in devitrite, $Na_2Ca_3Si_6O_{16}$; (c) Y_2O_3 in yttrium iron garnet, $Y_3Fe_5O_{12}$?

6.2 Sketch the phase diagram for the system Al_2O_3–SiO_2 using the following information. Al_2O_3 and SiO_2 melt at 2060 and 1720 °C. One congruently melting compound, $Al_6Si_2O_{13}$, forms between Al_2O_3 and SiO_2 with a melting point of 1850 °C. Eutectics occur at $\sim 5\,mol\%\,Al_2O_3$, 1595 °C and $\sim 67\,mol\%\,Al_2O_3$, 1840 °C.

6.3 Explain with the aid of examples the difference between the concepts of phases and components. Under what conditions can a component be a phase?

6.4 Phase diagrams are drawn with the compositions represented usually as either weight per cent or mole per cent. Derive expressions for converting from one to the other for a binary AB system.

6.5 Sketch a phase diagram for a system A–B that has the following features. Three binary compounds are present A_2B, AB and AB_2. Both A_2B and AB_2 melt congruently. AB melts incongruently to give A_2B and liquid. AB also has a lower limit of stability.

6.6 For the $MgAl_2O_4$–Al_2O_3 phase diagram (Fig. 5.31) describe the reactions that would be expected to occur, under equilibrium conditions, on cooling a liquid of composition 40 mol% MgO, 60% Al_2O_3. Using rapid cooling rates, how might the product(s) differ?

6.7 The system Mg_2SiO_4–Zn_2SiO_4 is a simple eutectic system in which the two end-member phases form limited ranges of solid solution. Sketch a probable phase diagram for this system. How would you determine experimentally: (i) the compositions of the solid solution limits; (ii) the mechanism of solid solution formation in each case; (iii) the eutectic temperature.

6.8 Pure iron undergoes the $\alpha \rightleftharpoons \gamma$ transformation at 910 °C. The effect of added carbon is to reduce the transformation temperature from 910 to 723 °C. Sketch the general appearance of the Fe-rich end of the Fe–C phase diagram using this information.

7.1 Explain why polyacetylene is an electrical conductor whereas polyethylene is not.

7.2 What kind of materials may be added to polypyrrole to make it (a) n-type, (b) p-type?

7.3 Using the atomic coordinates given in Fig. 7.6, draw the unit cell contents of $YBa_2 Cu_3O_7$ as a projection on (a) the ac plane and (b) the bc plane. Calculate the Cu–O, Ba–O and Y–O distances using the cell dimensions given in Fig. 7.6.

7.4 Sketch the energy band structure of silicon with a band gap of 1.1 eV. What elements would you add to silicon to make it p-type? Sketch the resulting energy level structure for acceptor levels that are located 0.01 eV above the top of the valence band. What fraction of the acceptor levels would be occupied at room temperature? Assume that the probability of excitation is proportional to $\exp(-E/kT)$. If the impurity concentration is 10^{-4} atom per cent, what is the carrier density due to the impurities? What would be the intrinsic carrier concentration at room temperature, in the absence of impurities? At what temperature is the intrinsic density equal to that caused by the impurities?

7.5 Repeat the above problem but for germanium with a band gap of 0.7 eV.

7.6 For use in practical semiconductor devices, why is it desirable to have materials with a large band gap between valence and conduction band but a small gap between valence/conduction band and impurity levels?

7.7 What effect if any, do you expect small amounts of the following impurities to have on the conductivity of NaCl crystals: (a) KCl; (b) NaBr; (c) $CaCl_2$; (d) AgCl; (e) Na_2O?

7.8 What effect, if any, do you expect small amounts of the following impurities to have on the conductivity of AgCl crystals: (a) AgBr; (b) $ZnCl_2$; (c) Ag_2O?

7.9 Using the conductivity data for NaCl given in Fig. 7.18, estimate (a) the enthalpy of cation vacancy migration, (b) the enthalpy of Schottky defect formation. Your results should be similar to those given in Table 7.2.

7.10 A particular solid has a conductivity of $10^{-5} ohm^{-1} cm^{-1}$ at room temperature. Suggest ways for determining the charge carrier responsible for conduction. How could you distinguish conduction by (a) electrons, (b) Na^+ ions, (c) O^{2-} ions?

7.11 Using the conductivity data for different ion exchanged β-aluminas given in Fig. 7.26, calculate the pre-exponential factor A and the activation energy E for each.

7.12 With the aid of Fig. 7.33, explain how the Na/S cell would differ in its mode of operation or usefulness if operated at (a) $200\,°C$, (b) $300\,°C$.

7.13 The Na/S cell is being developed for possible applications in electrically powered buses and cars. For such applications, what advantages and disadvantages do you think the Na/S cell would have in comparison with conventional Pb/acid batteries?

7.14 Suggest a design for a device containing a solid electrolyte that would be sensitive to fluorine gas.

7.15 What are the differences between paraelectric (i.e. dielectric), ferroelectric, ferrielectric and antiferroelectric materials?

7.16 If the following substances were placed between the plates of a capacitor, approximately what values would you expect for their apparent dielectric constant: (a) Ar gas; (b) water; (c) ice; (d) pure single-crystal silicon; (e) pure single-crystal KBr; (f) single crystal of KBr doped with $CaBr_2$; (g) Na β-alumina; (h) $BaTiO_3$?

7.17 Which, if any, of the following crystals might you expect to exhibit piezoelectricity: (a) NaCl; (b) CaF_2; (c) CsCl; (d) ZnS, wurtzite; (e) NiAs; (f) TiO_2, rutile?

7.18 Comment on the validity of the following statement: 'Pyroelectric substances are those that develop a net spontaneous polarization on heating.'

7.19 What are dielectric losses, what are they caused by and how might they be minimized in materials that are to be used as electrical insulators?

8.1 Indicate how you would distinguish between paramagnetic, ferromagnetic and antiferromagnetic behaviour using a Gouy balance.

8.2 Vanadium monoxide is diamagnetic and a good conductor of electricity whereas nickel oxide is paramagnetic/antiferromagnetic and a poor conductor of electricity. Account for these observations.

8.3 Account for the magnetic behaviour of the following spinels: (a) $ZnFe_2O_4$ is antiferromagnetic; (b) $MgFe_2O_4$ is ferrimangetic and its magnetic moment increases with temperature; (c) $MnFe_2O_4$ is ferrimagnetic and its magnetic moment is independent of temperature.

8.4 Explain why magnetic materials that are to be used in information storage should have either square or rectangular shaped hysteresis loops.

8.5 Why cannot pure iron metal be used in transformer cores?

8.6 Show that the following magnetic susceptibility data for Ni fit the Curie–Weiss law:

$T(K)$	800	900	1000	1100	1200
$\chi \times 10^{-5}$	3.3	2.1	1.55	1.2	1.0

Evaluate T_c or θ and C.

8.7 The potassium halides are all transparent to visible light. Calculate the wavelengths at which they become opaque. Band gap data are given in Table 2.17.

8.8 What are the general requirements for a solid material that is to be used as a laser source?

8.9 Anti-Stokes phosphors emit light of shorter wavelength than that used for excitation. Explain why the law of conservation of energy is not violated.

Index

Page references in bold type indicate major sections